The IMA Volumes in Mathematics and its Applications

Volume 150

For other titles published in this series, go to
www.springer.com/series/811

Institute for Mathematics and its Applications (IMA)

The **Institute for Mathematics and its Applications** was established by a grant from the National Science Foundation to the University of Minnesota in 1982. The primary mission of the IMA is to foster research of a truly interdisciplinary nature, establishing links between mathematics of the highest caliber and important scientific and technological problems from other disciplines and industries. To this end, the IMA organizes a wide variety of programs, ranging from short intense workshops in areas of exceptional interest and opportunity to extensive thematic programs lasting a year. IMA Volumes are used to communicate results of these programs that we believe are of particular value to the broader scientific community.

The full list of IMA books can be found at the Web site of the Institute for Mathematics and its Applications:

http://www.ima.umn.edu/springer/volumes.html.

Presentation materials from the IMA talks are available at

http://www.ima.umn.edu/talks/.

Video library is at http://www.ima.umn.edu/videos/.

Fadil Santosa, Director of the IMA

* * * * * * * * * *

IMA ANNUAL PROGRAMS

1982–1983	Statistical and Continuum Approaches to Phase Transition
1983–1984	Mathematical Models for the Economics of Decentralized Resource Allocation
1984–1985	Continuum Physics and Partial Differential Equations
1985–1986	Stochastic Differential Equations and Their Applications
1986–1987	Scientific Computation
1987–1988	Applied Combinatorics
1988–1989	Nonlinear Waves
1989–1990	Dynamical Systems and Their Applications
1990–1991	Phase Transitions and Free Boundaries
1991–1992	Applied Linear Algebra
1992–1993	Control Theory and its Applications
1993–1994	Emerging Applications of Probability
1994–1995	Waves and Scattering
1995–1996	Mathematical Methods in Material Science
1996–1997	Mathematics of High Performance Computing

Continued at the back

Craig John Benham · Stephen Harvey · Wilma K. Olson · De Witt L. Sumners · David Swigon
Editors

Mathematics of DNA Structure, Function and Interactions

Springer

Editors
Craig John Benham
Department of Mathematics
University of California
451 E. Health Sciences Drive
Davis CA 95616
GBSF
USA
http://genomics.ucdavis.edu/benham/benham.php

Stephen Harvey
Department of Chemistry and Biochemistry
Georgia Institute of Technology
Atlanta GA 30332-0400
USA
http://www.chemistry.gatech.edu/faculty/Harvey/

Wilma K. Olson
Department of Chemistry and Chemical Biology
Rutgers University
610 Taylor Rd.
Piscataway NJ 08854-8087
USA
http://rutchem.rutgers.edu/content_dynamic/faculty/wilma_k_olson.shtml

De Witt L. Sumners
Department of Mathematics
Florida State University
Tallahassee FL 32306-4510
208 Love Bldg.
USA
http://www.math.fsu.edu/~sumners/

David Swigon
Department of Mathematics
University of Pittsburgh
301 Thackeray Hall
Pittsburgh PA 15260
USA
http://www.math.pitt.edu/~swigon/

ISSN 0940-6573
ISBN 978-1-4419-0669-4 e-ISBN 978-1-4419-0670-0
DOI 10.1007/978-1-4419-0670-0
Springer Dordrecht Heidelberg London New York

Library of Congress Control Number: 2009931719

Mathematics Subject Classification (2000): 92-06, 92B99, 92C40, 92C05

Printed on acid-free paper

Springer is part of Springer Science+Business Media (www.springer.com)

FOREWORD

This IMA Volume in Mathematics and its Applications

MATHEMATICS OF DNA STRUCTURE, FUNCTION, AND INTERACTIONS

contains papers presented at a highly successful one-week workshop held on September 16-21, 2007 on the same title. The event was an integral part of the 2007–2008 IMA Thematic Year on "Mathematics of Molecular and Cellular Biology." We are grateful to all the participants for making this workshop a very productive and stimulating event.

We owe special thanks to Craig John Benham (Davis Genome Center, University of California, Davis), Stephen Harvey (Department of Chemistry and Biochemistry, Georgia Institute of Technology), Wilma K. Olson (Department of Chemistry and Chemical Biology, Rutgers University), De Witt L. Sumners (Department of Mathematics, Florida State University), and David Swigon (Department of Mathematics University of Pittsburgh) for their superb role as workshop organizers and editors of these proceedings.

We take this opportunity to thank the National Science Foundation for its support of the IMA.

Series Editors

Fadil Santosa, Director of the IMA

Markus Keel, Deputy Director of the IMA

PREFACE

Propelled by the success of the sequencing of the human and many related genomes, molecular and cellular biology has delivered significant scientific breakthroughs. Mathematics (broadly defined) continues to play a major role in this effort, helping to discover the secrets of life by working collaboratively with bench biologists, chemists and physicists. The critical need, which has already begun, is the development of a quantitative body of theory for biology. This development of theory is expected to have the same impact on biology as it did on the sciences of physics, chemistry and engineering in the 20th century. People with strong backgrounds in both biology and the mathematical sciences are creating this quantitative body of theory. Because of its outstanding record of interdisciplinary research and training, the IMA was an ideal venue for the 2007-2008 IMA thematic year on Mathematics of Molecular and Cellular Biology. This volume is dedicated to the memory of Nicholas Cozzarelli, a dynamic leader who fostered research and training at the interface between mathematics and molecular biology. Nick was the founding director of the Program in Mathematics and Molecular Biology (PMMB), a national research and training consortium in existence from 1987-2007. Two of the editors of this volume (Olson and Sumners) were members of PMMB, and one of the editors (Swigon) was a PMMB Fellow. Seven of the thirty-one authors of papers in this volume were PMMB Fellows, an indication of the influence of Nick Cozzarelli on research at the mathematics/molecular biology interface. The kickoff event for the IMA thematic year was the IMA tutorial on Mathematics of Nucleic Acids, and the following 6-day IMA workshop Mathematics of DNA Structure, Function and Interactions, held during September 15-21, 2007 in Minneapolis. The workshop consisted of 32 talks and 17 posters, and enjoyed participation by 120 interdisciplinary scientists, a mix of mathematicians, biologists, chemists, physicists and engineers. This volume consists of a remembrance of Nick Cozzarelli by two past members of his Berkeley molecular biology laboratory, and 15 papers contributed by speakers at the tutorial and workshop. It contains of some of the state-of-the-art in mathematical approaches to DNA as of September 2007. A short description of the articles in the volume follows. For a more complete idea of the content of each article, please see the introductions to each article.

1. Nick Cozzarelli: A personal remembrance by Stephen D. Levene and Lynn Zechiedrich. Steve and Lynn were postdocs in the Cozzarelli lab during the period 1989-1997. This remembrance is very perceptive in the description of Cozzarelli as a blast-ahead interdisciplinary scientist, and recounts a hilarious incident at the lab in which Nick accepts an unexpected NIH merit award over the phone.

2. Mathematical methods in DNA topology: Applications to chromosome organization and site-specific recombination, by Javier Arsuaga, Yuanan Diao, and Mariel Vazquez. This paper explores some of the uses of knot theory and 3-dimensional manifold topology to model chromosome organization and the binding and mechanism of site-specific DNA recombination enzymes. The paper reviews both theoretical and computational topological methods.

3. Conformational statistics of DNA and diffusion equations on the Euclidean group by Gregory S. Chirikjian. Using wormlike chain models for DNA, this paper studies the problem of determining the probability density of end-to-end chain position and orientation. Solutions are obtained by solving the Fokker-Planck equation that describes a diffusion process on the Euclidean motion group.

4. Perspectives on DNA looping, by Laura Finzi. This paper presents a survey of the field of DNA looping, with emphasis on three repressor systems lac, gal and phage lambda. The paper concentrates on the insight gained on transcriptionally-relevant DNA looping mechanisms by single-molecule approaches.

5. Differences between positively and negatively supercoiled DNA that topoisomerases may distinguish, by Jonathan M. Fogg, Daniel J. Catanese, Jr. Graham Randall, Michelle C. Swick, and Lynn Zechiedrich. This article presents a new biological perspective on DNA supercoiling, including a review of the functional importance and practical issues encountered in laboratory work. It provides hints of the features of DNA structure and energetics that topoisomerases may utilize in controlling the supercoiled state of DNA.

6. Calibration of tethered particle motion experiments, by Lin Han, Bertrand Lui, Seth Blumberg, John F. Beausang, Philip C. Nelson, and Rob Phillips. The Tethered Particle Motion (TPM) method has been used to observe and characterize a variety of protein-DNA interactions including DNA looping and transcription. This paper describes a detailed calibration of TPM magnitude as a function of DNA length and particle size, exploring how experimental parameters such as acquisition time and exposure time affect the apparent motion of the tethered particle

7. Difference topology: Analysis of high-order DNA-protein assemblies, by Makkuni Jayaram and Rasika Harshey. This paper studies Difference topology, a method for deciphering the DNA topology within DNA-protein complexes that are not readily amenable to standard structural analyses. The logic is to trap the crossings formed by distinct DNA segments by tying them into knots or links by site-specific DNA inversion and deletion, respectively, carried out by a recombinase. The number of such crossings can then be counted by analytical methods such as gel electrophoresis or electron microscopy.

8. Useful intrusions of DNA topology into experiments on protein-DNA geometry, by Jason D. Kahn, James R. Jenssen, and Vasavi Vittal.

This paper studies the use of small DNA minicircles to characterize protein-induced DNA bending and twisting. In every case studied, topological characterization of minicircle synthesis or properties has led to unexpected geometric or mechanistic conclusions.

9. Topological analysis of DNA-protein complexes, by Soojeong Kim and Isabel K. Darcy. Tangles have been used to model protein-bound DNA. The protein is represented by a 3D ball and the protein-bound DNA is represented by the strings embedded in the 3D ball. This paper reviews tangle analysis of protein-DNA complexes involving three or four segments of DNA.

10. Closing the loop on protein-DNA interactions: Interplay between shape and flexibility in nucleoprotein assemblies having implications for biological regulation, by Stephen D. Levene and Yongli Zhang. The formation of DNA loops by proteins bound at distant sites along a single molecule is an essential mechanistic aspect of many biological processes including gene regulation, DNA replication, and recombination. This paper describes a rigorous theory for DNA loop formation that connects the global mechanical and geometric properties of both DNA and protein, with applications to the problem of loop-mediated gene repression in vivo by lac repressor.

11. Four-way helical junctions in DNA molecules, by David M.J. Lilley. Four-way (Holliday) junctions are branch points in DNA where four helices are interconnected by the mutual exchange of strands. This paper presents a short review focusing on recent developments in understanding the structure and dynamics of DNA four-way junctions.

12. Micromechanics of single supercoiled DNA molecules, by John F. Marko. This paper reviews the theory of the mechanical response of single DNA molecules under stretching and twisting stresses. Using established results for the semiflexible polymer including the effect of torsional stress, and for the free energy of plectonemic supercoils, a theory of coexisting plectonemic and extended DNA is constructed and shown to produce phenomena observed experimentally.

13. Flexibility of nucleosomes on topologically constrained DNA, by Andrei Sivolob, Christophe Lavelle and Ariel Prunell. This paper reviews results on nucleosome conformational flexibility, its molecular mechanism and its functional relevance. The initial approach combined both empirical measurement and theoretical simulation of the topological properties of single particles reconstituted on DNA minicircles.

14. The mathematics of DNA structure, mechanics, and dynamics, by David Swigon. A brief review is given of the main concepts, ideas, and results in the fields of DNA topology, elasticity, mechanics and statistical mechanics. Discussion includes the notions of the linking number, writhe, and twist of closed DNA, elastic rod models, sequence-dependent base-pair level models, statistical models such as helical worm-like chain and freely jointed chain, and dynamical simulation procedures.

15. Paradox regained: A topological coupling of nucleosomal DNA wrapping and chromatin fibre coiling, by Andrew Travers. The folding and unfolding of the chromatin fibre is a fundamental control point for the regulation of eukaryotic transcription. This paper presents a novel solution to the so-called linking number paradox problem and shows that this solution implies that the chromatin fibre acts a tunable coil.

16. Statistical-mechanical analysis of enzymatic topological transformations in DNA molecule, by Alexander Vologodskii. This paper reviews computational approaches to the analysis of action of two classes of DNA enzymes: topoisomerase and recombinase. Comparing the simulated distribution with corresponding experimental data serves as a model test. The major principles and assumptions of the approach, which is based on the simulation of an equilibrium set of DNA conformations, are discussed.

On behalf of the editors, I would like to thank the authors of papers for contributing to this volume, and for their cooperation in the editorial process. Special thanks go to Patricia V. Brick and Dzung N. Nguyen for their excellent assistance in preparing papers for the volume publisher.

Craig John Benham
Department of Mathematics
University of California, Davis
http://genomics.ucdavis.edu/benham/benham.php

Stephen Harvey
Department of Chemistry and Biochemistry
Georgia Institute of Technology
http://www.chemistry.gatech.edu/faculty/Harvey/

Wilma K. Olson
Department of Chemistry and Chemical Biology
Rutgers University
http://rutchem.rutgers.edu/content_dynamic/faculty/wilma_k_olson.shtml

De Witt L. Sumners
Department of Mathematics
Florida State University
http://www.math.fsu.edu/~sumners/

David Swigon
Department of Mathematics
University of Pittsburgh
http://www.math.pitt.edu/~swigon/

CONTENTS

NICK COZZARELLI: A PERSONAL REMEMBRANCE

STEPHEN D. LEVENE* AND LYNN ZECHIEDRICH†

In the weeks following Nick Cozzarelli's untimely passing two years ago, much was written about his fundamental contributions to molecular biology [1–6]. It is not our intent here to recapitulate an account of his groundbreaking scientific contributions or his outstanding service to the scientific community, which were covered well previously. Instead, we offer

*Departments of Molecular & Cell Biology and Physics, University of Texas at Dallas, Richardson, TX 75080.

†Departments of Molecular Virology & Microbiology and Biochemistry & Molecular Biology, Baylor College of Medicine, Houston, TX 77030.

C.J. Benham et al. (eds.), *Mathematics of DNA Structure, Function and Interactions*,
The IMA Volumes in Mathematics and its Applications 150,
DOI 10.1007/978-1-4419-0670-0_1, © Springer Science+Business Media, LLC 2009

a personal portrayal of Nick as seen through the eyes of those who worked in his laboratory. As postdocs in the Cozzarelli laboratory spanning the years between 1989 and 1997 (S.D.L. 1989–92; L.Z. 1990–97), we were privileged to witness firsthand a period when the scope of Nick's interests and the range of techniques he would apply to problems underwent a dramatic expansion.

Nick called himself a biochemist, but one of the many things that made him unique was his ability to readily grasp and apply mathematical and physical concepts to problems involving DNA. Unlike many classically trained biochemists of his generation, Nick was as much at home discussing science with mathematicians and physicists as with colleagues who came from backgrounds similar to his. This was a major attraction of Nick's program for students and postdocs, who came from as diverse a collection of disciplines as one could imagine in a biochemistry laboratory. It is with this perspective that we dedicate this volume to the memory of Nick Cozzarelli.

Steve recalls that when he first arrived in Berkeley Nick's laboratory space was located in Stanley Hall, also known historically as the "virus lab." The building was a vestige of the 1950s, but has since been replaced by a state-of-the-art bioengineering and biophysics building of the same name. The Cozzarelli laboratory at that time was spacious, but remarkable for its lack of any recent renovation. It became clear shortly after one's arrival that the low-tech/high-tech dichotomy reflected in the ambience of the laboratory space mirrored Nick's approach to scientific problems. He was fearless in making use of new technology, sometimes long before the underlying principles became understood (such as the analysis of DNA topology by gel electrophoresis) or before equipment was readily available (such as postdoctoral colleague Junghuei Chen's improvised "PCR" set up, which consisted of a series of beakers over Bunsen burners, a timer and a pair of hands. At 2-minute intervals Junghuei would alternately plunge microcentrifuge tubes into hot- and cold-water baths – it worked beautifully). At the same time, Nick recognized the power of computer simulation and modeling before many other biochemists did and made extensive use of computation to verify or predict experimental outcomes.

Casual acquaintances would characterize Nick as an extrovert; he was wonderful to meet and had something interesting to say on almost any topic to everyone. When Lynn interviewed with him, he took her to lunch at the Berkeley Art Museum on campus. A famous abstract painter was giving a lecture on his art, which was installed then at the museum. When the artist finished his lecture, he asked for any questions and Nick was the first to raise his hand and they ended up having a long, animated conversation about the value of art in scientific publications. Indeed, Nick considered the art of illustration extremely important, which often led to the exchange of multiple figure drafts in the course of preparing manuscripts. This is just one example of many things that Nick passed along to his trainees.

To those who knew him well, "extrovert" was only part of the complicated equation that was Nick. "Intense" was another part. In conversation, Nick listened intently and always identified and grasped the most important points. With near-brutal execution, he would expose any weakness in the argument or the data, usually during laboratory meetings. There are many stories of how, over the years, his students and postdocs would respond to his direct approach. Tears were one response. A quick exit and slam of the door were another. But for the most part, his trainees prepared better and thought harder about their results; learning in the process how to identify weaknesses in an approach. The intensity of Nick's criticism was never personal, even if sometimes it could feel that way. The focus was always on asking the best questions and answering them conclusively.

Nick approached everything he did in exactly the same way – as a quest for perfection. He recognized and appreciated excellence in all realms-sports, food, music, art. He was a bibliophile. Nick said once that the day he had to stop reading literature for science would be the day he quit science. That explained how he kept up on a wide range of topics and could discuss any of them intelligently and passionately. One of Nick's most cherished possessions was his and Linda's Japanese garden, which was stocked with many rare plant specimens. His extraordinary attention to detail was apparent in this beautiful garden, and this trait, too, infused his approach to science.

He was acutely aware of his own faults and limitations; without hesitation, he sought feedback from those around him and could accept criticism as well as he could deliver it. Because of his directness, these interactions seemed natural. Lynn recalls an incident when Nick emerged from his office and blurted out, "Lynn, am I a sexist pig?" After a long pause, his color changed to pale. He hung his head, "Oh no, it's true." Her response: "You don't mean to be, but there are some things that you do and say that could be perceived as sexist by some people." After a long pause and with a dejected expression, Nick replied: "Please tell me." It takes a very strong person to look directly at his potential weaknesses. He listened carefully and thoughtfully to her comments, thanked her for her honesty, and committed himself to change. He truly had not realized how some of what he did or said might look or sound. The ability to adapt, and accept criticism without judgment or rancor was an important ingredient of Nick's success and accounts in large measure for the stunningly successful collaborations that he was involved in throughout his career.

With his directness and intensity Nick could be labeled, perhaps appropriately, as mercurial. However, Nick would always express his feelings and then move on. He was the exact opposite of the passive-aggressive stereotype, who maintains a pleasant facade, but is privately critical. If he was unhappy about something he would confront you (sometimes raising his voice), but this was usually followed up with praise behind your back. He was remarkably supportive when it counted.

Nick always appreciated and showed his appreciation to the members of his laboratory for the work that they did. He was generous in crediting others for ideas and data. This was a brilliant training move, one that empowered trainees and helped them to take ownership of their projects, thereby instilling confidence. However, it also speaks to the fact that Nick was never one to get caught up in the scramble for credit, who came first or did what. He put results and scientific goals above all else. Not that if you crossed the line he wouldn't let you know it. He once reviewed a manuscript submitted for publication and the colleague "forgot" to cite one of the ideas that Nick was most proud of: the "poison" hypothesis for the mechanism of quinolone gyrase inhibitors. He immediately picked up the phone and asked the scientist, "What the hell do you think you're doing ignoring our manuscript?" Of course, when the paper was published, Nick and Ken Kreuzer's work was properly credited.

When Nick was talking on the phone or working on a manuscript or grant in the office with his door closed, he was grumpy if you interrupted him. So, everyone in the laboratory was loathe to take a phone call on the laboratory extension for Nick because that meant you would be the one to have to interrupt him. One day Lynn picked up. "I need to speak with Dr. Nicholas Cozzarelli please." She said, "This is not his office phone number. Please call him there." The man responded, "I've been trying to call that number for hours; I need to speak with him- is he there?" Sighing, she said, "Yes, but he must be on the phone or busy if he is not answering his phone (trying hard to communicate that interrupting him was not a good idea)." The man was insistent so she grudgingly knocked on Nick's door. "TELL HIM TO CALL ME ON THIS LINE!" was Nick's response. "Nick, I told him that, but he is very insistent that you come to the phone." The door whipped open and she could see the frustration roil in his eyes as he stomped to the laboratory phone in his house slippers.

There was no missing Nick's half of the ensuing conversation:

> "HELLO?" he shouted into the mouthpiece.
> "YES, I KNOW ABOUT THAT PROGRAM AND IT STINKS. I THINK IT IS A TERRIBLE IDEA AND I HAVE TOLD EVERYONE I COULD..."
> "What?" "Oh, thank you very much... yes, I gratefully accept."

He hung up the phone and turned to Lynn and then-postdoc Roland Kanaar. "I just won one of those NIH merit awards." They took him out for a beer to celebrate a great honor - 10 years of NIH support without having to write a competing renewal.

There are many qualities involved in good mentorship, one of which is to know when to coach and when to let go. Nick had extraordinary intuition in this regard; a gift for knowing how to deliver encouragement at the right moment and in the right way, but giving students and postdocs the freedom to follow their own paths. He was sensitive to individual styles and recognized that science is an endeavor where "one size" does not fit all. The importance of having confidence in one's own abilities and the conviction of one's ideas were stressed; ever the optimist, Nick always looked at the positive first while maintaining a healthy level of skepticism.

One thing that his trainees and close friends would probably all agree on is that Nick was humble and considered himself lucky. "Me? Really" was his response to being told he had been elected into the National Academy of Sciences. Luck was something he always talked about in scientific contexts and we always had the sense of Nick's appreciation for all he was able to accomplish in a career that seems much too short. But it's our view that Nick created his own luck through passion, dedication, integrity, and a candid view of the world that sadly seems to belong to a different era. We miss the values that Nick espoused just as we miss him as a friend, colleague and mentor.

REFERENCES

[1] B. Alberts, *The Nick Cozzarelli I Knew.* Proc. Natl. Acad. Sci. **103**: 6077, 2006.

[2] J.C. Wang, *Nicholas Cozzarelli 1938–2006.* Nat. Struct. Mol. Biol. **13**: 469, 2006.

[3] R. Kanaar and D. Sherratt, *Nicholas (Nick) R. Cozzarelli 1938–2006.* Cell **125**: 415–417, 2006.

[4] C. Bustamante, *Nicholas R. Cozzarelli (1938–2006).* ACS Chem. Biol. **4**: 123–125, 2006.

[5] D. Kennedy, *Nick Cozzarelli.* Science **312**: 159, 2006.

[6] Preface and Dedication to Nicholas Cozzarelli, 1938–2006. Ann. Rev. Biochem. **76**, 2007.

MATHEMATICAL METHODS IN DNA TOPOLOGY: APPLICATIONS TO CHROMOSOME ORGANIZATION AND SITE-SPECIFIC RECOMBINATION

JAVIER ARSUAGA*, YUANAN DIAO†, AND MARIEL VAZQUEZ‡

Abstract. In recent years, knot theory and low-dimensional topology have been effectively used to study the topology and geometry of DNA under different spatial constraints, and to solve the topological mechanisms of enzymes such as site-specific recombinases and topoisomerases. Through continuous collaboration and close interaction with experimental biologists, many problems approached and the solutions proposed remain relevant to the biological community, while being mathematically and computationally interesting. In this paper, we illustrate the use of mathematical and computational methods in a variety of DNA topology problems. This is by no means an exhaustive description of techniques and applications, but is rather intended to introduce the reader to the exciting applications of topology to the study of DNA. Many more examples will be found throughout this book.

Key words. DNA knots, bacteriophage P4, DNA packing, random knots, site-specific recombination, Xer, tangles.

AMS(MOS) subject classifications. Primary 57M25, secondary 92B99.

Motivation. DNA presents high levels of condensation in all organisms. Volume reduction, defined as the ratio between the volume occupied by a given genome and the volume occupied by a random walk of the same length as the genome, ranges from 10^2 in *Escherichia coli* to 10^4 in humans[50].

These large condensation values lead to questions such as how the DNA is packed inside the eukaryotic cell nucleus, the prokaryotic cell, as well as inside other organisms such as DNA viruses. The complexity of the packing problem is magnified when one considers that the DNA molecule needs to be readily available to multiple biological processes essential to the proper functioning of the organism, such as DNA replication, transcription, recombination and repair. The cell has evolved tools to remove unwanted DNA entanglement and solve other topological problems, such as DNA un- (or over-)winding, knotting or linking, and formation of multimers, that may interfere with its functions. DNA topology, the study of geometrical

*Department of Mathematics, San Francisco State University, San Francisco, CA 94132, USA (jarsuaga@sfsu.edu). This research was supported in part by the Institute for Mathematics and its Applications (IMA). J. Arsuaga is partially supported by NIH grant NIH-2S06GM52588-12.

†Department of Mathematics and Statistics, University of North Carolina at Charlotte, Charlotte, NC 28223, USA (ydiao@uncc.edu). Y. Diao is partially supported by NSF grant DMS-0712958.

‡Department of Mathematics, San Francisco State University, San Francisco, CA 94132, USA (mariel@math.sfsu.edu). M. Vazquez is partially supported by the Institute for Mathematics and its Applications (IMA), and by NIH grants 2S06GM052588 and 20123550 U56 Mentoring core.

C.J. Benham et al. (eds.), *Mathematics of DNA Structure, Function and Interactions*, The IMA Volumes in Mathematics and its Applications 150,
DOI 10.1007/978-1-4419-0670-0_2,

(supercoiling) and topological (knotting) properties of circular DNA, provides the necessary experimental and computational techniques to describe and quantify these problems and their solutions.

The paper is divided into two parts. In each part we present an important problem in DNA topology, and the mathematical and computational tools used to address it.

In Part I we discuss the formation of knots in bacteriophages and its implications for phage packing geometry. Bacteriophages are viruses that propagate in bacteria. Most dsDNA bacteriophages pack their genome in a similar way inside the capsid, a proteinic enclosure with icosahedral symmetry. In the 1980s Liu and colleagues found that DNA extracted from bacteriophages P4 and P2 capsids was mostly knotted [63, 64]. The origin of these knots and whether they contained any information about the organization of the DNA inside the capsid remained unexplored. Here we will describe our current knowledge on how these DNA knots are formed, in particular we will focus on different mathematical models that have been proposed to explain their formation. We will also emphasize how this problem has been amenable to interdisciplinary studies and has generated new mathematics [2, 7, 66].

Part II deals with the resolution of topological obstructions arising during replication of the E. coli chromosome. The bacterial chromosome, a $4.6Mbp$ double-stranded DNA circle, is condensed 10^3 times inside the nucleoid. The two DNA strands are wrapped around each other an average of $420,000$ times in the supercoiled bacterial chromosome and therefore the DNA double-helix must be unwound in order to be copied. Interwinding of newly replicated sister chromosomes in a partially replicated chromosome forms precatenanes, which become catenanes (links) upon completion of replication. Without careful management by cellular machinery, replication of the bacterial chromosome would lead to two sister molecules highly linked together. The cell must solve the topological problem of separating the two linked sister chromosomes to ensure proper segregation at cell division. Unlinking of replication catenanes is mainly achieved by the type II topoisomerase Topo IV (reviewed in [41, 81]).

Furthermore, stalled or broken replication forks are repaired by homologous recombination. Occasionally crossing-over by homologous recombination generates DNA dimers, which may be knotted [84]. The dimers are resolved by Xer recombination. The Xer system consists of enzymes XerC and XerD, which act cooperatively and co-localize at the septum with the protein FtsK. FtsK plays an essential role in dimer resolution, coordinates chromosome segregation and cell division (reviewed in [9]). Recent experimental evidence shows that XerCD-FtsK recombination can unlink catenanes formed by site-specific recombination *in vitro* [52], as well as catenanes formed by replication *in vivo* [47]. Here we will review the tangle method for site-specific recombination. We will illustrate the method with applications to Xer recombination. The analysis will lead to several possible topological pathways followed by the enzymes. The question is posed as to whether the different pathways are simple planar projections of the same 3-dimensional topological mechanism.

Part I. DNA Knotting in Bacteriophages.

In this part of the paper we present the problem of DNA knotting in bacteriophage P4 as well as the various tools from the theory of random knotting used to approach this problem. Bacteriophage P4 knots are formed by random cyclization. In section 1 we introduce the problem of random cyclization of DNA in free solution. We discuss several computational methods currently used to simulate this process, as well as the corresponding analytical results to estimate the knotting probability of a random polygonal curve in $\Re^3$. This work is used as a framework to study the problem of DNA knotting in bacteriophages. Section 2 deals with cyclization of DNA in confined volumes. First, we review some of the experimental results on DNA knots found in the bacteriophage P4 system. This is followed by the description of three computational models and how these models have been used to address the biological problem. In section 3 we discuss the limitations of these approaches and future directions.

1. Cyclization of DNA molecules in free solution.

1.1. Experimental studies on random cyclization of DNA molecules. Random cyclization of long linear DNA molecules with sticky (i.e. complementary) ends produces knots with non-trivial probability. This knotting probability was independently estimated by Rybenkov et al. [78] and by Shaw and Wang [83]. Both groups showed that the formation of these knots depends on the length of the DNA molecule and on the ionic conditions of the solution (i.e. the effective diameter of the DNA molecule). In [78] it was found that the knotting probability for P4 DNA molecules circularized in solution was 3% and that the trefoil was the prevalent knot population followed by smaller amounts of the four crossing knot and even smaller amounts of the five crossing knots. Monte-Carlo simulations of idealized polymer chains (e.g [58, 66]) and analytical results [34] support these experimental results as explained below.

1.2. Simulations of Gaussian and equilateral random polygons without confinement. The wormlike chain is the most accurate polymer model for simulating DNA in solution. However other models such as the equilateral random polygon (ERP) or the Gaussian random polygon are good for estimating properties of long DNA molecules in the bulk and at the same time are more amenable to the development of rigorous analytical results. Several algorithms have been proposed for generating samples of equilateral random polygons. These include the crankshaft algorithm [58, 68], the hedgehog algorithm [58] and the pairwise rotation

TABLE 1

Summary results concerning various random polygons. Results in parenthesis are numerical results and – is the case when the numerical results vary because of the choices of parameters. GP_n stands for a Gaussian random polygon of n edges, EP_n stands for an equilateral random polygon of n edges, CEP_n stands for an equilateral random polygon of n edges within a confined space (usually a sphere with a predetermined radius), R_n stands for a uniform random polygon of n edges and SP_n for spooling random model.

	mean ACN	leading coeff.	knotting prob
GP_n	$O(n \ln n)$	$1/2\pi$	$\geq 1 - e^{-n^\epsilon} \to 1$
EP_n	$O(n \ln n)$	$3/16$	$\geq 1 - e^{-n^\epsilon} \to 1$
CEP_n	$O(n^2)$	$--$	$(\geq 1 - e^{-n^\epsilon} \to 1)$
R_n	$O(n^2)$	$(.115)$	$(\geq 1 - e^{-n^\epsilon} \to 1)$
SP_n	$O(n^2)$	$--$	$(\geq 1 - e^{-n^\epsilon} \to 1)$

algorithm [73]. The crankshaft algorithm is fairly popular among some researchers. In this algorithm two vertices of the polygon are selected at random, dividing the polygon into two subchains. One of the two subchains is selected at random (with equal probabilities for each subchain), and the selected subchain is rotated through a random angle around the axis connecting the two end vertices of the subchain. In the hedgehog algorithm an ERP is first generated and at each step two vectors are selected at random, rotated with respect to their sum and placed back in the polygon. The crankshaft algorithm generates an ergodic Markov chain in the space of all ERPs of fixed length [68]. However the main drawback with this algorithm is that the correlation between samples is very high and therefore many configurations need to be generated in between any two samples in the Markov chain. The hedgehog algorithm on the other hand generates independent samples of polygons however it is unknown whether it is ergodic or not. These algorithms have helped to estimate some of the biologically relevant properties, such as the knotting probability and the mean of the Average Crossing Number distribution (ACN), for equilateral polygons as shown in Table 1. The leading coefficient 3/16 in the equilateral random polygon case means the mean ACN of equilateral random polygons of length n is of the form $(3/16)n \ln n + O(n)$. This number is $1/2\pi$ for the case of Gaussian random polygons.

1.3. Analytical results for knotting of polygons without confinement: Gaussian and equilateral random polygon models. A Gaussian random vector $X = (x, y, z)$ is a random point whose coordinates x, y and z are independent standard normal random variables (with mean $= 0$ and variance $= 1$). The pdf (probability density function) of X is the joint pdf of x, y and z, which is

$$f(X) = \left(\frac{1}{\sqrt{2\pi}}\right)^3 e^{-\frac{x^2+y^2+z^2}{2}} = \left(\frac{1}{\sqrt{2\pi}}\right)^3 e^{-\frac{|X|^2}{2}}.$$

A Gaussian random walk of n steps (denoted by GW_n) consists of $n+1$ consecutive points $X_0 = (0,0,0) = O$, X_1, X_2, ... , X_n such that $Y_{k+1} = X_{k+1} - X_k$ $(k = 0, 1, ..., n-1)$ are independent Gaussian random vectors. It follows that the joint pdf for all the vertices is

$$f(X_1, X_2, ..., X_n) = \left(\frac{1}{\sqrt{2\pi}}\right)^{3n} e^{-\frac{1}{2}(|Y_1|^2+|Y_2|^2+\cdots+|Y_n|^2)}$$
$$= \left(\frac{1}{\sqrt{2\pi}}\right)^{3n} e^{-\frac{1}{2}(|X_1|^2+|X_2-X_1|^2+\cdots+|X_n-X_{n-1}|^2)}.$$

A Gaussian random polygon GP_n is a conditioned GW_n of n steps such that the last vertex X_n coincides with the starting point $X_0 = O$. Thus, if we let $g_n(X_n)$ be the pdf of X_n for a GW_n,

then the joint pdf of X_1, X_2, ..., X_{n-1} of a GP_n is

$$h(X_1, X_2, ..., X_n) = f(X_1, X_2, ..., X_n)/g_n(O).$$

The one advantage of the Gaussian random polygons (over other random polygon models) is that the joint probability density function of its vertices is of an explicitly nice form. This enabled the derivation of the following result concerning the knotting probability of a GP_n [34].

THEOREM 1.1. *[34] Let $\mathcal{K}$ be any knot type, then there exists a positive constant ϵ such that GP_n contains $\mathcal{K}$ as a connected sum component with a probability at least* $1 - \exp(-n^\epsilon)$ *provided that n is large enough.*

One can obtain a similar result for equilateral random polygons.

Suppose Y_1, Y_2, ... , Y_n are n independent random vectors uniformly distributed on S^2. An equilateral random walk of n steps, denoted by EW_n, is defined as the sequence of points in the three dimensional space $\mathbf{R}^3$: $X_0 = O$, $X_k = Y_1 + Y_2 + \cdots + Y_k$, $k = 1, 2, ..., n$. Each X_k is called a vertex of the EW_n and the line segment joining X_k and X_{k+1} is called an edge of EW_n (which is of unit length). Notice that the coordinates of each point are not independent from each other due to the fact that the distance between consecutive points in the polymer needs to be one. If the last vertex X_n of EW_n is fixed, then we have a conditioned random walk $EW_n|X_n$. In particular, EW_n becomes a polygon if $X_n = O$. In this case, it is called an equilateral random polygon and is denoted by EP_n. The joint probability density function $f(X_1, X_2, ..., X_n)$ of the vertices of an EW_n is $f(X_1, X_2, ..., X_n) = \varphi(U_1)\varphi(U_2)\cdots\varphi(U_n) = \varphi(X_1)\varphi(X_2 - X_1)\cdots\varphi(X_n - X_{n-1})$. Where $\varphi(U_i)$ is the density function of selecting a random point over the surface of the sphere.

Let X_k be the k-th vertex of an EW_n $(n \geq k > 1)$, its density function is defined by

$$f_k(X_k) = \int\int\cdots\int \varphi(X_1)\varphi(X_2 - X_1)\cdots \varphi(X_k - X_{k-1})dX_1dX_2\cdots dX_{k-1} \quad (1.1)$$

and it has the closed form $f_k(X_k) = \frac{1}{2\pi^2 r} \int_0^\infty x \sin rx \left(\frac{\sin x}{x}\right)^k dx$ [74]. In the case of EP_n, the density function of the vertex X_k can be approximated by a Gaussian distribution, as given in the following lemma [30, 33, 34].

LEMMA 1.1. *Let X_k be the k-th vertex of an EP_n and let h_k be its density function, then*

$$h_k(X_k) = \left(\sqrt{\frac{3}{2\pi\sigma_{nk}^2}}\right)^3 \exp\left(-\frac{3|X_k|^2}{2\sigma_{nk}^2}\right) + O\left(\frac{1}{k^{5/2}} + \frac{1}{(n-k)^{5/2}}\right), \quad (1.2)$$

where $\sigma_{nk}^2 = \frac{k(n-k)}{n}$.

In other words the density of the k step of an EP_n can be approximated by a Gaussian distribution with mean 0 and a standard deviation that depends on The vertex number k and on the distance from the vertex to the origin (or first vertex in the polygon).

This lemma provided the key link to apply the methods used in [34] for the Gaussian random polygons to the equilateral random polygons, which leads to the following theorem.

THEOREM 1.2. *[30] Let $\mathcal{K}$ be any knot type, then there exists a positive constant ϵ such that EP_n contains $\mathcal{K}$ as a connected sum component with a probability at least $1 - \exp(-n^\epsilon)$, provided that n is large enough.*

Numerical studies on EP_n suggest a scaling law of $1-\exp(-n/a)$ with $a = 244 \pm 5$ (see [66] and references therein).

The above two theorems imply that a long GP_n or EP_n contains many connected sum components (with a high probability), which makes it highly unlikely for the polygon to be achiral. This is stated in the following corollary. However, this only provides reason for the long GP_n and EP_n to favor chiral knots than achiral ones. For relatively short polygons, this is not clear.

COROLLARY 1.1. *[30, 34] There exists some constant $\theta > 0$ such that the probability that a GP_n or an EP_n is a chiral knot is at least $1 - \frac{1}{n^\theta}$.*

The determination of the knot type of a circular molecule can tell us its topological (minimum) crossing number, i.e., the minimum number of crossings one will see no matter how this molecule is artificially stretched, twisted, or bent. However, the average crossing number (ACN), defined as the average of crossing numbers over all orthogonal projections of the molecule, is a more natural geometric measure of the molecule entanglement as it refers to the actual number of crossings that can be perceived while observing a non-perturbed trajectory of a given molecule [55]. Furthermore, it is believed that DNA knots migrate in gel eletrophoresis accordingly with their ACN [99].

The following theorems are presented in [31, 32] and establish the $O(n \ln n)$ behavior of the mean ACN for the Gaussian and equilateral random polygons as illustrated in Table 1.

THEOREM 1.3. *Let χ_n be the ACN of an equilateral random walk of n steps; then*

$$E(\chi_n) = \frac{3}{16} n \ln n + O(n).$$

On the other hand, if χ'_n is the ACN of a Gaussian random polygon of n steps, then

$$E(\chi'_n) = \frac{1}{2\pi} n \ln n + O(n).$$

2. Cyclization of DNA molecules in confined volumes: DNA knotting in bacteriophage P4 capsids.

2.1. Experimental studies on DNA knots in bacteriophages. In dsDNA bacteriophages the volume of the bacteriophage genome is reduced 100 times inside the capsid [53]. This volume reduction imposes severe physical constraints on the DNA molecule. For instance the DNA molecule is under (at least) 50 atmospheres of pressure [42, 93] and at a concentration of 800mg/ml [56]. Despite these conditions the dsDNA molecule is believed to preserve its double helical structure [8] and not to have sequence-specific associations with the protein capsid. A number of models have been proposed to describe the organization of the viral chromosome under such extreme conditions of condensation. These include coaxial and concentric spooling models [4, 20, 35, 76, 82], coaxial models [10], toroidal models [51, 72], and liquid-crystaline models [61].

Bacteriophage P4 is an icosahedral phage of radius $r = 180$Å and a linear dsDNA genome of 11.5 kb ($l = 120 \times 10^3$Å). The genome is flanked by two 16*bp* long single stranded complementary sequences of DNA called *cos* sites. During phage morphogenesis a protein enclosure called capsid is assembled first. This is followed by the packing of a single linear DNA molecule into the capsid through the portal vertex. Infective viruses keep at least one of their *cos* sites attached near the portal [21]. This attachment prevents the two *cos* ends from meeting within the capsid and circularizing the chromosome. However in the experiments performed by Liu and colleagues it was found that most of the DNA molecules phenol extracted from bacteriophage P4 are circular and non-trivially knotted [63, 64].

Recent work [5, 6, 91, 92] reproduced and extended the results of Liu et al. Figure 1 shows a two dimensional gel of DNA knots from bacteriophage P4 in which different conditions are used in each dimension [91]. In this figure the top spot corresponds to the unknotted molecule followed, along a bell-shaped curve, by the trefoil knot, the figure eight (four-crossing) knot, and so on. The spot ahead of the bell is the linear chain. The most remarkable fact about this distribution is that about 95% of the DNA molecules are knotted and only about 2% are knots between 3 and 10 crossing knots.

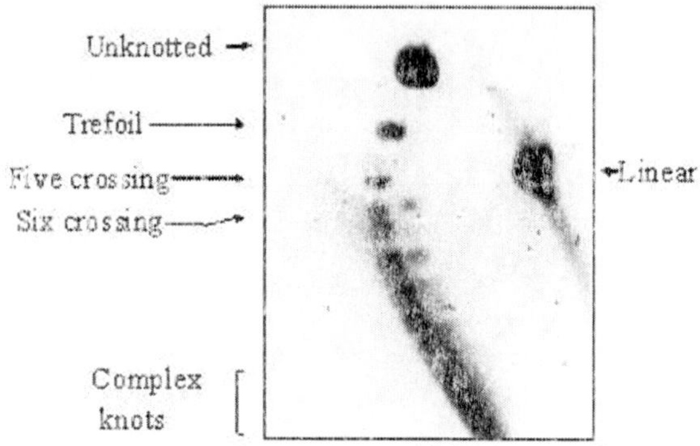

FIG. 1. *Two dimensional gel of knots extracted from bacteriophage P4.*

Furthermore the large majority of the population consist of knots with 30 crossings or more [5]. These results are in high contrast with those obtained by random cyclization of P4 DNA molecules in free solution (see Section 1.1 [78]) and suggested that knots extracted from bacteriophage P4 are formed inside the phage capsid and therefore may be used as reporters for chromosome organization in P4. Despite the small percentage of knots with less than 8 crossings (i.e. those that can be separated by gel electrophoresis) two important properties of the knot distribution were revealed. First the four crossing knot is mostly absent and second the torus knots 5_1 and the 7_1 are more probable than the twist knots 5_2 and 7_2 (contrary to what is expected in free solution). The theoretical work described next aims at explaining these experimental results. In this review we will focus only on the problems of knotting probability and complexity.

2.2. Random knotting within a confined space. A simple approach to study the knotting probability and complexity of P4 knots is by generating ensembles of random polygons inside different convex volumes. Next we describe three models: the Confined Equilateral Random Polygon (CEP_n), the Uniform Random Polygon (URP_n) and the Random Spooling (SP_n).

The Confined Equilateral Random Polygon.

In this model we consider ERPs confined to spheres of certain radius r and use CEP_n to denote such a polygon of length n. Figure 2 shows an example of such polygon.

Unfortunately, the extra condition that confines the polygon to a sphere of radius r completely invalidates the approximation formula given in Lemma 1.2 for the vertex X_k of CEP_n. Intuitively, a CEP_n would be more likely to be knotted than an EP_n. Indeed, this is confirmed by numerical studies [5, 65, 66] which were pioneered by Michels and Wiegel [65].

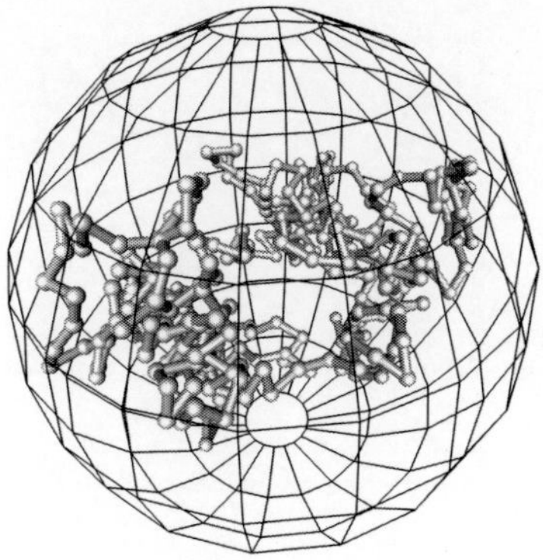

FIG. 2. *An equilateral random polygon inside a sphere.*

In their studies molecular dynamics algorithms were used to sample closed polygons and the knotting probability was computed. Michels and Wiegel found that the knotting probability of a polygon inside a sphere increases with respect to that in $\Re^3$ following a $\exp(N^\alpha/r^3)$ law with $\alpha = 2.28$. In more recent work [5, 66] large ensembles of CEP_n were generated by the crankshaft algorithm. In [66] the scaling law proposed by Michels and Wiegel was confirmed and the coefficient $\alpha = 2.15 \pm 0.04$ improved.

In [5], the combination of experimental and theoretical results led to propose that the effect of the confinement during the random cyclization process of the DNA molecule is one of the key drivers in the formation of knots in the P4 system. This argument has been extended by D. Smith′ group to explain the knotting of chains in confined volumes [75]

The Uniform Random Polygon.

Developing analytical results for CEP_n is a very difficult problem. An alternative model was proposed in [68] as a way to study the random behavior of circular DNA molecules packed in phage capsids that may provide clues about showing some of these analytical results. For $i = 1, 2, ..., n$, let $U_i = (u_{i1}, u_{i2}, u_{i3})$ be a three-dimensional random point that is uniformly distributed in the unit cube C^3 (or in a unit ball) such that U_1, U_2, ..., U_n are independent. Let e_i (called the i-th edge) be the line segment joining U_i and U_{i+1}, then the edges e_1, e_2, ..., e_n define a *uniform random polygon* R_n in the confined space (either the cube or the sphere), where e_n is the line segment joining U_n and U_1. A polygon of length n is denoted by URP_n.

While the knotting probability of an R_n has not been analytically determined (even in the case of $n \to \infty$), a numerical study carried out in [3] provided convincing data that the knotting probability of an R_n quickly approaches 1 as n approaches infinity. Figure 4 is the plot of the percentage of URPs with non-trivial determinant (i.e. those whose Alexander poly-

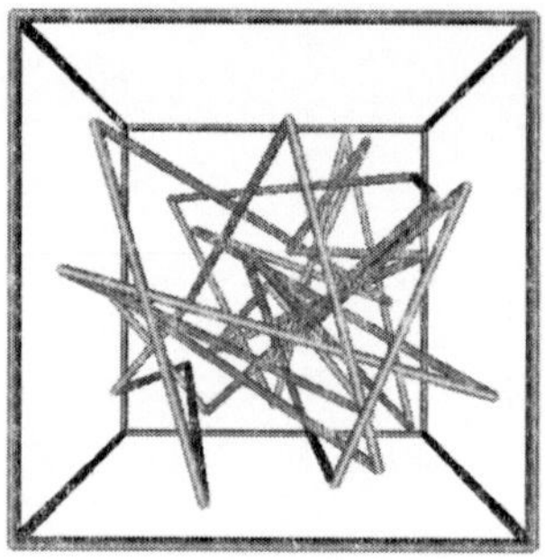

FIG. 3. *A uniform random polygon confined in the unit cube.*

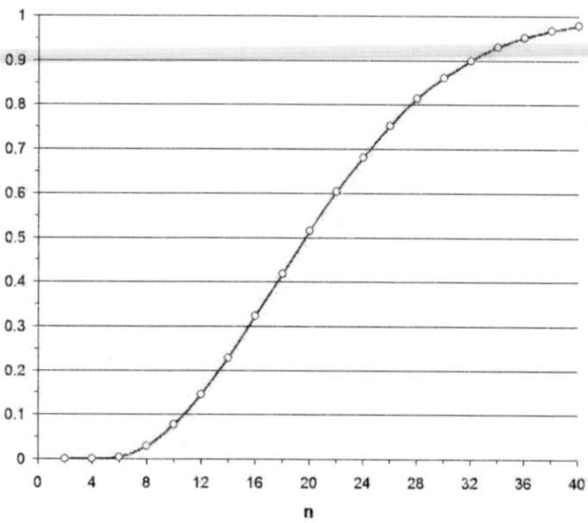

FIG. 4. *The lower bound of knotting probability for URPs up to 40 segments.*

nomial evaluated at $t = -1$ is non-trivial). Since the trivial knot has a trivial determinant, the results give a lower bound of the knotting probability. Figure 4 below is the plot of the data. The fitting curve used here is $1 - \exp(-0.000082n^3)$, although this is not to be expected as a general rule since the trivial knot probability of an R_n is at least of order $\exp(-n \ln n)$ as shown in [3].

It turns out that the mean ACN behavior for an R_n is much easier to determine, both analytically and numerically than for CEPs. Consider a uniform random polygon R_n with n edges e_1, e_2, ..., e_n in that consecutive order. Let $a(e_i, e_j)$ be the average crossing number between e_i and e_j, then the ACN of R_n is

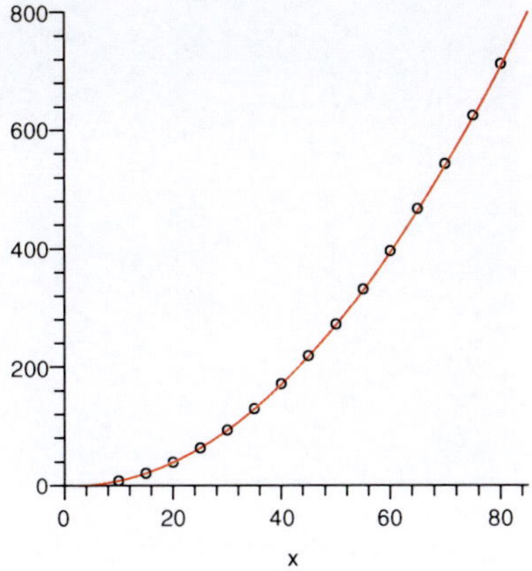

FIG. 5. *The mean ACN of uniform random polygons up to 80 vertices.*

$$\chi_n = \frac{1}{2}\sum_{i=1}^{n} \sum_{j\neq i-1,i,i+1} a(e_i, e_j).$$

It follows that the expected value of the average crossing number of R_n is

$$E(\chi_n) = \frac{1}{2}\sum_{i=1}^{n} \sum_{j\neq i-1,i,i+1} E(a(e_i, e_j)) = p(n-3)n.$$

This establishes the $O(n^2)$ behavior of the mean ACN of an R_n as shown in Table 1. Numerical studies in [2] produced the following near perfect fit using $E(\chi_n) \approx 0.115(n-3)n$.

There have not been enough numerical studies on the knot types of the CEP_n's [5, 66] and R_n's [67] to indicate their bias against achiral knots, even though this is generally expected for long random polygons since achiral knots are much rarer than the achiral ones within large knots.

The Random Spooling model.

The last model we discuss is the *random spooling model.* This model incorporates features from the random knotting models (described above) into the spooling and toroidal models [4, 10, 20, 35, 51, 72, 76, 82]. In the standard spooling model DNA fibers spool around an axis forming coaxial spherical layers. In [59] the knot type of molecular dynamics generated spooling conformations was studied and it was found that most of these conformations were unknotted. These results together with the wide distributions of knots that are observed in P4 suggested that current theoretical models of DNA packing disregard the effect of random fluctuations which

FIG. 6. *The random spooling model.*

in fact may play an important role in the packing of the viral chromosome. We recently proposed [7] that fibers follow spooling trajectories and at the same time they intermingle, as illustrated in the figure 6. This intermingling between fibers of different coaxial layers increases the knotting probability.

Some initial simulation and analytical results have been published [7]. For instance we have estimated the complexity of the average crossing number in the direction of the spooling axis as stated in the next theorem.

THEOREM 2.1. *[7] Let P_n^s be a spooling random polygon, then the average number of crossings in its projection to the xy-plane perpendicular to its center axis is of the order of $O(n^2)$.*

Although the knotting probability has not yet been shown to increase to 1 as suggested by the numerical results shown in Figure 7 a relationship between the writhe of the projection along the spooling axis and the knot type has been proven. The following theorem is a consequence of a theorem due to Morton[69]

THEOREM 2.2. *[7] Let $w(D_n)$ be the writhe of the projection in the direction of spooling axis and $\sigma(P_n^s)$ the number of times the spool goes around its axis. If $|w(D_n)| \geq \sigma(P_n^s)$, then P_n^s is a non-trivial knot. Furthermore, in this case P_n^s cannot be an achiral knot.*

This theorem shows that spooling conformations with high writhe are knotted. This agrees well with some of our results that relate DNA knotting and writhing in bacteriophage P4 [6, 11] and suggests that high writhe may also play an important role in the formation of knots in P4.

3. Conclusions. Here we have discussed the problem of knotting by random cyclization in free solution and in confined volumes. In both cases we have presented experimental, analytical and computational results. By comparing our experimental results with those obtained in free solution we

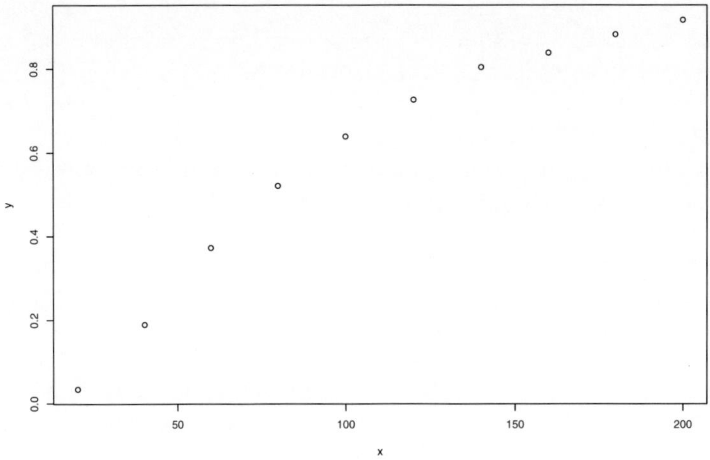

FIG. 7. *Knotting probability as a function of the length of the chain for the random spooling model.*

concluded that knotting in bacteriophage P4 occurs before, or very soon after, the disruption of the capsid and therefore P4 knots can be used as reporters of DNA packing. The large amount of knotting is still a feature that is not fully explained by current mathematical models. In this review we have presented three random knotting models: the confined equilateral polygon, the uniform random polygon and the spooling random polygon. All these models present consistent results however they do not reach the high levels of complexity found in bacteriophages. This is specially true if more accurate representations of the DNA molecule are taken into account. Nevertheless some information about the biological system has been extracted from these theoretical models. For instance our current simulations results suggest that DNA knotting in P4 is mainly driven by the confinement imposed by the capsid during the cyclization reaction, and perhaps also by possible biases introduced by the arrangement of the viral chromosome [6, 11]. Importantly none of the current idealized models proposed in the literature account for the formation of knots inside the capsid and previous simulations results failed to do so [59] thus suggesting that they may not reflect some important properties of the DNA packing. The random spooling model is our first attempt to address this issue. It remains to be seen if such models can reproduce the knot distributions observed experimentally. New experimental results have recently obtained for P4 deletion mutants whose genomes range between 5 and 8 *kb* [92]. These experiments hold a great promise for unveiling the properties that drive knotting in phage capsids as well as some of the essential features of the viral packing.

Part II. Enzymes that change the topology of DNA.

The cell has evolved a set of tools to remove unwanted DNA entanglement and solve other topological problems. Noteworthy are enzymes that change the topology of DNA such as site-specific recombinases and topoisomerases. For example, replication of the circular E. coli genome produces two catenated (linked) circles that cannot be separated without breaking the DNA chain; the type II topoisomerase topoIV plays an important role in unlinking the newly replicated genomes, thus ensuring proper segregation at cell division. Similarly, circular chromosome dimers arise occasionally due to crossing over by homologous recombination of newly replicated chromosomes. The site-specific recombination system XerCD coupled with the multifunctional protein FtsK resolves the chromosome dimers, thus allowing for stable inheritance. In Section 4 we introduce the biology of site-specific recombination. In Section 5 we briefly review the mathematics used in the tangle method, and the method itself, including the main assumptions. We finish the review by summarizing the tangle analysis of Xer recombination at *psi* and of XerCD-FtsK recombination at *dif*. We emphasize the 3-dimensional interpretation of the data.

4. Site-specific recombination. *Site-specific recombinases* are enzymes able to change the genetic code of their DNA substrate. They mediate important biological processes such as inversion of DNA segments, transposition of a DNA segment from one location to another along the genome, integration and excision of viral DNA into and out of its host genome, and resolution of multimeric DNA molecules [48, 79].

Site-specific recombination occurs in two steps. First, during *synapse* two short DNA segments of specific sequence are brought together by the site-specific recombinase(s) and any necessary accessory proteins. These segments are called *recombination sites*. Second, during *strand-exchange*, each recombination site is cleaved, the lose ends recombined and the recombined pieces are rejoined. The DNA sequence in a recombination site is usually non-palindromic, thus allowing to define a site orientation. When the DNA substrate consists of circular DNA molecules, the two recombination sites may occur in a single DNA circle or in two separate circles. In the first case, for *intramolecular recombination*, if the sites induce the same orientation on the circle they are said to be in *direct repeat*, otherwise they are in *inverted repeat* (Figure 8). Relative orientation of the sites is harder to characterize in the case of *intermolecular recombination* (i.e. two sites on separate DNA circles). In the simple case of $T(2,n)$ torus links with 4 or more crossings the sites are said to be in *parallel* or *anti-parallel* orientation with respect to each other (Figure 8).

Based on sequence homology and strand-passage mechanism, site-specific recombinases are divided into two families: serine recombinases and tyrosine recombinases [60, 71]. After synapse formation, the serine

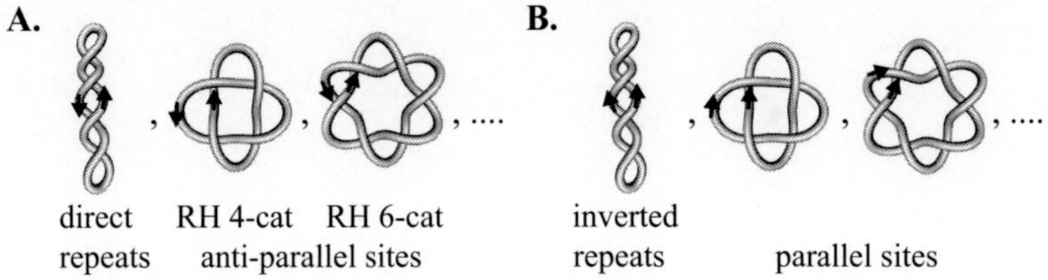

FIG. 8. *Site orientation. A. Negatively supercoiled circle with sites in direct repeat, right-hand 4-crossing torus link (RH 4-cat) with anti-parallel sites, right-hand 6-crossing torus link (RH 6-cat) with anti-parallel sites. B. Negatively supercoiled circle with sites in inverted repeats, RH 4-cat and RH 6-cat with parallel sites.*

recombinases introduce one double-stranded break on each recombination site and act by rotation of the synaptic complex around a dyad axis. The enzymes may iterate this process two or more times before releasing the sites, resulting in a *processive recombination* reaction. Enzymes in this family include the Tn3 resolvase and Gin of bacteriophage Mu [54, 87, 100]. Tyrosine recombinases go through a Holliday Junction intermediate performing single-stranded cleavage in two steps. Tyrosine recombinases are often represented by λ-Int from bacteriophage λ and include enzymes such as XerCD, Flp and Cre [71]. There is little evidence of the ability of tyrosine recombinases to act processively due to the fact that the HJ forces the synapse to reform at each step. However in Gourlay and Colloms [45] and in Grainge *et al.* [47], XerCD recombination is consistent with iterative cleavage and strand-exchange.

Site-specific recombinases can change the topology of circular DNA substrates [24, 87, 100]. Such changes can be quantified experimentally via gel electrophoresis and electron microscopy (*e.g.* [25]) . The experimental data can then be subjected to quantitative and mechanistic analysis to answer questions of DNA bending and strand-exchange. Close scrutiny of the data aided by geometrical models is used by biologists to understand the molecular mechanism of the enzyme(*e.g.* [87, 100]). Mathematicians can greatly improve the reliability of the results using rigorous mathematical tools. The most famous example of these is the tangle method proposed by Ernst and Sumners [39], which has been extensively used in the topological analysis of site-specific recombination reactions. Other more contemporary approaches include the topological characterization of site-specific recombination products ([13, 14, 15, 16]), the classification of 3-string tangles in order to generalize the 2-string tangle model [18, 19, 36], and novel methods to solve 2- and 3-string tangle equations for tangles that are not necessarily rational or Montesinos (*i.e.* sum of rational) [27, 28, 57].

5. The tangle method. The tangle method, based on knot theory and low-dimensional topology, is a mathematical method in which the ac-

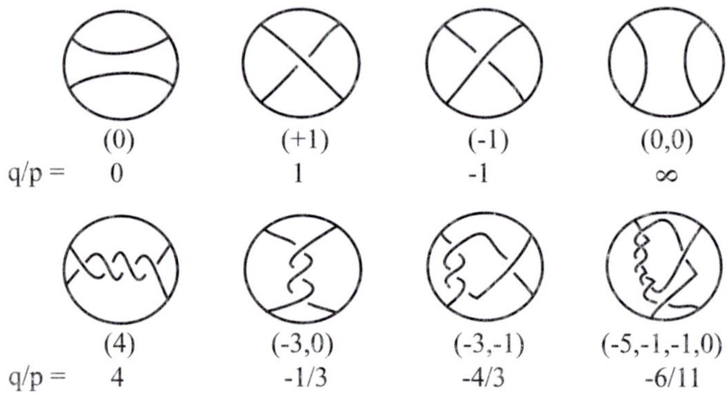

FIG. 9. *Rational tangles. The figure shows eight rational tangles, their Conway vectors and their corresponding classifying rational numbers (Conway Numbers). The top row illustrates the 4 trivial tangles.*

tion of site-specific recombinases on circular DNA substrates is modeled using 2 string tangles. Explaining the method requires a few basic definitions.

5.1. Tangles and knots. The following is a brief overview, more detailed definitions can be found in [17, 70, 77]. A 2-string tangle is an ordered pair (B, t), where B is a fixed oriented 3-ball in S^3 and t is a pair of non-oriented disjoint arcs properly embedded in B, whose endpoints lie on the bounding sphere. For each (B, t) there is an orientation-preserving homeomorphism

$$\Phi : (B, t) \to (D, t_\Phi)$$

that maps B onto the unit 3-ball D, and t onto two straight arcs t_Φ in D connecting the preferred equatorial points NE with SE, and NW with SW. The endpoints of t are mapped to the 4 special equatorial points {NW, NE, SE, SW}. Notice that in order to compare tangles defined in different 3-balls, we shall define a 2-string tangle more precisely as the triplet (B, t, Φ). In this way we may consider, without loss of generality, all tangles as defined on the unit 3-ball D with strings attached to the 4 special equatorial points. Two tangles (D, t_1) and (D, t_2) are equivalent if there is an ambient isotopy that takes t_1 to t_2 while fixing the endpoints.

A tangle diagram is a planar representation of the 3-dimensional tangle in the 3-ball, and may be obtained by projecting the arcs onto the equatorial disk (illustrated in Figure 9).

There are three different types of tangles: rational, locally knotted and prime. A tangle is rational if the strings can be continuously deformed to lie on the bounding 2-sphere [23]. Several rational tangles are illustrated in Figure 9. A tangle (D, t) is locally knotted if there is a 2-sphere S inside D

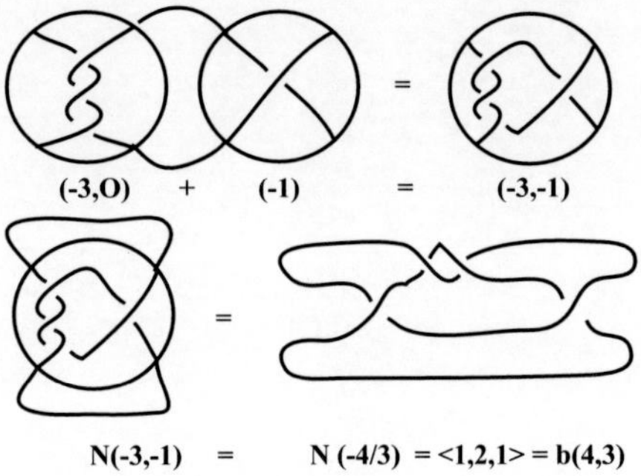

FIG. 10. *Tangle sum and numerator closure. The top diagram illustrates the sum of two rational tangles, yielding in this case another rational tangle with Conway vector* $(-3,-1)$. *The bottom diagram shows the numerator closure of* $(-3,-1)$ *which results in the 4-plat* $b(4,3)$ *described by the vector* $\langle 1,2,1\rangle$

that intersects one of the two arcs of t transversely in two points, and such that the 3-ball bounded by S contains t as a knotted arc with endpoints on S. If a tangle is neither rational nor locally knotted, then it is called a prime tangle [62].

There is a $1-1$ correspondence between equivalence classes of rational tangles and the extended rational numbers [23, 43]. Each equivalence class of rational tangles can be represented by its Conway symbol, a classifying vector $(a_1, a_2, , a_n)$ of integer entries such that $|a_1| > 1$, all entries are nonzero except possibly for a_n, and all entries have the same sign. This scheme applies to all but four exceptional rational tangles called trivial tangles. A unique extended rational number can be obtained from the classifying vector of a rational tangle via the following continuous fraction calculation :

$$\frac{q}{p} = a_n + \frac{1}{a_{n-1} + \frac{1}{a_{n-2} + \frac{1}{\cdots a_2 + \frac{1}{a_1}}}}$$

where $\frac{q}{p} \in \mathbf{Q} \cup \{\frac{1}{0}\}$, $q \in \mathbf{Z}$ and $\gcd(p,q) = 1$. Rational numbers associated to a few different rational tangles are shown in Figure 9.

Two tangle operations, the sum and the numerator closure, are illustrated in Figure 10. The numerator closure converts a tangle into a knot or link. In particular, it relates rational tangles with the family of 4-plats.

A 4-plat is a knot or link that can be obtained by braiding 4 strings and capping off the ends as illustrated in Figure 10 (bottom right) [17]. To each 4-plat can be associated a classifying vector $\langle c_1, c_2, ..., c_{2k+1}\rangle$ (Conway vector), such that $c_i > 0$ for all i. Two 4-plats $\langle c_1, c_2, ..., c_{2k+1}\rangle$ and $\langle d_1, d_2, ..., d_{2k+1}\rangle$ are equivalent if and only if $c_i = d_i$ or $c_i = d_{2k+1-i}$ for all $i = 1, ..., 2k+1$.

From the Conway vector, we can obtain a classifying rational number for each 4-plat through a continued fraction:

$$\frac{\beta}{\alpha} = \frac{1}{c_1 + \frac{1}{c_2 + \cdots + \frac{1}{c_{2k+1}}}}$$

where $0 < \beta < \alpha$. Hence, we can also denote a 4-plat knot by its *Conway symbol* $b(\alpha, \beta)$. Furthermore, by the Classification Theorem of 4-plats [17], we have that $b(\alpha, \beta)$ and $b(\alpha', \beta')$ are equivalent non-oriented links if and only if $\alpha = \alpha'$ and $\beta^{\pm 1} = \beta' \bmod(\alpha)$. If A is a rational tangle, then the result of taking its numerator closure $N(A)$ is a 4-plat (Figure 10). If A and B are rational tangles, then $N(A + B)$ is a 4-plat [17, 39].

5.2. The tangle method. The tangle method, first proposed by Ernst and Sumners in 1990 [39], and reviewed in [70, 90, 101], uses tangles to model the changes in topology of the synaptic complex before and after recombination. The method relies on knowledge on the topology of substrate and products, and on a few justified assumptions.

The first assumption is that the enzymatic mechanism is constant and independent of the geometry and topology of the DNA substrate. Changes in the DNA substrate due to binding and stand-exchange occur inside a tangle E, which contains the recombination sites and any other DNA bound by the enzyme(s). Topological differences between substrates are detected in the tangle O_f, the exterior of E, and are to remain unchanged during recombination.

The second assumption is that E can be partitioned into two tangles O_b and P. O_b contains any DNA bound by the enzymes which is unchanged by recombination. P contains the core regions of the recombination sites, *i.e.* the locations of cleavage and strand-exchange. Site-specific recombination is modeled by tangle surgery where P is converted into R. Most serine recombinases, and some tyrosine recombinases such as XerCD, display topological selectivity (*i.e.* they distinguish between sites in different orientations and in one or two molecules) and specificity (*i.e.* the topology of the product is uniquely determined by the substrate topology) [22, 86].

When the enzyme has topological specificity then, given a substrate of fixed topology, the mechanisms of binding and strand-exchange are constant. In this case the assumptions of the tangle method imply that the tangles O_b, P and R are constant (*i.e.* enzyme-specific). Solving for the topological mechanism of the enzyme when substrate and product of recombination are known is equivalent to knowing who O_b, P and R are. A site-specific recombination event where a substrate of specific topology (K_1) is converted into a product of specific topology (K_2) is modeled as a system of tangle equations:

$$N(O_f + O_b + P) = K_1 = N(O + P)$$
$$N(O_f + O_b + R) = K_2 = N(O + R),$$

where $O = O_f + O_b$ is the outside tangle consisting of all DNA (bound and unbound) which is not changed by strand-exchange.

A processive recombination event where two or more rounds of recombination take place in a single reaction is modeled as a system of two or more tangle equations.

$$\begin{aligned} N(O+P) &= K_0 \\ N(O+R) &= K_1 \\ N(O+2R) &= K_2 \\ \cdots &= \cdots \\ N(O+nR) &= K_n. \end{aligned}$$

Here K_0 is the substrate and K_i (i = 1, 2, ..., n) is the product of successive rounds of recombination. When substrate and products of recombination are 4-plats, systems of tangle equations can be solved for tangles that are rational or sums of rational tangles [39, 40]. Using tools from low-dimensional topology (e.g. Dehn surgery, cyclic surgery theorem) O, P and R can sometimes be shown to be rational or sums of two rational tangles [26, 39, 40, 97, 98]. In the absence of processive recombination the corresponding systems of two tangle equations often admit infinitely many computable solutions. When this is the case, reasonable assumptions can be made on P and R to limit the number of solutions to a small finite number.

Assumptions on P and R.

P is defined as the ball containing the core regions of the recombination sites (*i.e.* where the breakage and rejoining takes place). Usually these regions are very short DNA segments (e.g. 28bp for Tn4430 [95], 32 bp for XerCD [89]) and are thus unlikely to cause tangling inside P. Therefore, P can be any of the four trivial tangles with 0 or 1 crossings. We chose $P = (0)$. Any geometrical complexity induced on the DNA substrate by enzymatic binding is trapped in the O tangle (and more specifically in O_b).

Serine recombinases act by rotation of the sites around a dyad axis. If $P = (0)$ with parallel sites then $R = (k)$ for some integer k. Subsequent rounds of recombination in a processive event are modeled by $2R = (2k)$, $3R = (3k)$ and so on.

Members of the tyrosine family of site-specific recombinases (such as XerCD, Int, Cre, Flp, TnpI) catalyze recombination through a Holiday junction intermediate (HJ) [60, 71]. If $P = (0)$ with parallel sites and the enzyme recombines via a HJ, then $R = (+1)$ or (-1). If $P = (0)$ with antiparallel sites then $R = (0, 0)$. Should two or more rounds of recombination occur iteratively, then $R = (k)$ for $P = (0)$ parallel. In the case of anti-parallel sites the progression of R is harder to visualize (Shimokawa *et al.* preliminary report [85]).

3D considerations of the tangle model.

The orientation of the recombination sites is inherited into the tangle P. The two recombination sites in $P = (0)$ are in *parallel alignment* if both arrows point in the same direction in the tangle diagram, otherwise they are in *antiparallel alignment.*

However, the concept of parallel and antiparallel alignment represents a local geometric property of the recombination sites and is well-defined only in the tangle diagram, which corresponds to a planar projection of the 3-dimensional tangle. Unless the two sites are strictly coplanar, we can always obtain for the same 3-D tangle a planar projection with parallel alignment of the sites and another planar projection with antiparallel sites.

To take into account the most general situation, we will assume that the recombination sites are not coplanar in 3-dimensional space, and hence they can be in parallel or antiparallel alignment in the tangle diagram, based on the direction in which the projection is taken. Biological evidence suggests that enzymes in the tyrosine family present a pseudo-planar conformation at the synapse where the sites are presented in anti-parallel alignment [44, 94]. However co-planarity, in a strict mathematical sense, is unlikely to occur in nature (*cf.* argument and references in [97]).

The assumption of non-coplanarity followed by the assumptions on P implies that, without loss of generality, P can be assumed to be (0) with parallel sites.

5.3. Solving the tangle equations and TangleSolve. In sum, the tangle method models a site-specific recombination event as a system of two or more tangle equations on 3 unknowns O, P and R. If $P = (0)$, $R = (k)$ for some integer k, or $R = (0,0)$, then the system can be solved for O rational or sum of rational tangles, and for the integer k [37, 38, 39, 40]. Systems of tangle equations corresponding to processive and non-processive recombination have been studied extensively, and in many cases all possible solutions have been characterized [13, 14, 15, 16, 26, 37, 38, 39, 40, 46, 90, 96, 97, 98, 101]. Computing the solutions is not mathematically challenging but can be very tedious. In Saka and Vazquez [80], we introduced TangleSolve, a user-friendly computer implementation of the tangle method. TangleSolve is a java stand-alone program and web-based applet which offers a user-friendly interface for analyzing and visualizing recombination mechanisms. The program and documentation can be accessed from http://bio.math.berkeley.edu/TangleSolve/. This program is also reviewed and illustrated in Zheng *et al.* [101]. Based on the assumptions outlined in the previous section, TangleSolve finds rational and sums of rational tangle solutions to systems of equations arising from processive and non-processive recombination. TangleSolve computes only solutions that are rational or sum of rational tangles. In some cases, using tools from low-dimensional topology, it can be proven that all possible solutions to a system of tangle equations are rational or sums of two rational tangles. Such instances are highlighted in TangleSolve.

It is our goal to make this program available to the wider scientific community of mathematicians, molecular biologists and computational biologists. Therefore the calculations rest heavily on a graphical interface. Mathematical notation, or deep knowledge of the tangle method are not required to insert substrate and product topologies and compute enzymatic mechanisms.

TopoICE-R is another computer implementation of the tangle model which is available through KnotPlot [29]. TangleSolve and TopoICE-R have complementary features, noteworthy TopoICE-R provides a 3D rendition of the tangle equations and benefits from all the capabilities within Knotplot . TangleSolve has the ability to compute solutions for processive recombination reactions. Also the two applications use different sets of assumptions.

6. Tangle analysis of Xer recombination.

6.1. Dimer resolution at *psi* sites. In Colloms *et al.* [22] it was shown that, when acting on unknotted DNA circles with two *psi* recombination sites in direct repeats, XerC/XerD yield products of unique topology b(4,3) (the right-hand 4-crossing torus catenane) and anti-parallel sites (Figure 8). The reaction requires two accessory proteins PepA and ArgR [12], and there is experimental evidence that the sites wrap around the accessory proteins approximately three times prior to recombination [1]. This reaction can be written as the following system of tangle equations:

$$N(O+P)=b(1,1),\ N(O+R)=b(4,3)$$

Using results on Dehn surgeries on strongly invertible Knots [49], one can show that O is rational [26, 97] and therefore all solutions to the XerCD-*psi* system of equations can be computed. If no assumptions are made on P and R, there are infinitely many rational solutions to the XerCD-*psi* system of equations, most of which are too complex to be biologically reasonable [26]. In [97] the assumptions on P and R stated in section 4.2 were used to extract biologically relevant solutions. Under these assumptions the solution set was dramatically reduced and relevant solutions were computed using tangle calculus [97]. After introducing experimental information on the relative orientation of the sites in the recombination product (RH 4-cat with anti-parallel sites), the following are the only two solutions for $P=(0)$ with parallel alignment, R integral, and O minimal:

$$S_1:\ O=(-3,0),\ R=(-1);$$
$$S_2:\ O=(-5,0),\ R=(+1).$$

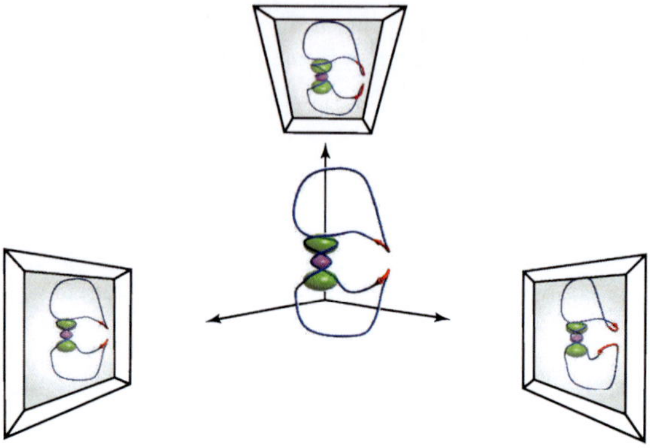

FIG. 11. *The center figure is a cartoon representation of the DNA substrate wrapped around the accessory proteins, and the two* psi *sites brought together in a pseudo-planar configuration by recombinases XerC/XerD. The surrounding images correspond to different planar projections of the center configuration. Each of these planar diagrams corresponds to one of the three tangle solutions for the Xer equations. From left to right these are:* $P = (0)$ *parallel and* $O = (-3, 0)$*;* $P = (0)$ *parallel and* $O = (-5, 0)$*; and* $P = (0)$ *anti-parallel and* $O = (-4, 0)$.

If P is assumed to be (0) with anti-parallel sites then, since XerC/XerD are tyrosine recombinases and recombine through a Holiday junction intermediate, the corresponding R is $(0, 0)$. In this case tangle calculus yields a unique solution:

$$S_3 : \ O = (-4, 0), \ R = (0, 0).$$

Solutions S_1, S_2 and S_3 are illustrated in Figure 11 left, top and right diagrams, respectively. The figure proposes a possible spatial relationship between the three solutions as outlined below.

It is worth noting that given a solution $S = \{O, P, R\}$ where O is rational, $P = (0)$ and $R = (+1)$ or $R = (-1)$, TangleSolve also displays a solution S' equivalent to S. S' is obtained by rotating the synapse so that $P = (0)$ becomes to $P = (0, 0)$, and by letting O have minimal number of crossings [80]. Interestingly, performing this simple transformation on S_1 and on S_2 yields the same solution $S' = \{P = (0, 0), O = (4), R = (0)\}$, where the sites in P are in anti-parallel alignment. Rotating S' in 3-space so that $P = (0)$ anti-parallel reveals S' to be the same as S_3.

In [97] we noted that there is a geometrical equivalence, obtained by rigid motion, between S_1 and S_3, and between S_2 and S_3. This equivalence suggests that a unique 3D representation of the synaptic complex, and topological mechanism of the enzymes, can be interpreted as three different tangle solutions when viewed from different spatial directions (see Figure 11). In [97] we presented a 3D cartoon model (as in Figure 11) and a 3D molecular model of XerCD/DNA which realized these spatial equivalences.

The molecular model was based on x-ray crystallographical data of the accessory proteins PepA and Arg R[88], and of the Cre/DNA complex [44, 94]. Cre is a tyrosine recombinase which shares high degree of homology with Xer.

This study indicates a limitation of the tangle model and suggests the need to consider equivalence classes of planar tangle diagrams related by 3D rigid motion (rotations and translations).

6.2. Unlinking by XerCD-FtsK. In [52] it was shown that XerCD-FtsK recombination at *dif* sites can unlink catenanes produced by λ-Int recombination with both parallel and anti-parallel sites (*cf* Figure 8). These results led to the hypothesis that *in vivo* XerCD-FtsK recombination at *dif* may work with topoIV to unlink catenanes produced by DNA replication. To test this hypothesis, supercoiled catenated plasmids with a *dif* site produced *in vivo* by replication in topoIV-deficient cells were incubated *in vitro* with XerCD-FtsK$_{50C}$. FtsK$_{50C}$ is a biochemically active form of FtsK [52]. In addition to catenanes with 2–14 crossings, a few dimeric knots were also extracted from the cells. The reaction, which was ATP dependent, efficiently produced unlinked circles. Experimental data suggested a stepwise reaction where crossings would be removed one at a time, thus converting catenanes into knots, into catenanes, iteratively until the two free circles would be obtained. A control experiment was done to demonstrate that XerCD-Ftsk$_{50C}$ recombination could convert knotted dimers of the type predicted above to free circles. These substrates were RH torus knots with two directly repeated *dif* sites produced by Cre-*loxP* recombination.

We used the tangle method to confirm that the recombination mechanism most consistent with the experimental data is one of stepwise unlinking where RH torus catenanes with parallel sites are converted to RH torus knots with directly repeated sites, and such knots are converted to RH torus catenanes until the reaction stops at two open circles. These results are summarized in Figure 12 and in [47]. The details of the mathematical analysis will be reported elsewhere [85].

Briefly, systems of tangle equations corresponding to the experimental data of Grainge et al. [47] were posed and solved using TangleSolve for tangles that are rational or sums of rational. For example, all possible systems of two equations converting a substrate of type RH 6-crossing torus catenane (6-cat) with parallel sites into a product with 5 or less crossings were considered. Under the tangle method assumptions outlined in section 4.2, only three solutions were found, all of which produced the RH 5-crossing torus knot with directly repeated sites. The solutions are shown in Figure 12. Iterating the mechanism shown in Figure 12.C. recombines the RH 6-cat into a RH 5-torus knot, into RH 4-cat, to + trefoil, to 2-cat, to unknot, to free circles, consistent with the experimental data (Figure 13).

7. Conclusion. In sections 3-5 we reviewed the tangle method for site-specific recombination, including certain computer implementations,

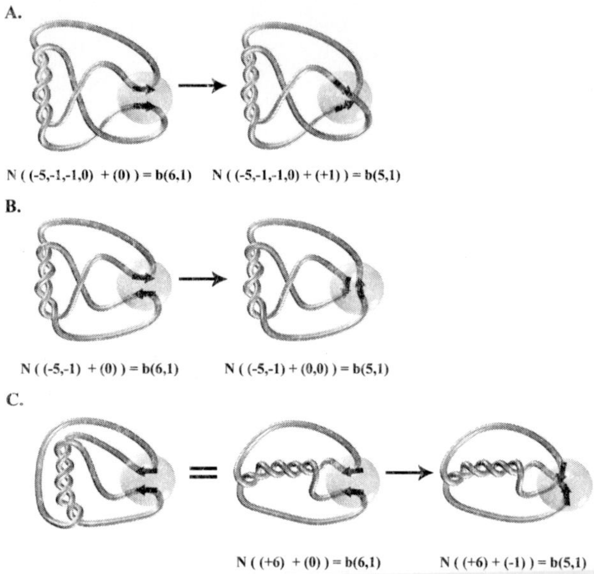

FIG. 12. *Solutions to the XerCD-FtsK tangle equations. Assuming a substrate of type RH 6-cat with parallel sites and a product knot or link with 5 or less crossings, we used TangleSolve to compute solutions that are rational or sum of rational tangles. We found that there are three biologically meaningful solutions, they are all rational and they produce the RH 5-crossing torus knot with sites in direct repeats. Panes A, B and C illustrate the three solutions. Each pane includes the synaptic complex before and after recombination, and shows how the P tangle is recombined into R. The three solutions are equivalent by rigid motion in 3 dimensions (results to be reported elsewhere [85]). Furthermore, iterating solution C results in the predicted gradual stepwise unlinking of the 6cat.*

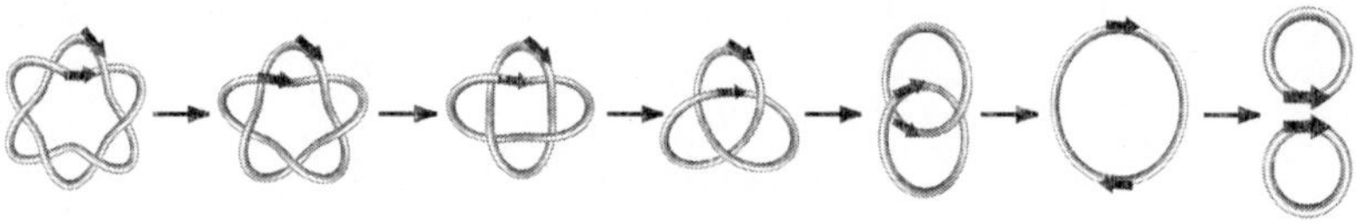

FIG. 13. *Stepwise unlinking by XerCD*-dif-*FtsK recombination.*

and we illustrated the method with the tangle analysis of Xer recombination. The tangle method is a powerful mathematical tool which uses contemporary pure mathematics to solve important biological questions. We highlighted the power of the method, as well as its limitations. In particular we pointed to the need to interpret the tangle results as 3-dimensional objects which may be related by rigid motions and thus equivalent from a biological (and geometrical) point of view.

Acknowledgements. The authors thank the following people: Jung Hun Koh for his help generating the spooling conformation, Rob Scharein for his assistance in generating figures using Knotplot, David Sherratt, Ian Grainge, Sean Colloms and Jonathan Bath for valuable discussions on the biology of Xer recombination, Kai Ishihara and Koya Shimokawa for their insights on the mathematical analysis of Xer. Special thanks go to De Witt L. Sumners for guiding the authors into this field. The three authors are very grateful to Nick Cozzarelli. All of them were fellows of the Program in Mathematics and Molecular Biology (PMMB). This program fostered the authors' interdisciplinary interest. Y. Diao was inspired by the joint work of Nick Cozzarelli and De Witt L. Sumners on random knot theory. J. Arsuaga was a postdoctoral fellow with N. Cozzarelli and is grateful for the invaluable training and for his career advice. M. Vazquez is most grateful to Nick Cozzarelli for his vision of a world where topology could help biology (and vice versa), and for letting this vision materialize by promoting close interactions between mathematicians and molecular biologists.

REFERENCES

[1] C. Alén, D.J. Sherratt, and S.D. Colloms, *Direct interaction of aminopeptidase A with recombination site DNA in Xer site-specific recombination.*, EMBO J. **16** (1997), pp. 5188–5197.

[2] J. Arsuaga, T. Blackstone, Y. Diao, E. Karadayi, and M. Saito, *Linking of Uniform Random Polygons in Confined Spaces*, J. Physics A **40** (2007), pp. 1925–1936.

[3] J. Arsuaga, T. Blackstone, Y. Diao, E. Karadayi, and M. Saito, *Sampling Large Random Knots in a Confined Space*, J. Physics A **40** (2007), pp. 11697–11711.

[4] J. Arsuaga, R. Tan, M. Vazquez, D.W. Sumners, and S.C. Harvey, *Investigation of viral DNA packaging using molecular mechanics models*, Biophys. Chem. **101** (2002), pp. 475–484.

[5] J. Arsuaga, M. Vazquez, S. Trigueros, D.W. Sumners, and J. Roca, *Knotting probability of DNA molecules confined in restricted volumes: DNA knotting in phage capsids.* Proc. Natl. Acad. Sci. USA **99** (2002), pp. 5373–5377.

[6] J. Arsuaga, M. Vazquez, P. McGuirk, S. Trigueros, D.W. Sumners, and J. Roca, *DNA knots reveal a chiral organization of DNA in phage capsids*, Proc. Natl. Acad. Sci. USA **102** (2005), pp. 9165–9169.

[7] J. Arsuaga and Y. Diao, *DNA Knotting in Spooling Like Conformations in Bacteriophages*, Computational and Mathematical Methods in Medicine **9**(3) (2008), pp. 303–316.

[8] K. Aubrey, S. Casjens, and G. Thomas, *Secondary structure and interactions of the packaged dsDNA genome of bacteriophage P22 investigated by Raman difference spectroscopy*, Biochemistry **31** (1992), pp. 11835–11842.

[9] F.X. Barre and D.J. Sherratt, *Chromosome dimer resolution.* In The Bacterial Chromosome (Higgins, N.P., ed.), Washington, DC: ASM Press (2005), pp. 513–524.

[10] L. Black,W. Newcomb, J. Boring, and J. Brown, *Ion etching bacteriophage T4: support for a spiral-fold model of packaged DNA*, Proc. Natl. Acad. Sci. USA **82** (1985), pp. 7960–7964.

[11] T. Blackstone,P. McGuirck,C. Laing, M. Vazquez, J. Roca, and J. Arsuaga, *The role of writhe in DNA condensation*, Proceedings of International Workshop on Knot Theory for Scientific Objects. OCAMI Studies Volume **1** (2007). Osaka Municipal Universities Press; pp. 239–250.

[12] M. Bregu, D.J. Sherratt, and S.D. Colloms, *Accessory factors determine the order of strand exchange in Xer recombination at psi.*, EMBO J. **21** (2002), pp. 3888–3897.

[13] D. Buck and E. Flapan, *Predicting Knot or Catenane Type of Site-Specific Recombination Products*, J Molecular Biology **374**(5) (2007), pp. 1186-1199.

[14] D. Buck and E. Flapan, *A topological characterization of knots and links arising from site-specific recombination.*, J. Phys. A: Math. Gen. **40** (2007), pp. 12377–12395.

[15] D. Buck and C. Verjovsky-Marcotte, *Tangle-solutions for a family of DNA rearranging proteins.*, Math Proc Camb Phil Soc **139** (2005), pp. 59–80.

[16] D. Buck and C. Verjovsky-Marcotte, *Classification of Tangle Solutions for Integrases, A Protein Family that Changes DNA Topology.*, J. Knot. Theory Ramifications **16** (2007), pp. 969–995.

[17] G. Burde and H. Zieschang, *Knots.*, vol. 5, In de Gruytier Studies in Mathematics (Gabriel, P., ed.) Walter de Gruyter, Berlin., 1985.

[18] H. Cabrera Ibarra, *On the classification of rational 3-tangles*, J. Knot Theory Ramifications **12**(7) (2003), pp. 921-946.

[19] H. Cabrera Ibarra, *Results on the classification of rational 3-tangles*, J. Knot Theory Ramifications **13**(2) (2004), pp. 175–192.

[20] K. Cerritelli, N. Cheng, A. Rosenberg, C. McPherson, F. Booy, and A. Steven, *Encapsidated conformation of bacteriophage T7 DNA*, Cell **91** (1997), pp. 271–280.

[21] D.K. Chattoraj and R.B. Inman, *Location of DNA ends in P2, 186, P4 and lambda bacteriophage heads*, J. Mol. Biol. **87** (1974), pp. 11–22.

[22] S.D. Colloms, J. Bath, and D.J. Sherratt, *Topological selectivity in Xer site-specific recombination*, Cell **88** (1997), pp. 855–864.

[23] J.H. Conway, *An enumeration of knots and links, and some of their algebraic properties.*, Computational Problems in Abstract Algebra, Pergamon, Oxford, UK (1967), pp. 329–358.

[24] N.R. Cozzarelli, M.A. Kraznow, S.P. Gerrard, and J.H. White, *A topological treatment of recombination and topoisomerases.*, Cold Spring Harbor Symp. Quant. Biol. **49** (1984), pp. 383–400.

[25] N.J. Crisona, R.L. Weinberg, B.J. Peter, D.W. Sumners, and N.R. Cozzarelli, *The topological mechanism of phage lambda integrase.*, J. Mol. Biol. **289** (1999), pp. 747–775.

[26] I. Darcy, *Biological distances on DNA knots and links: applications to Xer recombination.*, J. Knot Theory Ramification **10** (2001), pp. 269–294.

[27] I.K. Darcy, J. Chang, N. Druivenga, C. McKinney, R.K. Medikonduri, S. Mills, J. Navarra-Madsen, A. Ponnusamy, J. Sweet, and T. Thompson, sl Coloring the Mu transpososome., BMC Bioinformatics **7** (2006), pp. 435.

[28] I.K. Darcy, J. Luecke, and M. Vazquez *Tangle analysis of difference topology experiments: applications to a Mu-DNA protein complex*, IMA preprint series, (2008), https://www.ima.umn.edu/preprints/oct2007/2177.pdf.

[29] I.K. Darcy and R.G. Scharein, *TopoICE-R: 3D visualization modeling the topology of DNA recombination.*, Bioinformatics **22**(14) (2006), pp. 1790–1791.

[30] Y. Diao, *The Knotting of Equilateral Polygons in $\mathbf{R}^3$*, Journal of Knot Theory and its Ramifications, **4**(2) (1995), pp. 189–196.

[31] Y. Diao, A. Dobay, R.B. Kusner, K. Millet, and A. Stasiak, *The Average Crossing Number of Equilateral Random Polygons* J. Physics A **36**(46) (2003), pp. 11561–11574.

[32] Y. Diao and C. Ernst, *The Average Crossing Number of Gaussian Random Walks and Polygons*, Physical and numerical models in knot theory, J.A. Calvo, K.C. Millett, E.J. Rawdon, and A. Stasiak, editors, Series on Knots and Everything **36** (2005), World Scientific, pp. 275–292.

[33] Y. Diao, J. Nardo, and Y. Sun, *Global Knotting in Equilateral Random Polygons*; Journal of Knot Theory and its Ramifications, **10**(4) (2001), pp. 597–607.

[34] Y. Diao, N. Pippenger, and D.W. Sumners, *On Random Knots*, Journal of Knot Theory and its Ramifications, **3**(3) (1994), pp. 419–429.

[35] W.C. Earnshaw and S.R. Casjens, *DNA packaging by the double-stranded DNA bacteriophages*, Cell **21** (1980), pp. 319–331.

[36] J. Emert and C. Ernst, *N-string tangles*, J. Knot. Theory Ramifications **9(8)**(2000), pp. 987–1004.

[37] C. Ernst, *Tangle equations*, J. Knot. Theory Ramifications **5**(1996), pp. 145–159.

[38] C. Ernst, *Tangle equations II*, J. Knot. Theory Ramifications **6** (1997), pp. 1–11.

[39] C. Ernst and D.W. Sumners, *A calculus for rational tangles: applications to DNA recombination*, Math. Proc. Cambridge Phil. Soc. **108** (1990), pp. 489–515.

[40] C. Ernst and D.W. Sumners, *Solving tangle equations arising in a DNA recombination model*, Math. Proc. Cambridge Phil. Soc. **126** (1999), pp. 23–36.

[41] O. Espeli and K.J. Marians,, *Untangling intracellular DNA topology*, Mol Microbiol **52** (2004), pp. 925–931.

[42] A. Evilevitch, L. Lavelle, C.M. Knobler, E. Raspaud, and W.M. Gelbart, *Osmotic pressure inhibition of DNA ejection from phage*, Proc. Natl. Acad. Sci. USA **100** (2003), pp. 9292–9295.

[43] J.R. Goldman and L.H. Kauffman, *Rational tangles*, Advan. Appl. Math. **18** (1997), pp. 300–332.

[44] D.N. Gopaul, F. Guo, and G.D. Van Duyne, *Structure of the Holliday junction intermediate in Cre–loxP site-specific recombination*, EMBO J. **17** (1998), pp. 4175–4187.

[45] S.C. Gourlay and S.D. Colloms, *Control of Cre recombination by regulatory elements from Xer recombination systems*, Mol. Microbiol. **52** (2004), pp. 53–65.

[46] I. Grainge, D. Buck, and M. Jayaram, *Geometry of site alignment during Int family recombination: antiparallel synapsis by the FLP recombinase*, J. Mol. Biol. **298** (2000), pp. 749–764.

[47] I. Grainge, M. Bregu, M. Vazquez, V. Sivanathan, S.C. Ip, and D.J. Sherratt, *Unlinking chromosomes catenated in vivo by site-specific recombination*, EMBO J **26**(19) (2007), pp. 4228–4238.

[48] B. Hallet and D.J. Sherratt, *Transposition and site-specific recombination adapting DNA cut-and paste mechanism to a variety of genetic rearrangements*, FEMS Microbiol. Rev. (1997), p. 21.

[49] M. Hirasawa and K. Shimokawa, *Dehn surgeries on strongly invertible knots which yield lens spaces*, Proc. Am. Math. Soc. **128** (2000), pp. 3445–3451.

[50] V.F. Holmes and N.R. Cozzarelli, *Closing the ring: links between SMC proteins and chromosome partitioning, condensation, and supercoiling*, Proc. Natl. Acad. Sci. USA **97** (2000), pp. 1322–1324.

[51] N. Hud, *Double-stranded DNA organization in bacteriophage heads: an alternative toroid-based model*, Biophys. J. **69** (1995), pp. 1355–1362.

[52] S.C. Ip, M. Bregu, F.X. Barre, and D.J. Sherratt, *Decatenation of DNA circles by FtsK-dependent Xer site-specific recombination.*, EMBO J **22** (2003), pp. 6399–6407.

[53] P.J. Jardine and D.L. Anderson, *DNA packaging in double-stranded DNA phages* The bacteriophages (2006), Ed. Richard Calendar, Oxford University Press, pp. 49–65.

[54] R. Kanaar, A. Klippel, E. Shekhtman, J. M. Dungan, R. Kahmann, and N.R. Cozzarelli, *Processive recombination by the phage Mu Gin system: implications for the mechanisms of DNA strand-exchange, DNA site alignment, and enhancer action*, Cell **62** (1990), pp. 353–366

[55] V. Katritch, Bednar, D. Michoud, R. G. Scharein, J. Dubochet, and A. Stasiak, *Geometry and physics of knots*, Nature **384** (1996), pp. 142–145.

[56] E. Kellenberger, E. Carlemalm, J. Sechaud, A. Ryter, and G. Haller, *Considerations on the condensation and the degree of compactness in non-eukaryotic DNA-containing plasmas*, In Bacterial Chromatin: Proceedings of the Symposium "Selected Topics on Chromatin Structure and Function" (eds. C. Gualerzi and C. L. Pon), Springer, Berlin (1986), pp. 11–25.

[57] S. Kim and I.K. Darcy, *Topological analysis of DNA-protein complexes*, Included in this volume, Mathematics of DNA Structure, Function and Interactions (eds C.J. Benham, S. Harvey, W.K. Olson, D.W. Sumners and D. Swigon), Springer Science + Business Media, LLC, New York, (2009).

[58] K.V. Klenin, A.V. Vologodskii, V.V. Anshelevich, A.M. Dykhne, and M.D. Frank-Kamenetskii, *Effect of Excluded Volume on Topological Properties of Circular DNA*, J. Biomolec. Str. and Dyn. **5** (1988), pp. 1173–1185.

[59] J.C. LaMarque, T.L. Le, and S.C. Harvey, *Packaging double-helical DNA into viral capsids*, Biopolymers **73** (2004), pp. 348–355.

[60] A. Landy,*Coming or going its another pretty picture for the lambda-Int family album*, Proc. Natl Acad. Sci. USA **96** (1999), pp. 7122–7124.

[61] J. Lepault, J. Dubochet, W. Baschong, and E. Kellenberger, *Organization of double-stranded DNA in bacteriophages: a study by cryo-electron microscopy of vitrified samples* EMBO J. **6** (1987), pp. 1507–1512.

[62] W.B.R. Lickorish, *Prime knots and tangles.*, Trans. Am. Math. Soc. **267** (1981), pp. 321–332.

[63] L.F. Liu, J.L. Davis, and R. Calendar, *Novel topologically knotted DNA from bacteriophage P4 capsids: studies with DNA topoisomerases*, Nucleic Acids Res. **9** (1981), pp. 3979–3989.

[64] L.F. Liu, L. Perkocha, R. Calendar, and J.C. Wang, *Knotted DNA from bacteriophage capsids*, Proc. Natl. Acad. Sci. USA **78** (1981), pp. 5498–5502.

[65] J.P.J. Michels and F.W. Wiegel, *On the topology of a polymer ring*, Proc. R. Soc. London Ser A **403** (1986), pp. 269–284.

[66] C. Micheletti, D. Marenduzzo, E. Orlandini, and D.W. Sumners, *Knotting of random ring polymers in confined spaces*, J. Chem. Phys. **124** (2006), pp. 064903.1–10.

[67] K. Millett, *Knotting of regular polygons in 3-space*, Random knotting and linking (Vancouver, BC, 1993), World Sci. Publishing, Singapore (1994), pp. 31–46.

[68] K. Millett, *Monte Carlo Explorations of Polygonal Knot Spaces*, Knots in Hellas'98 (Delphi), Ser. Knots Everything **24** (2000), World Scientific, pp. 306–334.

[69] H.R. Morton, *Seifert circles and knot polynomials*, Math. Proc. Cambridge Phil. Soc. **99** (1986), pp. 107–109.

[70] K. Murasugi, *Knot Theory, Its Applications (Translated by B. Kurpita)*, Birkhauser, Boston, MA. 1996.

[71] S.E. Nunes-Duby, H.J. Kwon, R.S.T. Tirumalai, T. Ellenberger, and A. Landy, *Similarities and differences among 105 members of the Int family of site-specific recombinases*, Nucl. Acids Res. **26** (1998), pp. 391–406.

[72] A.S. Petrov, M.B. Boz, and S.C. Harvey, *The conformation of double-stranded DNA inside bacteriophages depends on capsid size and shape*, J Struct Biol. **160** (2007) pp. 241–248.

[73] P. Plunkett, M. Piatek, A. Dobay, J.C. Kern, K. Millet, A. Stasiak, and E. Rawdon, *Total curvature and total torsion of knotted polymers*, Macromolecules **40** (2007), pp. 3860–3867.

[74] L. Rayleigh, *On the problems of random vibrations, and of random flights in one, two, or three dimensions*, Phil. Mag. S. 6. **37**(220) (1919), pp. 321–347.

[75] D. Raymer and D. Smith, *Spontaneous knotting of an agitated string* Proc. Natl. Acad. Sci

[76] K. Richards, R. Williams, and R. Calendar, *Mode of DNA packing within bacteriophage heads*, J. Mol. Biol. **78** (1973), pp. 255–259.

[77] D. Rolfsen, *Knots* Mathematics Lecture Series **7**, Publish or Perish, Berkeley, CA., 1976.

[78] V.V. Rybenkov, N.R. Cozzarelli, and A.V. Vologodskii, *Probability of DNA knotting and the effective diameter of the DNA double helix*, Proc. Natl. Acad. Sci. USA **90** (1993), pp. 5307–5311.

[79] P.D. Sadowski, *Site-specific genetic recombination: hops, flips, and flops*, FASEB J. **7** (1993), pp. 760–767.

[80] Y. Saka and M. Vazquez, *TangleSolve: topological analysis of site-specific recombination*, Bioinformatics **18** (2002), pp. 1011–1012.

[81] J.B. Schvartzman and A. Stasiak, *A topological view of the replicon*, EMBO Rep. **5**(3) (2004), 256–261.

[82] P. Serwer, *Arrangement of double-stranded DNA packaged in bacteriophage capsids: An alternative model*, J. Mol. Biol. **190** (1986), pp. 509–512.

[83] S.Y. Shaw and J.C. Wang, *Knotting of a DNA chain during ring closure*, Science **260** (1993), pp. 533–536.

[84] Arciszewska, L.K. and D.J. Sherratt *Site-specific recombination and circular chromosome segregation*, Philos. Trans. R. Soc. Lond. B. Biol. Sci. **347** (1995), pp. 37–42.

[85] K. Shimokawa, K. Ishihara, I. Grainge, D.J. Sherratt, and M. Vazquez, *DNA unlinking by site-specific recombination: topological analysis of XerCD-FtsK action*, Preliminary report.

[86] W.M. Stark and M.R. Boocock, *Topological selectivity in site-specific recombination*, In Mobile Genetic Elements (Sherratt, D. J., ed.), IRL Press at Oxford University, Oxford (1995), pp. 101–129.

[87] W.M. Stark, D.J. Sherratt, and M.R. Boocock, *Site-specific recombination by Tn3 resolvase: topological changes in the forward and reverse reactions*, Cell **58** (1989), pp. 779–790.

[88] N. Strater, D.J. Sherratt, and S.D. Colloms, *X-ray structure of aminopeptidase A from Escherichia coli and a model for the nucleoprotein complex in Xer site-specific recombination.*, EMBO J. **18** (1999), pp. 4513–4522.

[89] D.K. Summers and D.J. Sherratt, *Multimerization of high copy number plasmids causes instability: ColE1 encodes a determinant essential for plasmid monomerization and stability*, Cell **36** (1984), pp. 1097–1103.

[90] D.W. Sumners, C. Ernst, N.R. Cozzarelli, and S.J. Spengler, *Mathematical analysis of the mechanisms of DNA recombination using tangles*, Quarterly Reviews of Biophysics **28** (1995), pp. 253–313.

[91] S. Trigueros, J. Arsuaga, M. Vazquez, D.W. Sumners, and J. Roca, *Novel display of knotted DNA molecules by two dimensional gel electrophoresis*, Nucleic Acids Research **29** (2001), e67.

[92] S. Trigueros and J. Roca, *Production of highly knotted DNA by means of cosmid circularization inside phage capsids*, BMC Biotechnol **7**(1) (2007), pp. 94.

[93] S. Tzill, J.K. Kindt, W.M. Gelbart, and A. Ben-Shaul, *Forces and Pressures in DNA Packaging and Release from Viral Capsids*, Biophys. J. **84** (2003), pp. 1616–1627.

[94] G.D. Van Duyne, *A structural view of Cre-loxP site-specific recombination*, Annu. Rev. Biophys. Biomol. Struct. **30** (2001), pp. 87–104.

[95] V. Vanhooff, C. Galloy, H. Agaisse, D. Lereclus, B. Revet, and B. Hallet, *Self-Control in DNA site-specific recombination mediated by the tyrosine recombinase TnpI*, Molecular Microbiology **60(3)** (2006), pp. 617–629.

[96] M. Vazquez, *Tangle analysis of site-specific recombination: Gin and Xer systems*, PhD dissertation in mathematics, Florida State University, Tallahassee, FL, 2000.

[97] M. Vazquez, S.D. Colloms, and D.W. Sumners, *Tangle analysis of Xer recombination reveals only three solutions, all consistent with a single 3-dimensional topological pathway*, J. Mol. Biol. **346** (2005), pp. 493–504.

[98] M. Vazquez and D.W. Sumners, *Tangle analysis of Gin site-specific recombination*, Math. Proc. Cambridge Phil. Soc. **136** (2004), pp. 565–582.

[99] A.V. Vologodskii, N.J. Crisona, B. Laurie, P. Pieranski, V. Katritch, J. Dubochet, and A. Stasiak, *Sedimentation and electrophoretic migration of DNA knots and catenanes*, J. Mol. Biol. **278** (1998), pp. 1–3.

[100] S.A. Wasserman, J.M. Dungan, and N.R. Cozzarelli, *Discovery of a predicted DNA knot substantiates a model for site-specific recombination*, Science **229** (1985), pp. 171–174.

[101] W. Zheng, C. Galloy, B. Hallet, and M. Vazquez, *The tangle model for site-specific recombination: a computer interface and the TnpI-IRS recombination system*, Knot Theory for Specific Objects, OCAMI studies **1**(2) (2007), pp. 251–271.

CONFORMATIONAL STATISTICS OF DNA AND DIFFUSION EQUATIONS ON THE EUCLIDEAN GROUP

GREGORY S. CHIRIKJIAN*

Abstract. Semi-flexible (or wormlike) polymer chains such as DNA possess bending and torsional stiffness. Given a semi-flexible polymer structure that is subjected to Brownian motion forcing, the distribution of relative positions and orientations visited by the distal end of the chain relative to its proximal end provides important information about the molecule that can be linked to experimental observations. This probability density of end-to-end position and orientation can be obtained by solving a Fokker-Planck equation that describes a diffusion process on the Euclidean motion group. In this paper, methods for solving this diffusion equation are reviewed. The techniques presented are valid for chains of up to several persistence lengths in open environments, where the effects of excluded volume can be neglected.

1. Introduction. The theory of semiflexible/wormlike polymer chains originated more than fifty years ago [21, 35, 41]. Since then, the statistical mechanics of chains such as DNA has received substantial attention in the literature (see e.g., [29, 33, 34, 38, 40, 42, 43, 46, 51, 57, 60, 62–64, 68, 74, 77–80, 83]). In particular, semiflexible polymer theories based on diffusion processes can be found in [5, 8, 45, 53]. Excluded-volume effects in polymer solutions in general [25, 28], and for semi-flexible chains in particular [56] have been studied. The main approaches are the use of renormalization group concepts [23, 61] and mean field potentials [25, 71].

In theories of ring-closure probabilities, the probability density function describing the relative frequency of occurrence of positions and orientations of the distal end of the chain for given position and orientation of the proximal end play an important role [22, 23, 25, 28, 32, 81]. And a number of new theoretical models have been developed by the author's group for generating this quantity from given stiffness models [10–15].

Experimental measurements of DNA stiffness parameters have been reported in [4, 36, 54, 55, 66, 81]. Efforts to characterize integrals of the joint positional and orientational probability density function over many of its arguments can be found in [77, 44], and the whole distribution in the case of the helical wormlike chain can be found in [81]. DNA Elastic properties and experimental measurements of DNA elastic properties such as twist/stretch coupling have been reported in [9, 16, 37, 48, 49, 69, 75].

Elastic models of DNA mechanics has a long history [6, 7]. A number of recent studies on chiral and uncoupled end-constrained elastic rod models of DNA with circular cross-section have been presented [18, 70, 72, 19]. These models use classical elasticity theory of continuum filaments with or without self-contact constraints to model the stable conformations of DNA in plasmids, in chromosomes, and during transcription.

*Department of Mechanical Engineering, Johns Hopkins University, Baltimore, MD 21218, USA (gregc@jhu.edu).

C.J. Benham et al. (eds.), *Mathematics of DNA Structure, Function and Interactions*,
The IMA Volumes in Mathematics and its Applications 150,
DOI 10.1007/978-1-4419-0670-0_3,

In some works, Euler angles are used in parameterizing equations of the Kirchhoff elastic rod theory to obtain equilibrium conformations of DNA and determine its stability [82, 26, 27]. Also, the worm-like chain model has been used to model the equilibrium behavior of DNA [59]. More recent works involve the modeling of DNA as an anisotropic inextensible rod and also include the effect of electrostatic repulsion for describing the DNA loops bound to Lac repressor, etc. [2, 3]. Another recent work includes sequence-dependent elastic properties of DNA [20]. All of these aforementioned works are based on Kirchhoff's thin elastic rod theory [47]. This theory, as originally formulated, deals with non-chiral elastic rods with circular cross-section. Another example is the special Cosserat theory of rods [1], which can be viewed as an extension of Kirchhoff's theory in that it includes extensible and shearable rods. Several researchers in elasticity have employed this rod theory to describe the static and dynamic characteristics of rods. For example, Simo and Vu-Quoc formulated a finite element method using rod theory [65]. Dichmann et al. employed a Hamiltonian formulation using the special Cosserat theory of rods for the purpose of describing DNA [24]. Coleman et al. reviewed dynamical equations in the theories of Kirchhoff and Clebsch [17]. Steigmann and Faulkner derived the equations of classical rod theory using parameter-dependent variational approach [67]. Recently, Gonzalez and Maddocks devised a method to extract sequence-dependent parameters for a rigid base-pair DNA model from molecular dynamics simulation [30]. In their paper, they used a force moment balance equation from Kirchhoff's rod theory to extract stiffness and inertia parameters. Another recent work includes the application of Kirchhoff rod theory to marine cable loop formation and DNA loop formation [31]. Recently, Wiggins et al. developed a theory based on nonlinear elasticity, called kinkable wormlike chain model, for describing spontaneous kinking of polymers including DNA [76].

The goal of this paper is to review a new kind of statistical treatment of semiflexible chains developed by the author that starts with the solution of diffusion equations on the Euclidean motion group. These solutions are solved using the motion-group Fourier transform, in which the probability density function of interest is expanded in terms of irreducible unitary representations (IURs). Since the motion group is noncompact and noncommutative, the IURs are infinite dimensional, which poses some interesting mathematical and computational challenges that are addressed here.

The remainder of this paper is structured as follows: Section 2 reviews the concept of frame distributions for semiflexible chains and explains how they can be computed accurately using diffusion equations for chain lengths of up to several persistence lengths. In Section 3, analytical examples based on the mathematical formulation of this paper are presented. Section 4 presents conclusions.

2. Diffusion on the motion group. In this section the relationship between the stiffness matrix and referential shape of an elastic filament and diffusion equations on the group of rigid-body motions is established. Subsection 2.2 reviews the mathematical tools required to define coordinate-free derivatives of functions of motion. Subsection 2.3 formulates the problem of finding the probability density of position and orientation of the distal end of an elastica relative to its proximal end when it is subjected to Brownian motion forcing from the environment. This is a 'phantom' model in which the effects of excluded volume are not considered. Modifications of this model that include excluded volume effects are somewhat involved and are discussed in the author's recent work [14].

2.1. Group-theoretic properties of rigid-body motion. An arbitrary rigid-body motion can be viewed as the pair $g = (\mathbf{r}, R)$ where $R \in SO(3)$ (i.e., R is a 3×3 rotation matrix), and $\mathbf{r} \in I\!R^3$ is a translation vector in three dimensional space. The composition law is $g_1 \circ g_2 = (R_1\mathbf{r}_2 + \mathbf{r}_1, R_1R_2)$ and the inverse of each element g is $g^{-1} = (-R^T\mathbf{r}, R^T)$. The action of the motion g on a position vector $\mathbf{x} \in I\!R^3$ is $g \cdot \mathbf{x} = R\mathbf{x} + \mathbf{r}$. In other words, the motion group "acts on" points $\mathbf{x} \in I\!R^3$ (which can be viewed as position vectors consisting of Cartesian coordinates) by moving them according to the rule $\mathbf{x} \to R\mathbf{x} + \mathbf{r}$. (Note the distinction that $\circ$ is used between group elements and $\cdot$ is used between a group element and a vector.) Any g describes the positional and orientational relationship between two reference frames. It is sometimes convenient to refer to the result of a rigid-body motion at a particular time as a "pose," and to refer to a function of motion as a pose distribution.

The collection of all rigid-body motions is denoted in this paper as $G = E(3)$ (The proper Euclidean motion group in three space). Any $g \in G$ can be faithfully represented with a 4×4 *homogeneous transformation matrix* of the form:

$$H(g) = \begin{pmatrix} R & \mathbf{r} \\ \mathbf{0}^T & 1 \end{pmatrix}$$

in the sense that $H(g_1 \circ g_2) = H(g_1)H(g_2)$ (i.e., the matrix product of $H(g_1)$ and $H(g_2)$). Here $\mathbf{0}^T = [0, 0, 0]$ and 1 is the number one. The structure of this bottom row is preserved under multiplication by matrices of the same kind.

Henceforth no distinction is made between G and the set of all 4×4 homogeneous transformation matrices with operation of matrix multiplication. That is, g and $H(g)$ will be used interchangeably, and since the group operator can be viewed as matrix multiplication, it does not need to be written explicitly as $\circ$.

Given a one-parameter motion $g(t)$, we can define the six-dimensional velocity of the rigid-body motion as observed in the moving frame as the nontrivial entries in the matrix

$$g^{-1}\dot{g} = \begin{pmatrix} R^T\dot{R} & R^T\dot{\mathbf{r}} \\ \mathbf{0}^T & 0 \end{pmatrix}.$$

Here t can be thought of as time, and a dot denotes differentiation with respect to t.

Since $R^T\dot{R}$ is skew symmetric as a result of R being orthogonal, it only has three independent nonzero entries. These can be extracted and used to form the dual vector $\boldsymbol{\omega}(t)$, which is the angular velocity of the moving frame as seen in the moving frame. In some contexts it will be convenient to write this as $\boldsymbol{\omega}_r(t)$ to distinguish it from the dual vector of $\dot{R}R^T$, which we will call $\boldsymbol{\omega}_l(t)$. These are related as $\boldsymbol{\omega}_l(t) = R\boldsymbol{\omega}_r(t)$.

The independent information in the matrix $g^{-1}\dot{g}$ can be extracted and put in a six-dimensional vector defined as

$$\boldsymbol{\xi}(t) = (g^{-1}\dot{g})^{\vee} = \begin{pmatrix} \boldsymbol{\omega} \\ \\ R^T\dot{\mathbf{r}} \end{pmatrix}.$$

The opposite operation of $\vee$ is:

$$\hat{\boldsymbol{\xi}}(t) = g^{-1}\dot{g} = \sum_{i=1}^{6} \xi_i \tilde{X}_i$$

where

$$\tilde{X}_1 = \begin{pmatrix} 0 & 0 & 0 & 0 \\ 0 & 0 & -1 & 0 \\ 0 & 1 & 0 & 0 \\ 0 & 0 & 0 & 0 \end{pmatrix}; \quad \tilde{X}_2 = \begin{pmatrix} 0 & 0 & 1 & 0 \\ 0 & 0 & 0 & 0 \\ -1 & 0 & 0 & 0 \\ 0 & 0 & 0 & 0 \end{pmatrix};$$

$$\tilde{X}_3 = \begin{pmatrix} 0 & -1 & 0 & 0 \\ 1 & 0 & 0 & 0 \\ 0 & 0 & 0 & 0 \\ 0 & 0 & 0 & 0 \end{pmatrix}; \quad \tilde{X}_4 = \begin{pmatrix} 0 & 0 & 0 & 1 \\ 0 & 0 & 0 & 0 \\ 0 & 0 & 0 & 0 \\ 0 & 0 & 0 & 0 \end{pmatrix};$$

$$\tilde{X}_5 = \begin{pmatrix} 0 & 0 & 0 & 0 \\ 0 & 0 & 0 & 1 \\ 0 & 0 & 0 & 0 \\ 0 & 0 & 0 & 0 \end{pmatrix}; \quad \tilde{X}_6 = \begin{pmatrix} 0 & 0 & 0 & 0 \\ 0 & 0 & 0 & 0 \\ 0 & 0 & 0 & 1 \\ 0 & 0 & 0 & 0 \end{pmatrix}.$$

These correspond to infinitesimal rotations and translations about the 1, 2, and 3 axes and form a basis for the Lie algebra associated with G. Matrix exponentiation of any weighted sum of these basis elements produces elements of G. For example,

$$\exp(\theta\tilde{X}_3 + z\tilde{X}_6) = \begin{pmatrix} R(\mathbf{e}_3, \theta) & z\mathbf{e}_3 \\ \\ \mathbf{0}^T & 1 \end{pmatrix}$$

where

$$R(\mathbf{e}_3, \theta) = \begin{pmatrix} \cos\theta & -\sin\theta & 0 \\ \sin\theta & \cos\theta & 0 \\ 0 & 0 & 1 \end{pmatrix}.$$

Furthermore, for small values of θ and z, the matrix exponential is approximated well as

$$\exp(\theta \tilde{X}_3 + z\tilde{X}_6) \approx I + \theta \tilde{X}_3 + z\tilde{X}_6.$$

In the study of Lie algebras, the matrix commutator (Lie bracket), $[\tilde{X}_i, \tilde{X}_j] = \tilde{X}_i\tilde{X}_j - \tilde{X}_j\tilde{X}_i$, plays a central role. It is well known that

$$[\tilde{X}_i, \tilde{X}_j] = \sum_{k=1}^{6} C_{ij}^k \tilde{X}_k$$

where C_{ij}^k are called the structure constants of the Lie algebra.

2.2. Differentiating functions of motion. Given a smooth function $f(g)$ where $g \in G$ is described as a 4×4 rigid-body transformation matrix, the Lie derivatives from the right and left are defined as

$$\tilde{X}_i^r f(g) = \frac{df(g \circ \exp(t\tilde{X}_i))}{dt}\Big|_{t=0} \quad \text{and} \quad \tilde{X}_i^l f(g) = \frac{df(\exp(-t\tilde{X}_i) \circ g)}{dt}\Big|_{t=0} \tag{1}$$

for the motion group where $g \in G$ is described using 4×4 transformation matrices. The 'r' in the symbol $\tilde{X}_i^r$ is used to denote the position of $\exp(t\tilde{X}_i)$ on the 'right side' of g inside the function $f(\cdot)$, and similarly for 'l'. $\tilde{X}_i^r$ commutes with left shifts of the form $(L(g_1)f)(g) = f(g_1^{-1} \circ g)$, and $\tilde{X}_i^l$ commutes with right shifts of the form $(R(g_1)f)(g) = f(g \circ g_1)$.

It can be shown that for $G = E(3)$, [10–12]

$$\tilde{X}_i^r = \begin{cases} X_i^r & \text{for } i = 1,2,3 \\ (R^T \nabla_{\mathbf{r}})_{i-3} & \text{for } i = 4,5,6 \end{cases} \tag{2}$$

and

$$\tilde{X}_i^l = \begin{cases} X_i^l + \sum_{k=1}^{3} (\mathbf{r} \times \mathbf{e}_i) \cdot \mathbf{e}_k \dfrac{\partial}{\partial r_k} & \text{for } i = 1,2,3 \\ -\dfrac{\partial}{\partial r_{i-3}} & \text{for } i = 4,5,6. \end{cases} \tag{3}$$

Here $\nabla_q f = [\partial f/\partial q_1, \partial f/\partial q_2, \partial f/\partial q_3]^T$ and $X_i^r f = (\nabla_q f) \cdot (J_r^{-1}\mathbf{e}_i)$ is the $SO(3)$ differential operator with J_r being the Jacobian that relates angular velocity as seen in the moving frame to the rate of change of the rotational parameterization q as $\boldsymbol{\omega}_r = J_r\dot{\mathbf{q}}$. An analogous relationship holds for the left operators and angular velocity as observed from the inertial reference frame.

For example, if R is parameterized with ZXZ Euler angles so that $R = R(\alpha, \beta, \gamma)$, then the X_i^r are defined as

$$
\begin{aligned}
X_1^r &= -\cot\beta\sin\gamma\frac{\partial}{\partial\gamma} + \frac{\sin\gamma}{\sin\beta}\frac{\partial}{\partial\alpha} + \cos\gamma\frac{\partial}{\partial\beta};\\
X_2^r &= -\cot\beta\cos\gamma\frac{\partial}{\partial\gamma} + \frac{\cos\gamma}{\sin\beta}\frac{\partial}{\partial\alpha} - \sin\gamma\frac{\partial}{\partial\beta};\\
X_3^r &= \frac{\partial}{\partial\gamma}
\end{aligned}
\tag{4}
$$

when rotations are parameterized using the ZXZ Euler angles (α, β, γ).

2.3. The Fokker-Planck equation for a semi-flexible chain. A nonuniform extensible elastic filament with unstretched length L has elastic energy of the form

$$E_1 = \int_0^L F(\boldsymbol{\xi}(s), s)\, ds, \tag{5}$$

where

$$F(\boldsymbol{\xi}(s), s) = \frac{1}{2}[\boldsymbol{\xi}(s) - \boldsymbol{\xi}_0(s)]^T K(s)[\boldsymbol{\xi}(s) - \boldsymbol{\xi}_0(s)]$$

where $\boldsymbol{\xi}_0(s)$ defines the minimal energy conformation. Given $\boldsymbol{\xi}_0(s)$, it is possible to integrate the matrix differential equation

$$\dot{g}_0(s) = g_0(s)\,\hat{\boldsymbol{\xi}}_0(s)$$

subject to the initial condition $g(0) = e$ (the 4×4 identity matrix) for $s \in [0, L]$ to obtain the minimal energy conformation rooted at the identity. In the case when $\boldsymbol{\xi}_0(s)$ is a constant vector, this will be a helix (with circular arcs and line segments as special cases).

Note that the independent variable is now a curve parameter, s, rather than time, t. Here the curve parameter s is taken to be the arclength of the filament in its undeformed (referential) conformation $g_0(s)$.

Consider the equilibrium statistics of a stochastically forced elastic filament. Let the evolution of the probability density of relative pose of reference frames attached to a stochastically forced elastic filament at values of curve parameter 0 and s be denoted as $f(g; 0, s)$. Since it is a probability density, by definition

$$\int_G f(g; 0, s)dg = 1. \tag{6}$$

Clearly $f(g;0,s)$ must be related in some way to the equilibrium shape of the filament, its stiffness, and the strength of the Brownian motion forcing from the ambient solvent. And the strength of this noise should be related in some way to the temperature. In fact, since $f(g;0,s)$ is the function describing the distribution of poses for a filament at equilibrium, it can be represented exactly as a path integral [12, 39, 81]:

$$f(g;0,s) = \frac{1}{Z(s)} \int_{g(0)=e}^{g(s)=g} \exp\left[-\frac{1}{k_B T} \int_0^s F(\boldsymbol{\xi}(\sigma),\sigma)d\sigma\right] \mathcal{D}(s). \quad (7)$$

Conceptually, this adds the contribution of the integrand over all possible paths $g(\sigma) \in G$ for $\sigma \in [0,s]$ that satisfy the end constraints $g(0) = e$ (the identity, or "do nothing" motion corresponding to the proximal end of the filament) and $g(s) = g$ (the pose of the frame attached to the distal end of the segment). The constant k_B is Boltzmann's constant and T is temperature measured in degrees Kelvin. The integrand normalized by the partition function, $Z(s)$, is the Maxwell-Boltzmann distribution for the filament of length s. For conformations that are highly deformed relative to the referential shape (which is the shape that a filament would take as $T \to 0$), the contribution to the path integral will be very small. Also, in this non-inertial theory, the statistical properties of any segment are independent of those of other concatenated segments that make up a longer chain.

Equation 7 does not take into account the effects of excluded volume, which can be ignored for moderate values of filament length in the case when the DNA is not enclosed in a small compartment. The mathematical machinery associated with path integrals produces an evolution equation (i.e., a partial differential equation) for $f(g;0,s)$ of the form:

$$\frac{\partial f}{\partial s} = \frac{1}{2} \sum_{k,l=1}^{6} D_{lk}(s)\, \tilde{X}_l^r \tilde{X}_k^r f - \sum_{l=1}^{6} (\boldsymbol{\xi}_0(s) \cdot \mathbf{e}_l)\, \tilde{X}_l^r f \quad (8)$$

subject to the initial conditions

$$f(g;0,0) = \delta(g).$$

Here the diffusion matrix is related to the stiffness matrix as $D(s) = k_B T K^{-1}(s)$. This equation takes into account anisotropy and inhomogeneity of the elasticity (which has been observed in, e.g., [50]), as well as arbitrary minimal energy shape, and has essentially the same derivation as the homogeneous case presented in [10–12].

Under the extreme condition that $T \to 0$, no diffusion would take place, and $f(g;,0,s) \to \delta(g_0^{-1}(s) \circ g)$. For the biologically relevant case ($T \approx 300$), (8) can be solved using the harmonic analysis approach in [10–12]. If we make the shorthand notation $f_{s_1,s_2}(g) = f(g;s_1,s_2)$, then it will always be the case for $s_1 < s < s_2$ that

$$f_{s_1,s_2}(g) = (f_{s_1,s} * f_{s,s_2})(g) = \int_G f_{s_1,s}(h) f_{s,s_2}(h^{-1} \circ g) dh. \tag{9}$$

This is the convolution of two pose distributions. Here h is a dummy variable of integration, and the explicit form of the invariant integration measure, dh, is described in detail in [12]. While (9) will always hold for semiflexible phantom chains, for the homogenous rod there is the additional convenient properties that

$$f(g; s_1, s_2) = f(g; 0, s_2 - s_1) \quad \text{and} \quad f(g; s_2, s_1) = f(g^{-1}, s_1, s_2). \tag{10}$$

The first of these says that for a uniform chain the pose distribution only depends on the difference of arclength along the chain. The second provides a relationship between the pose distribution for a uniform chain resulting from taking the frame at s_1 to be fixed at the identity and recording the poses visited by s_2, and the distribution of frames that results when s_2 is fixed at the identity. However, neither of these nor (9) will hold when excluded-volume interactions are taken into account.

As a specific example of when $f(g; s_1, s_2) = f(g; 0, s_2 - s_1)$, if the chain is uniform, inextensible and shearless, we have the constant diffusion matrix

$$D = \begin{pmatrix} D_{11} & D_{12} & D_{13} & 0 & 0 & 0 \\ D_{12} & D_{22} & D_{23} & 0 & 0 & 0 \\ D_{13} & D_{23} & D_{33} & 0 & 0 & 0 \\ 0 & 0 & 0 & 0 & 0 & 0 \\ 0 & 0 & 0 & 0 & 0 & 0 \\ 0 & 0 & 0 & 0 & 0 & 0 \end{pmatrix} \tag{11}$$

and if the minimal energy conformation is an arc-length-parameterized helix, we have the constant vector

$$\boldsymbol{\xi}_0^T = [\omega_1, \omega_2, \omega_3, 0, 0, 1]^T.$$

In this case (8) is a *degenerate diffusion* on $E(3)$ with constant coefficients. Methods for solving such equations are presented in [10–12]. These methods use the concept of the noncommutative Fourier transform for the Euclidean group. This builds on the work of Miller [52] and Vilenkin [73].

3. Solving the motion-group Fokker-Planck equation. The diffusion equation in (8) can be solved in two different ways, depending on whether $L << 1$ or $L >> 1$ when $\|D\| \approx 1$. If $L << 1$, and if exponential coordinates are used to parameterize rigid-body motions, then (8) can be solved in closed form as a Gaussian distribution. When $L >> 1$, methods of noncommutative harmonic analysis become effective. These two different approaches are described in the following two subsections.

3.1. $L << 1$ **solution as a closed-form Gaussian distribution.** Using the fact that the exponential map

$$g(\mathbf{x}) = \exp\left(\sum_{i=1}^{6} x_i \tilde{X}_i\right)$$

can be used to parameterize a ball around the identity, then for short lengths of DNA, the differential operators simply become

$$\tilde{X}_i^r \approx \frac{\partial}{\partial x_i}.$$

This is because the 6×6 Jacobian matrix for the exponential paramaterization becomes the identity matrix as $\|\mathbf{x}\| \to 0$, which is the same as $g \to e$ (the group identity).

It therefore follows that in this parameterization, (8) becomes a diffusion equation with drift in $se(3)$ (which can be identified with $\mathbb{R}^6$). When $\boldsymbol{\xi}_0(s) = \boldsymbol{\xi}_0$ and $D(s) = D$ are constant, this diffusion equation has constant coefficients, and the solution can be identified immediately as the Gaussian distribution:

$$f(g(\mathbf{x}), s) = \frac{1}{(2\pi s)^3 |\det D|^{\frac{1}{2}}} \exp\left[-\frac{1}{2}(\mathbf{x} - s\boldsymbol{\xi}_0)^T D^{-1}(\mathbf{x} - s\boldsymbol{\xi}_0)\right]. \quad (12)$$

The two limitations of this solution are: (1) it breaks down for larger values of L since as $\|\mathbf{x}\|$ increases the motion-group differential operators become complicated expressions in terms of the exponential coordinates rather than simply partial derivatives; and (2) in the case of degenerate diffusions where $|\det D| = 0$ and D^{-1} does not exist (such as in (11)), this solution cannot be used. Solutions of the form (12) have been found to be useful in Robotics problems [58].

The next subsection reviews another solution technique.

3.2. $L >> 1$ **solution using noncommutative harmonic analysis.** Noncommutative Harmonic Analysis is an extension of Fourier analysis in which functions of group-valued argument are expanded in terms of irreducible unitary representations (IURs) for the group of interest. Recall that a unitary group representation is a matrix-valued function of group-valued argument, $U(g; \lambda)$, that has the properties

$$U(g_1 \circ g_2; \lambda) = U(g_1; \lambda)U(g_2; \lambda) \qquad U(e; \lambda) = I \qquad U(g^{-1}; \lambda) = U^*(g; \lambda)$$

where $*$ denotes the Hermitian conjugate of a matrix, and λ is a parameter (that can be discrete or continuous depending on whether the group is compact or not) used to enumerate inequivalent representations. These properties are fully analogous to those of $e^{in(\theta_1+\theta_2)} = e^{in\theta_1} \cdot e^{in\theta_2}$ and $|e^{in\theta}| = 1$ (with n taking the role of λ), while reflecting the fact that for noncommutative groups $g_1 \circ g_2 \neq g_2 \circ g_1$ in general. A representation $U(g;\lambda)$ is called *irreducible* if it cannot be block diagonalized under the same similarity transformation for all values of $g \in G$. Every possible representation matrix for a group can be constructed as similarity transformed versions of direct sums of IURs.

The details of how the matrices $U(g;\lambda)$ are constructed for the Euclidean motion group are reviewed in the following subsection, together with the definition and properties of the group Fourier transform, which is the tool that is used to solve (8) for $L >> 1$ in both the nondegenerate and degenerate cases.

3.2.1. Irreducible unitary representations of the Euclidean motion group. Let $g = (\mathbf{a}, A) \in G$. The unitary representations $U^s(\mathbf{a}, A; p)$, which act on functions $\phi(\mathbf{u}) \in L^2(S^2)$ with the usual inner product are defined by [52, 12]

$$(U^s(\mathbf{a}, A; p)\phi)(\mathbf{u}) = e^{-ip\mathbf{u}\cdot\mathbf{a}}\, \Delta_s(R_{\mathbf{u}}^{-1}\, A\, R_{A^{-1}\mathbf{u}})\, \phi(A^{-1}\mathbf{u}) \ , \qquad (13)$$

where $A \in SO(3)$, $R_{\mathbf{u}}$ is the rotation matrix which converts $(0,0,1)^T$ to any $\mathbf{u} \in S^2$, and $\Delta_s(R(\mathbf{e}_3, \theta)) = e^{is\theta}$ are representations of $SO(2)$ enumerated by $s = 0, \pm 1, \pm 2, \ldots.$

Each representation $U^s(\mathbf{a}, A; p)$, characterized by $0 \leq p < \infty$ and s, is irreducible (they, however, become reducible if we restrict G to $SO(3)$, i.e. when $|\mathbf{a}| = 0$). They are unitary, because $(U^s(\mathbf{a}, A; p)\phi_1, U^s(\mathbf{a}, A; p)\phi_2) = (\phi_1, \phi_2)$. The set of all such representations is also complete. In an appropriate basis for $L^2(S^2)$, these representation operators can be expressed as infinite dimensional matrices with elements indexed as $U^s_{l',m';l,m}(\mathbf{a}, A; p)$ [52, 12].

Representations (13), which are denoted below as $U^s(g;p)$, satisfy the homomorphism properties

$$U^s(g_1 \circ g_2; p) = U^s(g_1; p) \cdot U^s(g_2; p),$$

where $\circ$ is the motion group operation and $\cdot$ denotes composition of linear operators.

3.2.2. Fourier transform, convolution, inversion and marginalization. Given a complex-valued function $f(\mathbf{a}, A)$ on G, the Fourier transform is the matrix-valued function

$$\mathcal{F}^s\{f\}(p) = \hat{f}^s(p) = \int_G f(g) U^s(g^{-1}; p)\, dg$$

where $g = (\mathbf{a}, A) \in G$, $dg = dA\, d\mathbf{a}$, and $U(g;p)$ is the unitary matrix with elements $U^s_{l',m';l,m}(\mathbf{a}, A; p)$.

The matrix elements of the transform are given in terms of matrix elements as

$$\hat{f}^{s}_{l',m';l,m}(p) = \int_G f(\mathbf{a}, A)\, \overline{U^s_{l,m;l',m'}(\mathbf{a}, A; p)}\, dA\, d\mathbf{a} \tag{14}$$

where we have used the unitary property.

An important property of this Fourier transform is that it converts convolutions on the motion group to pairwise products in Fourier space as:

$$\mathcal{F}^s\{f_1 * f_2\}(p) = \hat{f}^s_2(p)\hat{f}^s_1(p). \tag{15}$$

The inverse Fourier transform recovers $f(g)$ from the set of Fourier transforms $\hat{f}^s(p)$ as [12]:

$$f(g) = \frac{1}{2\pi^2} \sum_{s=-\infty}^{\infty} \int_0^{\infty} \mathrm{Tr}(\hat{f}^s(p) U^s(g;p))\, p^2\, dp\ . \tag{16}$$

If only the distribution of end-to-end distance is sought, this can be obtained from $f(a\mathbf{u}, A)$ by integrating over all $A \in SO(3)$ and all $\mathbf{u} \in S^2$. Due to the structure of the matrix elements of $U^s(g;p)$, it can be shown that [10, 84, 85]:

$$\frac{a^2}{2\pi^2} \int_{S^2} \int_{SO(3)} f(a\mathbf{u}, A) d\mathbf{u} dA = \frac{2}{\pi} a \int_0^{\infty} \hat{f}^0_{0,0;0,0}(p) \sin(pa) p\, dp. \tag{17}$$

3.2.3. Operational properties. By the definition of the $SE(3)$-Fourier transform $\mathcal{F}[\cdot]$ and operators $\tilde{X}^r_i$ reviewed in earlier subsections of this section, one observes that

$$\mathcal{F}[\tilde{X}^r_i f] = \int_G \frac{d}{dt}\left(f(g \circ \exp(t\tilde{X}_i))\right)|_{t=0} U^s(g^{-1};p) dg. \tag{18}$$

Here g can be thought of as $H(g)$ and $\exp(t\tilde{X}_i)$ is an element of the subgroup of G generated by $\tilde{X}_i$, which for small values of t is approximated as $I + t\tilde{X}_i$. By performing the change of variables $h = g \circ \exp(t\tilde{X}_i)$ and using the homomorphism property of the representations $U^s(\cdot)$, one finds

$$\mathcal{F}[\tilde{X}^r_i f] = \int_G f(h) \frac{d}{dt}\left(U^s(\exp(t\tilde{X}_i) \circ h^{-1}; p)\right)|_{t=0} dh \tag{19}$$

$$= \frac{d}{dt}\left(U^s(\exp(t\tilde{X}_i); p)\right)|_{t=0} \int_G f(h) U^s(h^{-1}, p) dh. \tag{20}$$

By defining

$$u^s(\tilde{X}_i; p) = \frac{d}{dt}\left(U^s(\exp(t\tilde{X}_i); p)\right)|_{t=0},$$

we write

$$\mathcal{F}[\tilde{X}^r_i f] = u^s(\tilde{X}_i; p) \hat{f}^s(p).$$

Hence, (8) can be transformed to the infinite system of linear differential equations:

$$\frac{d\hat{f}^s}{dL} = \mathcal{B}^s \hat{f}^s, \tag{21}$$

where

$$\mathcal{B}^s = \frac{1}{2} \sum_{k,l=1}^{6} D_{lk}(s) u^s(\tilde{X}_l; p) u^s(\tilde{X}_k; p) - \sum_{l=1}^{6} (\boldsymbol{\xi}_0(s) \cdot \mathbf{e}_l) u^s(\tilde{X}_l; p).$$

In the case when $D_{lk}(s)$ and $\boldsymbol{\xi}_0(s) \cdot \mathbf{e}_l$ are constant, $f(\mathbf{a}, A; L)$ is then found by simply substituting $\hat{f}^s(p; L) = \exp(L\mathcal{B}^s)$ into the $SE(3)$ Fourier inversion formula (16). For the case when parameters vary with arc length, the system of ordinary differential equations in (21) can be integrated numerically. Generally this will be efficient because $\mathcal{B}^s$ is a sparse matrix.

For numerical results obtained using this methodology, see [10, 84, 85, 15]

4. Conclusions. Double helical DNA can be described as an anisotropic elastic filament that has a helical referential configuration. When this filament is subjected to Brownian motion forcing, reference frames attached to it move relative to each other. The statistics of this motion at equilibrium is descibed fully by a diffusion (Fokker-Planck) equation on the Euclidean motion group. The Fourier transform for the motion group has useful operational properties that converts this diffusion equation to a system of ordinary differential equations in a dual space, which can be solved. This dual space solution is then brought back to real space by applying the inverse Fourier transform for the motion group.

REFERENCES

[1] ANTMAN S.S., *Nonlinear Problems of Elasticity*, Springer-Verlag, New York, 1995.

[2] BALAEFF A., MAHADEVAN L., AND SCHULTEN K., "Modeling DNA loops using the theory of elasticity," E-print archive arXiv.org (`http://arxiv.org/abs/physics/0301006`, 2003).

[3] BALAEFF A., MAHADEVAN L., AND SCHULTEN K., "Structural basis for cooperative DNA binding by CAP and Lac Repressor," *Structure*, **12**: 123–132, 2004.

[4] BAUMANN C.G., SMITH S.B., BLOOMFIELD V.A., AND BUSTAMANTE C. "Ionic Effects on the Elasticity of Single DNA Molecules," *Proceedings of the National Academy of Sciences of the USA*, **94**(12): 6185–6190, 1997.

[5] BAWENDI M.G. AND KARL F.F., "A Wiener Integral Model for Stiff Polymer Chains," *Journal of Chemical Physics* **83**(5): 2491–2496, Sep. 1, 1985.

[6] BENHAM C.J., "Elastic Model of the Large-Scale Structure of Duplex DNA," *Biopolymers*, **18**(3): 609–23, 1979.

[7] BENHAM C.J. AND MIELKE S.P.,"DNA mechanics," *Annual Review of Biomedical Engineering*, **7**: 21–53, 2005

[8] BHATTACHARJEE S.M. AND MUTHUKUMAR M., "Statistical Mechanics of Solutions of Semiflexible Chains: A Pathe Integral Formulation," *Journal of Chemical Physics*, **86**(1): 411–418, Jan. 1, 1987.

[9] BUCHIAT C., WANG M.D., ALLEMAND J.F., STRICK T., BLOCK S.M., AND CROQUETTE V., "Estimating the Persistence Length of a Worm-like Chain Molecule from Force-Extension Measurements," *Biophysical Journal*, **76**: 409–413, Jan. 1999.

[10] CHIRIKJIAN G.S. AND WANG Y.F., "Conformational Statistics of Stiff Macromolecules as Solutions to PDEs on the Rotation and Motion Groups," *Physical Review E*, **62**(1): 880–892, July 2000.

[11] CHIRIKJIAN G.S. AND KYATKIN A.B.,"An Operational Calculus for the Euclidean Motion Group with Applications in Robotics and Polymer Science," *J. Fourier Analysis and Applications*, **6**(6): 583–606, December 2000.

[12] CHIRIKJIAN G.S. AND KYATKIN A.B., *Engineering Applications of Noncommutative Harmonic Analysis*, CRC Press, Boca Raton, FL 2001.

[13] CHIRIKJIAN G.S., *Stochastic Models, Information Theory, and Lie Groups*, Birkhäuser, 2009.

[14] CHIRIKJIAN G.S., "The Stochastic Elastica and Excluded-Volume Perturbations of DNA Conformational Ensembles," unpublished manuscript, 2008.

[15] CHIRIKJIAN G.S., "Conformational Statistics of Macromolecules Using Generalized Convolution," *Computational and Theoretical Polymer Science*, **11**: 143–153, February 2001.

[16] CLUZEL P., LEBRUN A., CHRISTOPH H., LAVERY R., VIOVY J.L., CHATENAY D., AND CARON F., "DNA: An Extensible Molecule," *Science*, **271**: 792, Feb 9, 1996.

[17] COLEMAN B.D., DILL E.H., LEMBO M., LU Z., AND TOBIAS I., "On the dynamics of rods in the theory of Kirchhoff and Clebsch," *Arch. Rational Mech. Anal.*, **121**: 339–359, 1993.

[18] COLEMAN B.D., TOBIAS I., AND SWIGON D., "Theory of the influence of end conditions on self-contact in DNA loops," *J. Chem. Phys.*, **103**: 9101–9109, 1995.

[19] COLEMAN B.D., SWIGON D., AND TOBIAS I., "Elastic stability of DNA configurations. II: Supercoiled plasmides with self-contact," *Phys. Rev. E*, **61**: 759–770, 2000.

[20] COLEMAN B.D., OLSON W.K., AND SWIGON D., "Theory of sequence-dependent DNA elasticity," *J. Chem. Phys.*, **118**: 7127–7140, 2003.

[21] DANIELS H.E., 'The Statistical Theory of Stiff Chains," *Proc. Roy. Soc.* (Edinburgh), A**63**: 290–311, 1952.

[22] DES CLOIZEAUX J. AND JANNINK G., *Polymers in Solution: Their Modelling and Structure*, Clarendon Press, Oxford, 1990.

[23] DE GENNES P.G., *Scaling Concepts in Polymer Physics*, Cornell University Press, 1979.

[24] DICHMANN D.J., LI Y., AND MADDOCKS J.H., "Hamiltonian Formulations and Symmetries in Rod Mechanics," in *Mathematical Approaches to Biomolecular Structure and Dynamics*, Mesirov J.P., Schulten K., and Summers D., eds., pp. 71–113, Springer-Verlag, New York, 1995.

[25] DOI M. AND EDWARDS S.F., *The Theory of Polymer Dynamics*, Clarendon Press, Oxford, 1986.

[26] FAIN B. AND RUDNICK J., "Conformations of closed DNA," *Phys. Rev. E*, **60**: 7239–7252, 1999.

[27] FAIN B. AND RUDNICK J., Östlund, S., "Conformations of linear DNA," *Phys. Rev. E*, **55**: 7364–7368, 1997.

[28] FLORY P.J., *Statistical Mechanics of Chain Molecules*, Wiley-Interscience, New York, 1969.

[29] GOBUSH W., YAMAKAWA H., STOCKMAYER W.H., AND MAGEE W.S., "Statistical Mechanics of Wormlike Chains. I. Asymptotic Behavior," *The Journal of Chemical Physics*, **57**(7): 2839–2843, Oct. 1972.

[30] GONZALEZ O. AND MADDOCKS J.H., "Extracting parameters for base-pair level models of DNA from molecular dynamics simulations," *Theor. Chem. Acc.*, **106**: 76–82, 2001.

[31] GOYAL S., PERKINS N.C., AND LEE C.L., "Nonlinear dynamics and loop formation in Kirchhoff rods with implications to the mechanics of DNA and cables," *J. Comp. Phys.*, **209**: 371–389, 2005.

[32] GROSBERG A.YU. AND KHOKHLOV A.R., *Statistical Physics of Macromolecules*, American Institute of Physics, New York, 1994.

[33] HA B.Y. AND THIRUMALAI D., "Semiflexible Chains under Tension," *Journal of Chemical Physics*, **106**(8): 4243–4247, 1997.

[34] HAGERMAN P.J., "Analysis of the Ring-Closure Probabilities of Isotropic Wormlike Chains: Application to Duplex DNA," *Biopolymers*, **24**: 1881–1897, 1985.

[35] HERMANS J.J. AND ULLMAN R., "The Statistics of Stiff Chains, with Applications to Light Scattering," *Physica*, **18**(11): 951–971, 1952.

[36] HOROWITZ D.S. AND WANG J.C., "Torsional Rigidity of DNA and Length Dependence of the Free Energy of DNA Supercoiling," *Journal of Molecular Biology*, **173**: 75–91, 1984.

[37] KAMEIN R.D., LUBENSKY T.C., NELSON P., AND O'HERN C.S., "Direct Determination of DNA Twist-Stretch Coupling," *Europhysics Letters*, **28**(3): 237–242, Apr. 20, 1997.

[38] KHOLODENKO A.L., "Statistical Mechanics of Semiflexible Polymers: Yesterday, Today and Tomorrow," *J. Chem. Soc. Farady Trans.*, **91**(16): 2473–2482, 1995.

[39] KLEINERT H., *Path Integrals in Quantum Mechanics, Statistics and Polymer Physics*, 2nd ed., World Scientific, Singapore, 1995.

[40] KLENIN K., MERLITZ H., AND LANGOWSKI J., "A Brownian Dynamics Program for the Simulation of Linear and Circular DNA and Other Wormlike Chain Polyelectrolytes," *Biophysical Journal*, **74**: 780–788, Feb. 1998.

[41] KRATKY O. AND POROD G., "Röntgenuntersuchung Gelöster Fadenmoleküle," *Recueil des Travaux Chimiques des Pays-Bas*, **68**(12): 1106–1122, 1949.

[42] KROY K. AND FREY E., "Force-Extension Relation and Plateau Modulus for Wormlike Chains," *Physical Review Letters*, **77**(2): 306–309, 1996.

[43] LAGOWSKI J.B., NOOLANDI J., AND NICKEL B., "Stiff Chain Model – functional Integral Approach," *Journal of Chemical Physics*, **95**(2): 1266–1269, 1991.

[44] LEVENE S.D. AND CROTHERS D.M., "Ring Closure Probabilities for DNA Fragments by Monte Carlo Simulation," *J. Mol. Biol.*, **189**: 61–72, 1986.

[45] LIVERPOOL T.B. AND EDWARDS S.F., "Probability Distribution of Wormlike Polymer Loops," *Journal of Chemical Physics*, **103**(15): 6716–6719, Oct. 15, 1995.

[46] LIVERPOOL T.B., GOLESTANIAN R., AND KREMER K., "Statistical Mechanics of Double-Stranded Semiflexible Polymers," *Physical Review Letters*, **80**(2): 405–408, 1998.

[47] LOVE A.E.H., *A Treatise on the Mathematical Theory of Elasticity*, Dover, New York, 1944.

[48] MARKO J.F. AND SIGGIA E.D., "Bending and Twisting Elasticity of DNA," *Macromolecules*, 1994, **27**: 981–988.

[49] MARKO J.F., "DNA Under High Tension: Overstretching, Undertwisting, and Relaxation Dynamics," *Physical Review E*, **57**(2): 2134–2149, Feb. 1998.

[50] MAROUN R.C. AND OLSON W.K., "Base sequence effects in double-helical DNA. 2. Configurational statistics of rodlike chains," *Biopolymers*, **27**: 561–584, 1988.

[51] MATSUTANI S., "Statistical Mechanics of no-stretching elastica in three-dimensional space," *J. Geometry and Physics*, **29**: 243–259, 1999.

[52] MILLER W., "Some Applications of the Representation Theory of the Euclidean Group in Three-Space," *Commun. Pure App. Math.*, **17**: 527–540, 1964.

[53] MIYAKE A., "Stiff-Chain Statistics in Relation to the Brownian Process," *Journal of the Physical Society of Japan*, **50**(5): 1676–1682, May 1981.

[54] MOROZ J.D. AND NELSON P., "Torsional directed walks, entropic elasticity, and DNA twist stiffness," *Proceedings of the National Academy of Sciences of the USA*, **94**(26): 14418–14422, 1997.

[55] MOROZ J.D. AND NELSON P., "Entropic elasticity of twist-storing polymers," *Macromolecules*, **31**(18): 6333–6347, 1998.

[56] NORISUYE T., TSUBOI A., AND TERAMOTO A., "Remarks on Excluded-Volume Effects in Semiflexible Polymer Solutions," *Polymer Journal*, **28**(4): 357–361, 1996.

[57] ODIJK T., "Stiff Chains and Filaments under Tension," *Macromolecules*, **28**(20): 7016–7018, 1995.

[58] PARK W., LIU Y., ZHOU Y., MOSES M., AND CHIRIKJIAN G.S., "Kinematic State Estimation and Motion Planning for Stochastic Nonholonomic Systems Using the Exponential Map," *Robotica*, **26**(4): 419–434. 2008.

[59] SCHIESSEL H., RUDNICK J., BRUINSMA R., AND GELBART W.M., "Organized condensation of worm-like chains," *Europhys. Lett.*, **51**: 237–243, 2000.

[60] SCHMIDT M. AND STOCKMAYER W.H., "Quasi-Elastic Light Scattering by Semiflexible Chains," *Macromolecules*, **17**(4): 509–514, 1984.

[61] SHÄFER L., *Excluded volume effects in polymer solutions, as explained by the renormalization group*, Springer, New York : 1999.

[62] SHI Y., HE S., AND HEARST J. E., "Statistical mechanics of the extensible and shearable elastic rod and of DNA," *Journal of Chemical Physics*, **105**(2): 714–731, July 1996.

[63] SHIMADA J. AND YAMAKAWA H., "Statistical Mechanics of DNA Topoisomers," *Journal of Molecular Biology*, **184**: 319–329, 1985.

[64] SHORE D. AND BALDWIN R. L., "Energetics of DNA Twisting," *Journal of Molecular Biology*, **170**: 957–981, 1983.

[65] SIMO J.C. AND VU-QUOC L., "A three dimensional finite-strain rod model. Part II: Computational aspects," *Comput. Meth. Appl. Mech. Engr.*, **58**: 79–116, 1986.

[66] SMITH S.B., FINZI L., AND BUSTAMANTE C., "Direct Mechanical Measurements of the Elasticity of Single DNA-Molecules by Using Magnetic Beads," *Science*, **258**: 1122–1126, Nov. 13, 1992.

[67] STEIGMANN D.J. AND FAULKNER M.G., "variational theory for spatial rods," *Arch. Rational Mech. Anal.*, **133**: 1–26, 1993.

[68] STEPANOW S., "Kramer Equation as a Model for Semiflexible Polymers," *Physical Review E*, **54**(3): R2209–R2211, 1996.

[69] STRICK T.R., ALLEMAND J.F., BENSIMON D., BENSIMON A., AND CROQUETTE V., "The Elasticity of a Single Supercoiled DNA Molecule," *Science*, **271**: 1835–1837, Mar. 29, 1996.

[70] SWIGON D., COLEMAN B.D., AND TOBIAS I., "The elastic rod model for DNA and its application to the tertiary structure of DNA minicircles in Mononucleosomes," *Biophys. J.*, **74**: 2515–2530, 1998.

[71] THIRUMALAI D. AND HA B.-Y., "Statistical Mechanics of Semiflexible Chains: A Mean Field Variational Approach," pp. 1–35 in *Theoretical and Mathematical Models in Polymer Research*, A. Grosberg, ed., Academic Press, 1998.

[72] TOBIAS I., SWIGON D., AND COLEMAN B.D., "Elastic stability of DNA configurations. I: General theory," *Phys. Rev. E*, **61**: 747–758, 2000.

[73] VILENKIN N.J., AKIM E.L., AND LEVIN A.A., *The Matrix Elements of Irreducible Unitary Representations of the Group of Euclidean Three-Dimensional Space Motions and Their Properties*, Dokl. Akad. Nauk SSSR **112**: 987–989, 1957 (in Russian); also Vilenkin N.J and Klimyk A.U., *Representation of Lie Groups and Special Functions*, Vols. 1–3, Kluwer Academic Publ., Dordrecht, Holland 1991.

[74] VOLOGODSKII A.V., ANSHELEVICH V.V., LUKASHIN A.V., AND FRANK-KAMENETSKII M.D., "Statistical mechanics of supercoils and the torsional stiffness of the DNA double helix," *Nature*, **280**: 294–298, July 1979.

[75] WANG M.D., YIN H., LANDICK R., GELLES J., AND BLOCK. S.M., "Stretching DNA with Optical Tweezers," *Biophysical Journal*, **72**: 1335–1346, Mar. 1997.

[76] WIGGINS P.A., PHILLIPS R., AND NELSON P.C., "Exact theory of kinkable elastic polymers," E-print archive arXiv.org (arXiv:cond-mat/0409003 v1, Aug. 31, 2004).

[77] WILHELM J. AND FREY E., "Radial Distribution Function of Semiflexible Polymers," *Physical Review Letters*, **77**(12): 2581–2584, Sept. 16, 1996.

[78] WINKLER R.G., HARNAU L., AND REINEKER P., "Distribution functions and dynamical properties of stiff macromolecules," *Macromolecular Theory and Simulation*, **6**: 1007–1035, 1997.

[79] WINKLER R.G., "Analytical Calculation of the Relaxation Dynamics of Partially Stretched Flexible Chain Molecules: Necessity of a Wormlike Chain Descrption," *Physical Review Letters*, **82**(9): 1843–1846, 1999.

[80] YAMAKAWA H. AND STOCKMAYER W.H., "Statistical Mechanics of Wormlike Chains. II. Excluded Volume Effects", *Journal of Chemical Physics*, **57**(7): 2843–2854, October 1, 1972.

[81] YAMAKAWA H., *Helical Wormlike Chains in Polymer Solutions*, Springer, 1997.

[82] ZANDI R. AND RUDNICK J., "Constrainst, histones, and 30-nm spiral," *Phys. Rev. E*, **64**, Art. No. 051918, 2001.

[83] ZHAO S.R., SUN C.P., AND ZHANG W.X., "Statistics of wormlike chains. I. Properties of a Single Chain", *Journal of Chemical Physics*, **106**(6): 2520–2529, February 8, 1997.

[84] ZHOU Y. AND CHIRIKJIAN G.S., "Conformational statistics of bent semiflexible polymers," *Journal of Chemical Physics*, **119**(9): 4962–4970, Sept. 1, 2003.

[85] ZHOU Y. AND CHIRIKJIAN G.S., "Conformational Statistics of Semi-Flexible Macromolecular Chains with Internal Joints," *Macromolecules*, **39**(5): 1950–1960, 2006.

PERSPECTIVES ON DNA LOOPING

LAURA FINZI*

1. Introduction. DNA looping is a ubiquitous regulatory mechanism which can be involved in DNA transcription, recombination, repair, etc. Here, I will focus on protein-mediated DNA looping as a mechanism of transcriptional regulation. Indeed, such topological change in DNA is known to repress and/or activate many prokaryotic and viral genes [1–4] and is believed to mediate interaction between promoters and enhancers as well as insulate them in eukaryotes [5–9].

Protein-mediated loop formation involves DNA-protein and protein-protein interactions. While, some biochemical and biophysical parameters that govern DNA-protein and protein-protein interactions (for example, salt, temperature and DNA bending elasticity) have been studied for a long time and are fairly well understood, other parameters that affect these interactions are not as well understood. In particular, the effect of: (i) the DNA scaffold on protein-protein interactions, (ii) protein non-specific binding to DNA, (iii) DNA torsional stress, (iv) tension on DNA are poorly characterized despite their essential role in facilitating or impeding loop formation.

The influence of these parameters on loop formation and, more generally, on DNA-bound protein-protein interactions has not been addressed experimentally for lack of suitable approaches. For the past 15 years or so, single-molecule approaches have developed to become reliable and powerful methods to investigate the mechanistic features of protein-mediated loop formation and breakdown. By considering one molecule at the time, single-molecule approaches avoid ensemble averaging and reveal transient intermediates and infrequent heterogeneous behavior that go undetected in bulk experiments.

Thus, in this article, I will concentrate on the insight gained on transcriptionally-relevant DNA looping mechanisms by single-molecule approaches. However, I would like to emphasize that there is no one perfect, all-problem solving technique, and that the most detailed understanding is always achieved by combining information from various approaches.

2. Overview, results, and discussion. In the following I will use three well known prokaryotic repressors of transcription to exemplify the kind of mechanistic problems into which single-molecule techniques have provided much insight and also to show the areas where the application of these techniques can still be refined.

*Department of Physics, Emory University, Atlanta, GA 30322-2430 (lfinzi@physics.emory.edu).

C.J. Benham et al. (eds.), *Mathematics of DNA Structure, Function and Interactions*,
The IMA Volumes in Mathematics and its Applications 150,
DOI 10.1007/978-1-4419-0670-0_4,

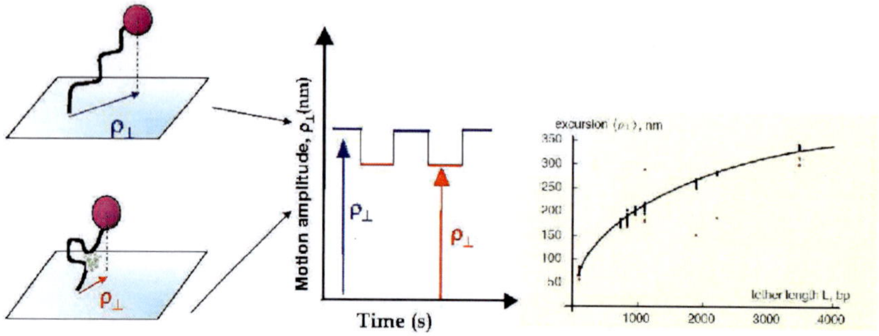

FIG. 1. *Schematic of a TPM experimental setup. The amplitude of the Brownian motion of the bead depends on the length of the DNA (black, curved line). If a protein (green complex) induces looping, the bead's tether shortens and reduces the bead's amplitude of motion (excursion). When the DNA molecule changes conformation, due to protein association and dissociation (left), the amplitude of the Brownian motion, ρ, will fluctuate, in time, between two levels, resembling a telegraphic signal (center). A calibration curve was obtained using DNA tethers of different length [12].*

***Lac repressor* and the: (i) real-time detection of small loop formation and breakdown, (ii) determination of kinetic and thermodymanic parameters of simple two-state systems and (iii) characterization of structural features that affect loop formation.**

(i) Real-time detection of small loop formation and breakdown. In a first, ground-breaking study, the tethered particle motion (TPM) technique was used to detect the dynamic loop formation and breakdown mediated by the lac repressor [10]. TPM is a simple, elegant technique that consists of observing through an optical microscope the thermal (Brownian) motion of a small bead tethered to the glass surface of a microscope flow-chamber by a single DNA molecule. The DNA tether is invisible, but the range of Brownian motion of the bead depends on its tether length. Thus protein-induced DNA conformational changes, such as looping, are revealed as changes in the Brownian motion of the bead over time and, in the simplest case, give rise to a telegraph-like signal (Figure 1) [11, 12].

The lac repressor protein efficiently regulates transcription of the genes responsible for the metabolism of lactose in *E. coli* by binding to three specific binding sites (operators): O_1, O_2, O_3 [13]. O_3 is 90 bp upstream of O_1, while O_2 is 402 bp downstream of O_1. In the presence of repressor, DNA containing two copies of the O_1 operator separated by 305 bp, gave a telegraph-like TPM signal which suggested the presence of two DNA configurations. The histogram displaying the frequency distribution of the TPM signal showed two peaks: one at larger and one at smaller amplitudes of Brownian motion (Figure 2H in [10]). This further supported the idea of two states in dynamic equilibrium: the looped and unlooped configurations.

It was therefore shown that TPM could be used to verify the DNA looping activity of a transcriptional regulator.

(ii) Kinetic and thermodymanic parameters. Establishing a threshold between the peaks of a bimodal histogram such as the one described above (see also Figure 5), it is possible to calculate the percentage of overall time spent in the looped and unlooped state, respectively. The ratio between these two percentages is related to thermodynamic parameters. TPM traces may also provide the rate constants for loop formation and breakdown. The histograms of the measured dwell times for each configuration can be fitted to this end. If a single or double exponential fit satisfactorily the histogram of the looped state dwell times, for example, the rate constant for the loop breakdown process is easily extracted (Figure 3 in [10]).

(iii) Structural features that affect loop formation. The simple TPM measurements are amenable to many implementations and to the characterization of many different parameters that can modulate transcriptional regulation via DNA loop formation. For example, one can study how (i) loop length or (ii) protein-induced DNA bends or (iii) intrinsic curvature in DNA affect the probability of loop formation when introduced at different positions within the loop or nearby the specific binding sites that secure it. Studies in these directions were presented recently by Zurla *et al.* [14] and by Goyal *et al.* [15]. In the simple case of the lac repressor protein, which is a dimer of dimers, each with a surface of interaction with DNA, TPM was used to study the effect of the hinge domain flexibility on loop formation using protein mutants [16]. In the case of other transcriptional factors, mutant proteins may be used to characterize the role of different amino acids in the protein-protein interaction surface that mediates loop closure, while mutations in the different surfaces of the proteins involved in loop closure can help understanding the role of different operators.

Atomic force microscopy (AFM) is an imaging technique where DNA and proteins can be deposited on a surface and imaged dry or in solution. Dry samples are used to visualize with high definition DNA-protein complexes and their configuration. Solution imaging is best for the characterization of dynamic behaviors but is more challenging. AFM images provide data complementary to TPM investigations since they help visualize and characterize the geometry, structure and, sometimes, stoichiometry [17, 18] of protein-mediated DNA loops. AFM imaging can further be used to characterize the relative occurrence of alternative structures and the effect on them of changes in the flexibility of DNA. The latter can indeed be modulated quite precisely by changing the density of positive charges on the deposition surface for an AFM sample [19].

***Gal repressor* and the: (i) effect of DNA supercoiling and (ii) alternate loop trajectories.**

(i) Effcct of DNA supercoiling. The gal repressor regulates transcription of the genes responsible for the metabolism of galactose in *E. coli.* It

does so by regulating initiation from the two promoters, $P1$ and $P2$, of the *gal* operon. It binds to operators, O_E and O_I, which encompass the promoters and are separated by 113 bp [20]. In this case, repression also requires DNA to be supercoiled and the presence of the nucleoid-associated protein, HU [21]. GalR belongs to the same family of proteins, has high sequence similarity and supposedly similar crystal structure to the lac repressor [22], therefore it was proposed that repression was, like in the case of its relative, the consequence of a DNA loop arising from the interaction between two operator-bound gal repressors which inactivates the promoters [23]. In 2003, two reports were published which demonstrated GalR/HU-mediated loop formation using single-molecule approaches. The first one succeeded in visualizing loops in supercoiled DNA minicircles containing O_E and O_I at a distance greater than the wild-type (wt) using AFM [24]. The authors were able to show that GalR/HU-mediated loops adopt an antiparallel and not a parallel geometry confirming theoretical predictions [25]. The second, implemented magnetic tweezers to monitor wt loop formation and breakdown in a supercoiled, linear fragment of DNA in solution [26]. Magnetic tweezers are an extension of TPM which allows manipulation of single DNA molecules. Most commonly, they consist of a pair of magnets placed on a mount above the microscope stage that can be both translated along or rotated around the optical axis of the microscope. The magnetic field generated by the magnets can both attract and rotationally trap a DNA-tethered paramagnetic bead. This results in stretching of the DNA tether, that can also be positively or negatively supercoiled by rotating the magnets [27, 28]. This work also yielded details about the mechanism of loop formation clarifying the role of DNA supercoiling and of the HU co-factor. The authors showed that negative supercoiling facilitates HU binding at the apex of the loop and that HU lowers the energy barrier for GalR-GalR dimer interaction and loop closure. In brief, it was shown that (i) looping occurs only in DNA substrates that have been negatively supercoiled beyond a given threshold and not in relaxed or positively supercoiled DNA, (ii) negative supercoiling denatures the preferred HU binding site at the apex of the loop and thus facilitates HU binding (HU binds preferentially to ssDNA), (iii) HU lowers the free energy for loop formation and stabilizes the GalR-mediated loop [26] by juxtaposing the two GalR operators as a consequence of its DNA bending.

It is interesting to note that, in the GalR case, DNA supercoiling does not facilitate loop formation by reducing the loop size and therefore reducing the penalty of entropy loss, but by changing the flexibility of DNA locally (ssDNA is more flexible than dsDNA) and by facilitating the binding of a co-factor (HU) which directly facilitates loop formation. This may be the case for several of the regulatory loops with size below or around the persistence length of DNA where the entropy loss due to loop formation is not so significant, while overcoming the bending rigidity of DNA is energetically very costly.

(ii) Alternate loop trajectories. A loop of DNA might, in principle, follow either a parallel or an anti-parallel trajectory [29], and the particular trajectory can be influenced by symmetry requirements in the protein-DNA and protein-protein interactions, flexibility of protein-protein interfaces, binding of architectural proteins, and length of the intervening DNA. In some systems, such as that of the gal operon, the scheme of protein-DNA and protein-protein interactions would allow more than one parallel and anti-parallel geometry [25, 29, 30]. Experimental [24, 26] and theoretical [25] analysis of the GalR/HU-mediated loop suggested that only loops with antiparallel geometry are formed, but the two alternate antiparallel trajectories (A1 and A2) could not be distinguished experimentally. However, recently only A1 was found to repress *gal* transcription [31]. Magnetic tweezers were used to test the two explanations proposed for this observation: (i) the A2 loop is thermodynamically unfavored such that it either does not form or forms with such thermodynamics and/or kinetics that it fails to repress transcription or (ii) the A2 and A1 loops are geometrically/topologically different (Lia *et al.*, submitted). It was found that *gal* DNA loops with A1 and A2 trajectories form with similar energies and probabilities, when DNA is underwound similarly to the situation found *in vivo* and in *in vitro* transcriptional assays. However, the DNA shortening due to loop formation (Δl, in Figure 2) was observed to be different for A1 and A2 at low values of force and its dependence on the tension applied to the molecule is shown in Figure 3. As one would expect, the amplitude of this shortening decreased under higher tension for all conditions. However, as the tension was increased, Δl decreased gradually in A2 DNA, while decreased abruptly at a tension around 0.9 pN for both heterodimer/HU-mediated and wt-GalR/HU-mediated A1 loops. Although, the force determination is accurate to only 10 percent, this abrupt transition is significant. Furthermore, at low forces, Δl associated with A1 loop formation is larger than that associated with the A2 loop by about 10 nm. Only at the highest applied force, the two loops give rise to a similar Δl, compatible with the expected length of the GalR-induced loop. The difference in Δl associated with A1 vs. A2 loop formation may indicate that formation of the A1 loop is accompanied by more compensatory negative writhe in the surrounding DNA segments. In a torsionally constrained DNA molecule the linking number is constant [32]. Thus, any change in the twist of one segment of the DNA due to the interaction with proteins (here GalR and HU) will generate an opposite change in the remaining segments. Such twist compensation has been observed in similar single molecule experiments for DNA operated on by the RSC chromatin remodeling factor which produces negatively supercoiled loops [33]. At higher forces, the writhe in the DNA is eliminated as DNA denatures and the Δl due to A1 loop formation shrinks to the value found for A2. This high force Δl should most closely correspond to loop formation alone. Since the Δl associated with A2 loop formation does not have the same force dependence, the A1 and A2 trajec-

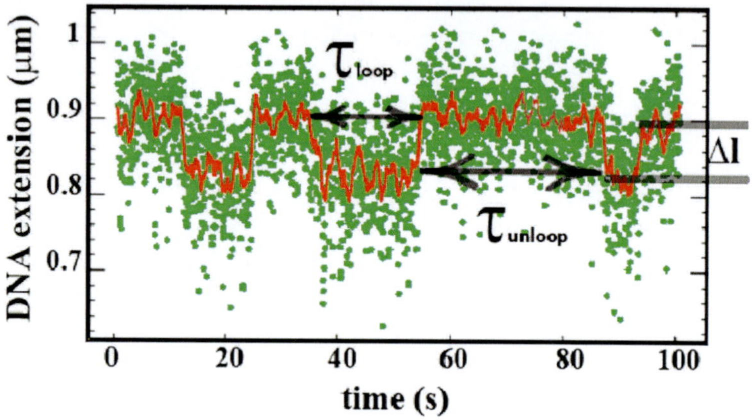

FIG. 2. *A typical telegraph-like signal observed in the A1 DNA molecule unwound by 3% (-0.03σ) and pulled with a force of 0.9 pN . The green dots are raw data and the red line is the averaged signal (1s). From the trace is possible to measure the transition time (τ_{loop} and τ_{unloop}) between the looped and unlooped state, as well the amplitude of the transition due to loop formation.*

tories may be more significantly different (i.e. topologically) in supercoiled DNA than previously indicated in relaxed DNA [25].

Compared to these two trajectories, wt GalR/HU-induced looping of wt DNA at two different values of forces [26], is most similar to that observed for A1 DNA acted upon by heterodimer GalR (Figure 3). This suggests that the wild-type repressosome most frequently involves loops with A1 trajectories.

Although, measurements should be performed on a wider range of forces, it is intriguing to think that the observations described above may point to a difference in torsion and its compensation as one of the discriminating factors between a transcriptionally active and inactive loop. In particular, one can envision using differences in compensatory writhe, like the ones detected by magnetic tweezers, as an indicator of topological differences among loops of the same size and with otherwise similar characteristics, but functionally different.

Lambda repressor **and the: (i) formation of large, multi-protein-mediated loops, (ii) effect of non-specific binding, (iii) effect of tension and supercoiling in large loops.**

(i) Formation of large, multi-protein-mediated loops. The lambda repressor or CI protein is both an activator and a repressor of transcription and is required for the maintenance of the lysogenic state in *E. coli* after infection by bacteriophage lambda [34]. During lysogeny, dimers of CI bind

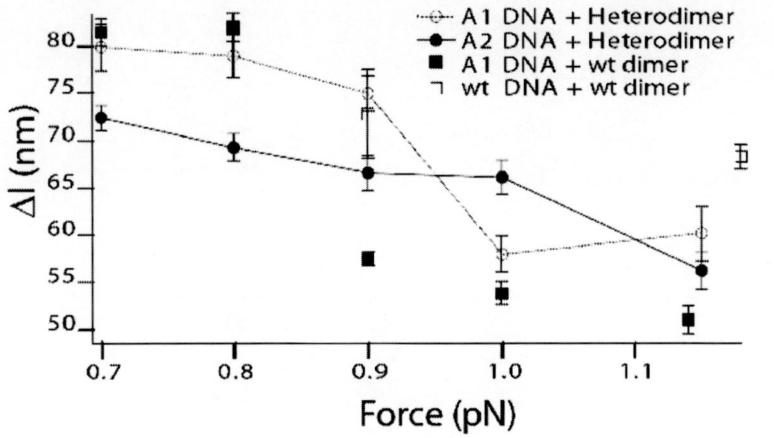

FIG. 3. *Force dependence of the change in DNA length (Δl) due to loop formation. Data referring to loop formation induced between the wt GalR operators by the wt GalR protein were included from a previous publication [26].*

to specific sites within the O_L and O_R control regions, located about 2,3 kbp apart on the phage genome (Figure 4A). Each control region contains three binding sites for CI, O_L1, O_L2, O_L3 and O_R1, O_R2, O_R3 [35–37]. Lambda CI has been a paradigm for the concept of cooperativity in gene regulation. First, CI binds to its operator sites as a dimer with an intrinsic affinity $O_L1 > O_R1 > O_L3 > O_L2 > O_R2 > O_R3$ [38, 39]. Pairs of dimers can interact when bound to adjacent or nearby operators (Figure 4A) forming tetramers. These cooperative interactions improve the specificity and strength of CI binding to O_R1 and O_R2, and O_L1 and O_L2, respectively ($O_{R/L}1 \sim O_{R/L}2 > O_{R/L}3$). Occupancy of O_R2 by CI also activates transcription of the CI gene from the P_{RM} promoter, giving rise to a positive CI auto-regulatory mechanism [40–42] (Figure 4A). This is needed to maintain the level of CI required for repression of the lytic genes, as described above. Finally, using the full λ regulatory region, Dodd and collaborators showed that CI is able to repress its own transcription from P_{RM} by binding to O_R3 at lysogenic concentrations [43]. This ensures that excessive production of CI does not interfere with efficient switching to the lytic stage, when necessary. Thus, they suggested that tetramers of CI, bound to $O_L1 - O_L2$ and $O_R1 - O_R2$, interact to form an octamer causing the intervening DNA sequence to loop (Figure 4B) [35]. This higher–order DNA structure juxtaposes O_L3 and O_R3 and, due to a local increase in DNA concentration and (again!) cooperativity [1], ensures CI occupancy of the weak O_R3 operator which results in the repression of P_{RM}. This model implies interaction

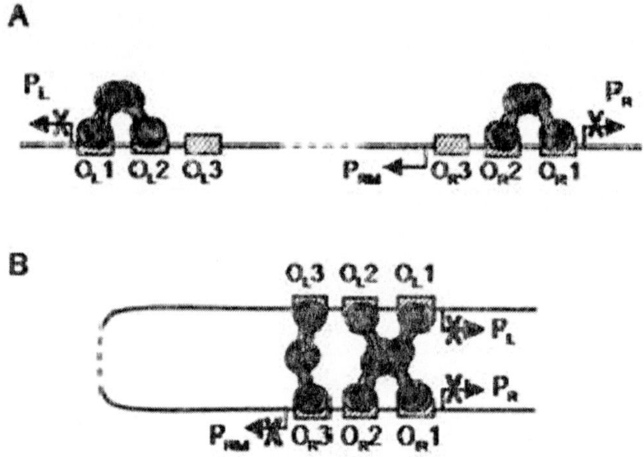

FIG. 4. *Model of CI regulation by long-range DNA looping proposed by Dodd & collaborators [35].*
A) CI dimers bound cooperatively at oR1 and oR2 repress transcription at pR while the dimer at oR2 also activates transcription from pRM. The dimers bound cooperatively at oL1 and oL2 repress transcription at pL.
B) Tetramers of CI bound at oL and oR interact forming an octameric complex and a 2.4 kbp DNA loop. This higher-order complex facilitates cooperative binding of another pair of CI dimers at oL3 and oR3, resulting in formation of another CI tetramer and repression of transcription from P_{RM}.

between the CI dimers bound at O_L3 and O_R3 and explains how CI can negatively auto-regulate its own expression when present at physiological concentrations. Using TPM, direct evidence for loop formation by CI was obtained [44]. However, in this case, given the presence of six specific operators, a large number of nucleoprotein complexes are possible increasing the complexity of the system. The superimposed frequency histograms of many TPM measurements clearly indicate that there is only one size of loop, at least given the spatial resolution of this kind of measurement (Figure 5). How then can one understand the mechanism of loop formation and test, for example, if the CI octamer precedes the CI dodecamers-mediated loop or if there is another mechanism of DNA-CI and CI-CI interactions that regulates the formation and breakdown of the loop? Answering these questions has entailed the application and refinement of several techniques of analysis. Comparison of the results obtained from each of these different approaches has further deepened our understanding of the interaction between CI and DNA.

First, we used a thermodynamic approach (C. Zurla *et al.*, in preparation). As mentioned above, from the TPM traces, one can measure the

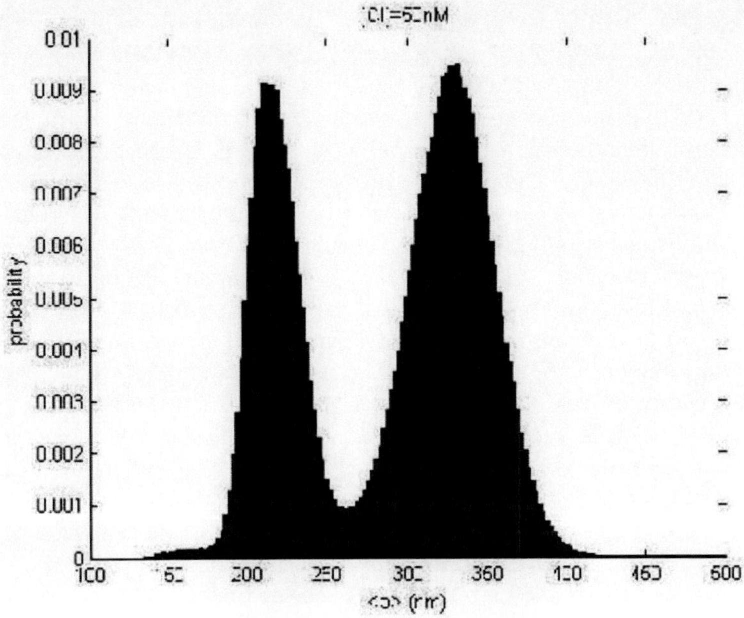

FIG. 5. *Probability distribution of the TPM signal observed for several tethered beads (about 40 for each condition) in the presence of 50 nM CI (nominal concentration).*

overall time spent in the looped state and in the unlooped state. The ratio between these two values (D_L/D_U) is equivalent to the relative probability of occurrence of the two kinds of configurations and we should consider that there are 64 possible unlooped and 49 looped configurations (Figure 6),

$$\frac{D_L}{D_U} = \frac{\sum_{i=1}^{49} p_{\text{looped},i}}{\sum_{j=1}^{64} p_{\text{unlooped},j}} \tag{1}$$

where $p_{\text{looped},i}$ and $p_{\text{unlooped},j}$ are the probability of occurrence of each looped or unlooped species, respectively. Also, from Equation 1, we know that this probability is related to the probability of occurrence of each looped/unlooped configuration. Such probability is given by:

$$p_i = \frac{[CI_2]^{s_i} e^{-\frac{\Delta G_i}{RT}}}{\sum_{j=1}^{113} [CI_2]^{s_j} e^{\frac{-\Delta G_j}{RT}}} \tag{2}$$

where $[CI_2]$ is the concentration of dimeric CI, s is the number of CI dimers in configuration i, and ΔG_i is the free energy for that particular configuration. This free energy is equal to the sum of three possible contributions: the free energy of binding to a specific operator, the free energy of cooperative interaction between adjacent CI dimers, and the looping free energy that incorporates DNA bending, configurational entropy, and protein-protein interactions. The values for the free energy of binding at each of the different six CI operators and those of cooperative binding are known [38, 39] and summarized in Table 1 in ref. [35]. On the other hand, the free energies of each specific kind of looping are unknown. However, we measured values for D_L/D_U at different CI concentrations (Figure 7) which, as expected, increase with CI concentration. Therefore, we can develop an expression in terms of CI concentration [35] and free energy, leaving looping free energies as fit parameters. Global fitting of the data by means of Equation 1 yields $\Delta G_{\text{oct}} = -0.6$ kcal/mol and $\Delta G_{\text{dod}} = -1.5$ kcal/mol for the free energy for loop formation mediated by CI octamers and CI dodecamers, respectively. The difference in ΔG between octamer- or dodecamer-mediated loop is not very large. This observation may suggest that, as in the case of many proteins which fluctuate between the folded and unfolded state at room temperature, the looped state mediated by CI fluctuates between the octamer and the dodecamer-mediated looped species. To test this hypothesis, C. Manzo *et al.* followed a kinetic approach which consists in using a multiscale analysis [45] after using a generalized likelihood ratio test that determines the location of a signal amplitude change point (cp) [46] and thus eliminates the limitations that arise from binning and thresholding (manuscript in preparation). In general, the complexity of the system demands the development of more sophisticated analyses than those that have been devised for two-state systems [10, 16].

(ii) Effect of non-specific binding. Another complicating feature of lambda CI is that, on the contrary of the gal or lac repressors, it displays a significant non-specific binding to DNA (Figure 8). Clearly, this form of interaction with DNA may affect CI-mediated loop formation and breakdown both *in vitro* and *in vivo* where CI has been estimated to be present at 30 dimers per prophage [37]. Single-molecule approaches can help in quantifying the effect of non-specific binding and we are presently conducting measurements similar to those described by Marko and Siggia [47] to characterize CI non-specific binding to lambda DNA. In general, non-specific binding is an important, common aspect of protein-DNA interactions which is seldom addressed properly and can be characterized in detail using single-molecule approaches.

(iii) Effect of tension and supercoiling in large loops. Magnetic tweezers are the tool of choice for investigating the effect of DNA tension and supercoiling on protein-mediated loop formation. Magnetic tweezers were used to manipulate the supercoiling level of the λ regulatory region. Detec-

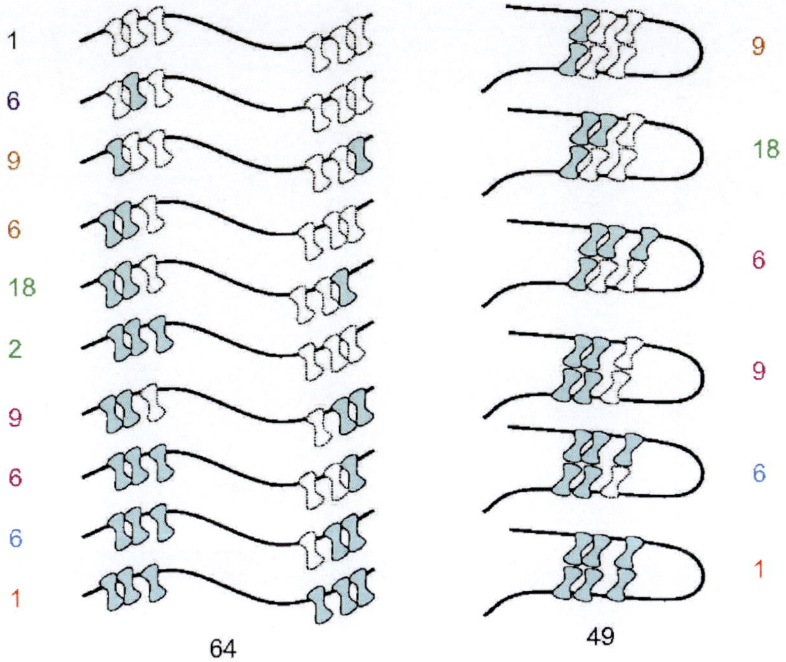

FIG. 6. *Pictorial representation of the 64 unlooped and 49 looped DNA configurations which may arise from interaction of CI with its specific binding sites. The numbers on the sides indicate the multiplicity of each configuration. The cyan shapes indicate CI dimers bound to lambda operators; the shapes with dotted contours indicate vacant positions for CI dimers.*

tion of CI-mediated looping in the magnetic tweezers set-up has proved to be challenging given the large size of the loop and the presence of tension in the DNA which interferes with the looping event. Indeed, at high forces, the DNA is mostly stretched and loops rarely form. Instead, at very low forces, DNA is basically a random coil and formation of the loop does not yield a detectable reduction in DNA length.

CI-induced wt-length loop formation in supercoiled DNA was observed nonetheless using a different approach, pioneered by Strick *et al.* [48]. DNA was supercoiled at low forces and then subjected to a sudden force jump. At low forces, twisting the DNA generates plectonemes and perhaps looping too. A sudden increase in force instantly eliminates the plectonemes extending all but the part of the DNA involved in the CI-mediated loop. A second abrupt lengthening of the DNA occurs when the CI-secured loop ruptures. The duration of the loop can thus be measured until it breaks down. The measurements can be repeated several times by lowering the

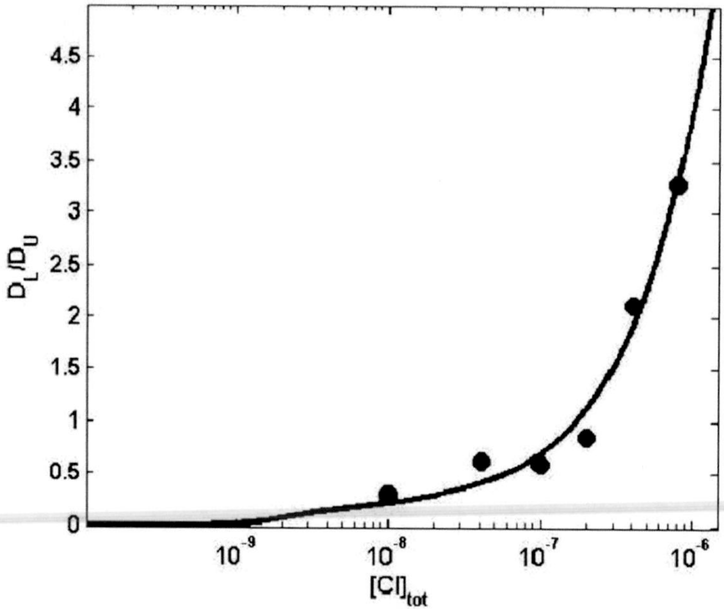

FIG. 7. *Experimental values for the relative probability of loop formation D_L/D_U obtained at different CI concentrations and global, thermodynamic fitting of these points obtained as described in the text.*

force and repeating the cycle. Figure 9 shows this type of measurement performed on a DNA molecule ~11,000 bp-long which had been underwound by 20 turns or ~ 2.4%.

These results clearly show that it will be possible to determine the lifetime of the wt loop state. These assays will not yield information on the unlooped state lifetime. Nonetheless, by introducing different levels of DNA twisting, these assays will allow characterization of the effect of DNA supercoiling on the loop lifetime, and for example, testing of the hypothesis that supercoiling may increase the local concentration of CI and facilitate loop formation at lower CI concentrations than those observed in relaxed DNA.

In order to obtain a telegraph-like signal from which to derive both the lifetime of the looped and unlooped configuration, the loop size should be shortened with respect to that of the wt. The left panel of Figure 10 shows the typical dependence of DNA length vs. number of turns of the magnets for a molecule 4900 bp-long where the O_L and O_R regions are separated by 393 bp only. The lower panel on the right of Figure 10 shows a typical

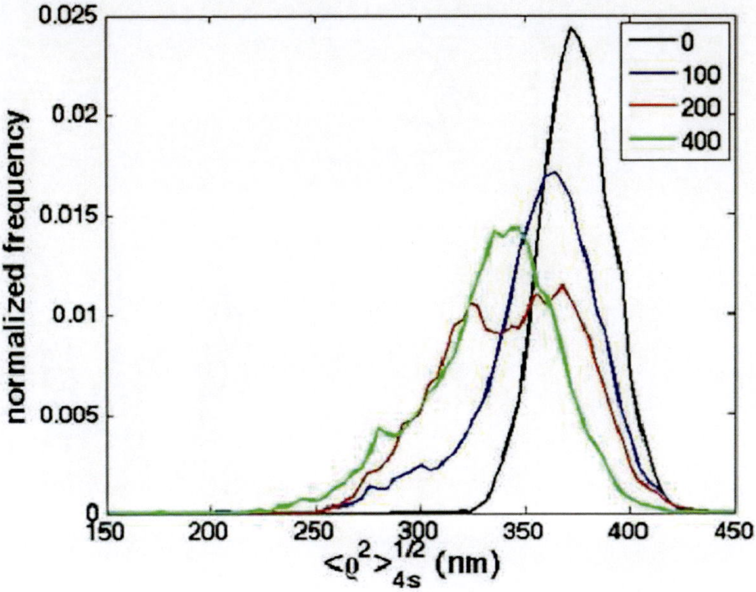

FIG. 8. *Frequency distribution histograms of the TPM signal measured for null DNA (DNA not containing any specific site for CI) at various concentrations of CI. Histograms are not filled to better reveal the shape and overlapping of distributions in different experimental conditions.*

trace obtained when CI is added to molecules unwound by 10 turns while subjected to a force of 0.3 pN. One can then characterize the effect of different values of DNA supercoiling and tension on the formation of loops of different size and eventually extrapolate the wt behavior. Initial results reported in Figure 11 show that a small increase in force ($\sim$ 0.1 pN) leads to a significant reduction of looping probability and an increase of negative supercoiling from $\sigma = -0.012$ to $\sigma = -0.024$ dramatically decreases the probability of the 393 bp-long loop formation. The observation that tension and unwinding do not favor loop formation is in agreement with theoretical predictions by Vologodskii [49, 50].

A detailed characterization of how DNA supercoiling and tension affect loop stability as a function of the size of the loop in the case of the lambda CI-mediated loop will also help predict the behavior of other loop-based regulatory systems and therefore will be of great general significance.

3. Conclusion. Single-molecule approaches have been applied to the characterization of several aspects of molecular biophysics. Here, I have briefly reviewed some applications to the characterization of the molecu-

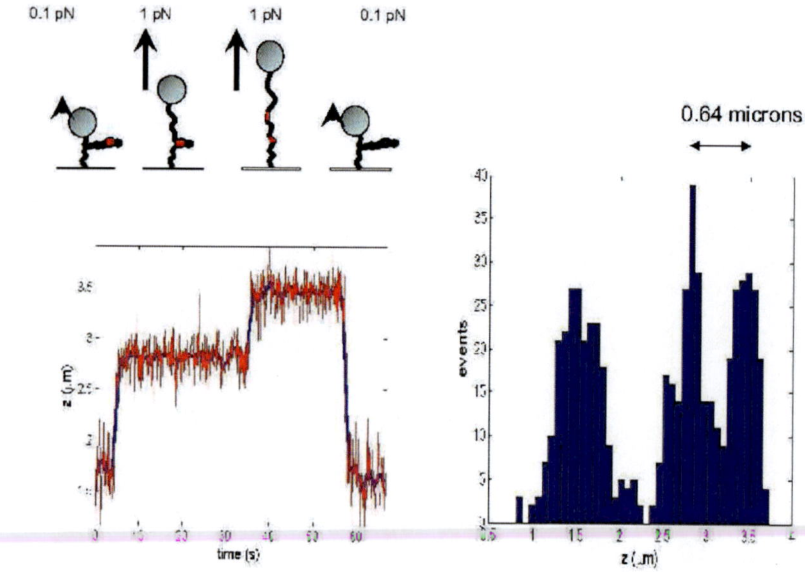

FIG. 9. *CI-mediated looping in supercoiled DNA. DNA end-to-end distance vs time (left panel). DNA end-to-end distance histogram (right panel). DNA molecules were unwound by approximately 3% in the presence of an applied force of 0.3 pN. The end-to-end distance of the DNA at this point is given by the low trace and by the leftmost peak in the histogram. The force is then suddenly increased to 1 pN which removes all plectonemes causing a rapid increase in DNA length. DNA is not yet completely extended because of the loop. After sometime, loop breakdown further increases DNA to its full extension (second plateau and rightmost peak. The distance between the rightmost and middle peaks corresponds to the length of the loop.*

lar mechanisms that regulate transcription, with particular emphasis on protein-induced DNA looping. Single-molecule techniques can be successfully employed to detect formation of small (hundreds of bp) as well as large (kbp) regulatory loops, to characterize the kinetics, thermodynamics and the mechanism of loop formation in exquisite detail, to determine the effect of structural and topological features as well as that of biochemical interactions and DNA tension. Finally they are the best suited to provide quantitative information pertinent to theoretical predictions of either a general or system-specific character advanced by several investigators, some of whom are contributing to this volume [25, 30, 51–66].

Acknowledgments. I am grateful to the talented past and present members of my lab for their contributions to the work presented here and to David Dunlap, Sankar Adhya, Phil Nelson and David Bensimon for their collaboration.

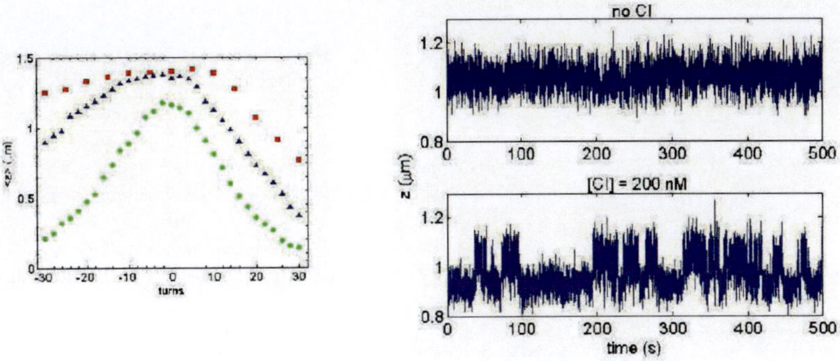

FIG. 10. *Left: Dependence of DNA length on number of turns of the magnet at three different forces for a DNA fragment containing the complete oL and oR regions separated by 393 bp. The overall fragment length was 4396 bp. Force is: 0.3 pN (green curve), 0.7 pN (blue curve) and 1 pN (red curve).*
Right: Dependence of DNA length versus time for the same DNA molecule stretched by 0.3 pN and negatively twisted by 10 turns. Top: control experiment in the absence of CI. Bottom: in the presence of 200 nM CI.

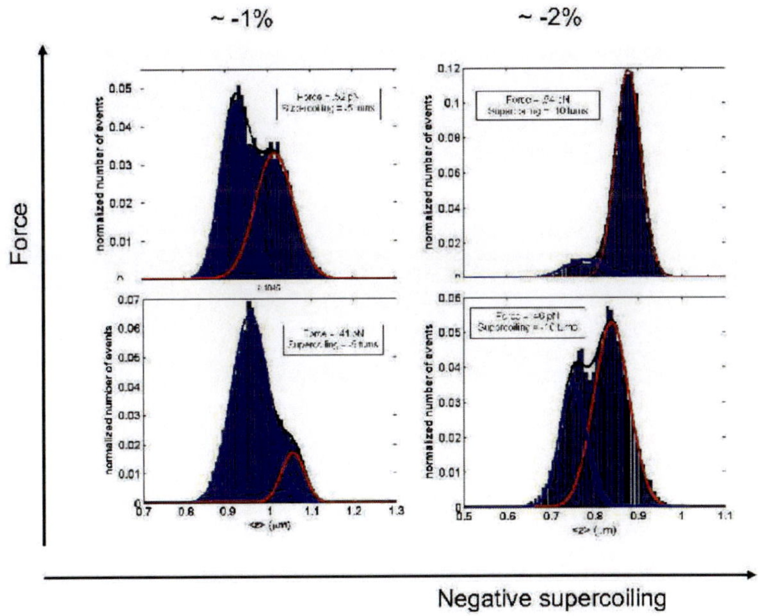

FIG. 11. *Histograms of the DNA length at different forces and level of supercoiling. The data show that a small increase in force (∼ 0.1 pN) leads to a significant reduction of the looped population and the increase of negative supercoiling from* $\sigma = -0.012$ *to* $\sigma = -0.024$ *dramatically decreases the probability of loop formation.*

REFERENCES

[1] Schleif R. (1992). Dna Looping. Annual Review Of Biochemistry **61**: 199–223.

[2] Dodd I.B., Shearwin K.B., and Sneppen K. (2007). Modelling transcriptional interference and DNA looping in gene regulation. Journal Of Molecular Biology **369**: 1200–1213.

[3] Dodd I.B., Shearwin K.E., and Egan J.B. (2005). Revisited gene regulation in bacteriophage lambda. Current Opinion In Genetics & Development **15**: 145–152.

[4] Matthews K.S. (1992). DNA LOOPING. Microbiological Reviews **56**: 123–136.

[5] Gaszner M. and Felsenfeld G. (2006). Insulators: exploiting transcriptional and epigenetic mechanisms. Nature Reviews Genetics **7**: 703–713.

[6] Ameres S.L., Drueppel L., Pfleiderer K., Schmidt A., Hillen W., and Berens C. (2005). Inducible DNA-loop formation blocks transcriptional activation by an SV40 enhancer. Embo Journal **24**: 358–367.

[7] Bondarenko V.A., Jiang Y.I., and Studitsky V.M. (2003). Rationally designed insulator-like elements can block enhancer action in vitro. Embo Journal **22**: 4728–4737.

[8] Moreau P., Hen R., Wasylyk B., Everett R., Gaub M.P., and Chambon P. (1981). The SV40-72 base repair repeat has a striking effect on gene-expression both in SV40 and other chimeric recombinants. Nucleic Acids Research **9**: 6047–6068.

[9] Tolhuis B., Palstra R.J., Splinter E., Grosveld F., and de Laat W. (2002). Looping and interaction between hypersensitive sites in the active beta-globin locus. Molecular Cell **10**: 1453–1465.

[10] Finzi L. and Gelles J. (1995). Measurement Of Lactose Repressor-Mediated Loop Formation And Breakdown In Single Dna-Molecules. Science **267**: 378–380.

[11] Finzi L. and Dunlap D. (2003). Single-molecule studies of DNA architectural changes induced by regulatory proteins. Methods Enzymol **370**: 369–378.

[12] Nelson P.C., Zurla C., Brogioli D., Beausang J.F., Finzi L., and Dunlap D. (2006). Tethered particle motion as a diagnostic of DNA tether length. Journal Of Physical Chemistry B **110**: 17260–17267.

[13] Lewis M. (2005). The lac repressor. Comptes Rendus Biologies **328**: 521–548.

[14] Zurla C., Samuely T., Bertoni G., Valle F., Dietler G., Finzi L., and Dunlap D.D. (2007). Integration host factor alters LacI-induced DNA looping. Biophysical Chemistry **128**: 245–252.

[15] Goyal S., Lillian T., Blumberg S., Meiners J.C., Meyhofer E., and Perkins N.C. (2007). Intrinsic curvature of DNA influences LacR-mediated looping. Biophysical Journal **93**: 4342–4359.

[16] Vanzi F., Broggio C., Sacconi L., and Pavone F.S. (2006). Lac repressor hinge flexibility and DNA looping: single molecule kinetics by tethered particle motion. Nucleic Acids Research **34**: 3409–3420.

[17] Wyman C., Grotkopp E., Bustamante C., and Nelson H.C.M. (1995). Determination Of Heat-Shock Transcription Factor-2 Stoichiometry At Looped Dna Complexes Using Scanning Force Microscopy. Embo Journal **14**: 117–123.

[18] Wyman C., Rombel I., North A.K., Bustamante C., and Kustu S. (1997). Unusual oligomerization required for activity of NtrC, a bacterial enhancer-binding protein. Science **275**: 1658–1661.

[19] Podesta A., Indrieri M., Brogioli D., Manning G.S., Milani P., Guerra R., Finzi L., and Dunlap D. (2005). Positively charged surfaces increase the flexibility of DNA. Biophysical Journal **89**: 2558–2563.

[20] Irani M.H., Orosz L., and Adhya S. (1983). A control element within a structural gene - the gal operon of Escherichia-coli. Cell **32**: 783–788.

[21] Aki T., Choy H.E., and Adhya S. (1996). Histone-like protein HU as a specific transcriptional regulator: Co-factor role in repression of gal transcription by GAL repressor. Genes to Cells **1**: 179–188.

[22] Geanacopoulos M., Vasmatzis G., Lewis D.E.A., Roy S., Lee B., and Adhya S. (1999). GalR mutants defective in repressosome formation. Genes & Development **13**: 1251–1262.

[23] Choy H.E., Park S.W., Parrack P., and Adhya S. (1995). Transcription Regulation By Inflexibility Of Promoter Dna In A Looped Complex. Proceedings Of The National Academy Of Sciences Of The United States Of America **92**: 7327–7331.

[24] Virnik K., Lyubchenko Y.L., Karymov M.A., Dahlgren P., Tolstorukov M.Y., Semsey S., Zhurkin V.B., and Adhya S. (2003). "Antiparallel" DNA loop in gal repressosome visualized by atomic force microscopy. Journal Of Molecular Biology **334**: 53–63.

[25] Geanacopoulos M., Vasmatzis G., Zhurkin V.B., and Adhya S. (2001). Gal repressosome contains an antiparallel DNA loop. Nature Structural Biology **8**: 432–436.

[26] Lia G., Bensimon D., Croquette V., Allemand J.F., Dunlap D., Lewis D.E.A., Adhya S.C., and Finzi L. (2003). Supercoiling and denaturation in Gal repressor/heat unstable nucleoid protein (HU)-mediated DNA looping. Proceedings Of The National Academy Of Sciences Of The United States Of America **100**: 11373–11377.

[27] Strick T.R., Allemand J.F., Bensimon D., Bensimon A., and Croquette V. (1996). The elasticity of a single supercoiled DNA molecule. Science **271**: 1835–1837.

[28] Strick T.R., Allemand J.F., Bensimon D., and Croquette V. (1998). Behavior of supercoiled DNA. Biophysical Journal **74**: 2016–2028.

[29] Semsey S., Virnik K., and Adhya S. (2005). A gamut of loops: meandering DNA. Trends In Biochemical Sciences **30**: 334–341.

[30] Mehta R.A. and Kahn J.D. (1999). Designed hyperstable lac repressor center dot DNA loop topologies suggest alternative loop geometries. Journal Of Molecular Biology **294**: 67–77.

[31] Semsey S., Tolstorukov M.Y., Virnik K., Zhurkin V.B., and Adhya S. (2004). DNA trajectory in the Gal repressosome. Genes & Development **18**: 1898–1907.

[32] White J.H. (1969). SELF-LINKING AND GAUSS-INTEGRAL IN HIGHER DIMENSIONS. American Journal of Mathematics **91**: 693–&.

[33] Lia G., Praly E., Ferreira H., Stockdale C., Tse-Dinh Y.C., Dunlap D., Croquette V., Bensimon D., and Owen-Hughes T. (2006). Direct observation of DNA distortion by the RSC complex. Mol Cell **21**: 417–425.

[34] Ptashne M.a.G.A. (2002). Genes and Signals (Cold Spring Harbor Laboratory Press).

[35] Dodd I.B., Shearwin K.E., Perkins A.J., Burr T., Hochschild A., and Egan J.B. (2004). Cooperativity in long-range gene regulation by the lambda CI repressor. Genes & Development **18**: 344–354.

[36] Maniatis T. and Ptashne M. (1973). Multiple Repressor Binding At Operators In Bacteriophage-Lambda - (Nuclease Protection Polynucleotide Sizing Pyrimidine Tracts Supercoils E-Coli). Proceedings Of The National Academy Of Sciences Of The United States Of America **70**: 1531–1535.

[37] Oppenheim A.B., Kobiler O., Stavans J., Court D.L., and Adhya S. (2005). Switches in bacteriophage lambda development. Annual Review Of Genetics **39**: 409–429.

[38] Koblan K.S. and Ackers G.K. (1992). Site-Specific Enthalpic Regulation Of Dna-Transcription At Bacteriophage-Lambda Or. Biochemistry **31**: 57–65.

[39] Senear D.F., Brenowitz M., Shea M.A., and Ackers G.K. (1986). Energetics Of Cooperative Protein Dna Interactions - Comparison Between Quantitative Deoxyribonuclease Footprint Titration And Filter Binding. Biochemistry **25**: 7344–7354.

[40] Jain D., Nickels B.E., Sun L., Hochschild A., and Darst S.A. (2004). Structure of a ternary transcription activation complex. Molecular Cell **13**: 45–53.

[41] Nickels B.E., Dove S.L., Murakami K.S., Darst S.A., and Hochschild A. (2002). Protein-protein and protein-DNA interactions of sigma(70) region 4 involved in transcription activation by lambda cl. Journal Of Molecular Biology **324**: 17–34.

[42] Revet B., von Wilcken-Bergmann B., Bessert H., Barker A., and Muller-Hill B. (1999). Four dimers of lambda repressor bound to two suitably spaced pairs of lambda operators form octamers and DNA loops over large distances. Current Biology **9**: 151–154.

[43] Dodd I.B., Perkins A.J., Tsemitsidis D., and Egan J.B. (2001). Octamerization of lambda CI repressor is needed for effective repression of P-RM and efficient switching from lysogeny. Genes & Development **15**: 3013–3022.

[44] Zurla C., Franzini A., Galli G., Dunlap D.D., Lewis D.E.A., Adhya S., and Finzi L. (2006). Novel tethered particle motion analysis of CI protein-mediated DNA looping in the regulation of bacteriophage lambda. Journal Of Physics-Condensed Matter **18**: S225–S234.

[45] Liebovitch L.S., Scheurle D., Rusek M., and Zochowski M. (2001). Fractal methods to analyze ion channel kinetics. Methods **24**: 359–375.

[46] Watkins L.P. and Yang H. (2005). Detection of intensity change points in time-resolved single-molecule measurements. Journal of Physical Chemistry B **109**: 617–628.

[47] Marko J.F. and Siggia E.D. (1997). Driving proteins off DNA using applied tension. Biophysical Journal **73**: 2173–2178.

[48] Strick T.R., Croquette V., and Bensimon D. (2000). Single-molecule analysis of DNA uncoiling by a type II topoisomerase. Nature **404**: 901–904.

[49] Vologodskii A.V. and Frankkamenetskii M.D. (1992). Modeling Supercoiled Dna. Methods In Enzymology **211**: 467–480.

[50] Vologodskii A.V., Levene S.D., Klenin K.V., Frankkamenetskii M., and Cozzarelli N.R. (1992). Conformational And Thermodynamic Properties Of Supercoiled Dna. Journal Of Molecular Biology **227**: 1224–1243.

[51] Marko J.F. (2007). Torque and dynamics of linking number relaxation in stretched supercoiled DNA. Physical Review E **76**.

[52] Vologodskii A., Do Q., Shiffeldrim N., and Smith C. (2005). Microscopic mechanisms of DNA flexibility. Biophysical Journal **88**: 59A–59A.

[53] Levene S.D., Abola P.M., Vologodskii A.V., and Cozzarelli N.R. (1992). Studies Of Loop Closure In Supercoiled Dna. Faseb Journal **6**: A222–A222.

[54] Zhang Y., McEwen A.E., Crothers D.M., and Levene S.D. (2006). Statistical-mechanical theory of DNA looping. Biophysical Journal **90**: 1903–1912.

[55] Zhang Y., McEwen A.E., Crothers D.M., and Levene S.D. (2007). Analysis of in-vivo LacR-mediated gene repression based on the mechanics of DNA looping. Biophysical Journal, 232A–232A.

[56] Du Q., Kotlyar A., and Vologodskii A. (2008). Kinking the double helix by bending deformation. Nucleic Acids Research **36**: 1120–1128.

[57] Polikanov Y.S., Bondarenko V.A., Tchernaenko V., Jiang Y.I., Lutter L.C., Vologodskii A., and Studitsky V.M. (2007). Probability of the site juxtaposition determines the rate of protein-mediated DNA looping. Biophysical Journal **93**: 2726–2731.

[58] Lillian T.D., Goyal S., Perkins N.C., Meiners J.C., and Kahn J.D. (2007). Computational rod theory predicts experimental characteristics of DNA looping by the Lac repressor. Biophysical Journal, 416A–417A.

[59] Mehta R.A. and Kahn J.D. (1999). Stability of designed Lac repressor-DNA loops. Faseb Journal **13**: A1371–A1371.

[60] Tolstorukov M.Y., Colasanti A.V., McCandlish D.M., Olson W.K., and Zhurkin V.B. (2007). A novel roll-and-slide mechanism of DNA folding in chromatin: Implications for nucleosome positioning. Journal of Molecular Biology **371**: 725–738.

[61] Matsumoto A. and Olson W.K. (2006). Predicted effects of local conformational coupling and external restraints on the torsional properties of single DNA molecules. Multiscale Modeling & Simulation **5**: 1227–1247.

[62] Swigon D., Coleman B.D., and Olson W.K. (2006). Modeling the Lac repressor-operator assembly: The influence of DNA looping on Lac repressor conformation. Proceedings of the National Academy of Sciences of the United States of America **103**: 9879–9884.

[63] Czapla L., Swigon D., and Olson W.K. (2006). Sequence-dependent effects in the cyclization of short DNA. Journal of Chemical Theory and Computation **2**: 685–695.

[64] Vilar J.M.G. (2006). Modularizing gene regulation. Molecular Systems Biology.

[65] Vilar J.M.G. and Saiz L. (2006). Multiprotein DNA looping. Physical Review Letters **96**.

[66] Villa E. and Schulten K. (2007). Multiscale simulations of the DNA loop topologies induced by the lac repressor. Biophysical Journal, 185A–185A.

DIFFERENCES BETWEEN POSITIVELY AND NEGATIVELY SUPERCOILED DNA THAT TOPOISOMERASES MAY DISTINGUISH

JONATHAN M. FOGG*, DANIEL J. CATANESE, JR.*, GRAHAM L. RANDALL[†], MICHELLE C. SWICK[‡], AND LYNN ZECHIEDRICH*[†‡§]

Abstract. In all living cells, DNA is homeostatically underwound relative to its lowest energy conformation, resulting in negative supercoiling. This underwinding of DNA is critical to the metabolism of DNA and, thus, is vital to cell survival. Enzymes called topoisomerases regulate and maintain the supercoiled state of DNA and are critical to the successful replication of the genome. These enzymes are major targets for drugs used in the treatment of bacterial infections and cancer. One puzzling phenomenon of the topoisomerase mechanism is how these enzymes, orders of magnitude smaller than their substrate, can search, recognize and act at a local level to affect global DNA topology. While the homeostatic state of DNA supercoiling in cells is negative, both positive and negative supercoils exist transiently. Because of the right-handed nature of the DNA helix, the positive and negative supercoils are not equivalent. Several computational and theoretical models have been developed in an effort to describe the features of both positively and negatively supercoiled DNA. These models have accurately predicted some of the phenomena observed *in vivo*. However, the over-simplifying assumptions cannot account for the different biological activities of positively and negatively supercoiled DNA. This review will discuss the models in place and the mathematical and energetic properties of this elegant molecule and the "machines that push it around."

1. Introduction. The classical double-helix structure of DNA, as deduced by Watson and Crick, is one of the most recognizable icons of modern scientific endeavor (Watson and Crick, 1953). Although some details are still debated (the exact number of base pairs per turn for example), it is almost universally accepted that DNA in its lowest energy state exists in the form known as B-DNA. Undoubtedly elegant, the antiparallel double-

*Department of Molecular Virology and Microbiology, Baylor College of Medicine, Houston, TX 77030.

JMF: Formerly funded by a post-doctoral fellowship from the Program in Mathematics and Molecular Biology at Florida State University, with funding from the Burroughs Wellcome Fund Interfaces Program; currently funded by National Institutes of Health (NIH) grant to LZ.

DJC: Formerly funded by the Pharmacoinformatics Training Program of the W.M. Keck Center for Computational and Structural Biology of the Gulf Coast Consortia; NIH T90 DK070109; currently funded by the Infection and Immunity NIH Grant T32 AI55413.

[†]Institutional Program in Structural and Computational Biology and Molecular Biophysics, Baylor College of Medicine, Houston, TX 77030.

GLR: Formerly funded by the Computational Biology and Medicine Program of the W.M. Keck Center for Computational and Structural Biology of the Gulf Coast Consortia, NLM T15 LM07093.

[‡]Interdepartmental Program in Cell and Molecular Biology, Baylor College of Medicine, Houston, TX 77030.

MCS: Funded by the Pharmacoinformatics Training Program of the W.M. Keck Center for Computational and Structural Biology of the Gulf Coast Consortia, NIH T90 DK070109.

[§]LZ: Funded by NIH Grant RO1 A1054830. (Corresponding Author, elz@bcm.edu.)

C.J. Benham et al. (eds.), *Mathematics of DNA Structure, Function and Interactions*, The IMA Volumes in Mathematics and its Applications 150, DOI 10.1007/978-1-4419-0670-0_5, © Springer Science+Business Media, LLC 2009

helical structure also profoundly influences the biological activity of DNA. The challenges the structure presents to its own replication were recognized by the mathematicians Pohl and Roberts (Pohl and Roberts, 1978). They argued that the intertwining of the two strands would prohibit their complete separation during replication and perhaps the side-by-side conformation, as postulated by Rodley *et al.* (1976), would better fulfill the requirements for a replicating molecule. Despite the seemingly insurmountable challenges, the helical structure of DNA offers many advantages. One of these is incredible stability, which maintains the integrity of the genetic material. Modulation of helical winding also provides a mechanism for regulating access to the genetic information. These mechanisms include underwinding, which facilitates strand separation, and overwinding, which inhibits strand separation.

The helical nature of DNA imparts interesting mechanical properties on the molecule. A full understanding of the biological function of DNA requires a detailed knowledge of these properties. Likewise, an appreciation of the utility of the double helical polymer requires consideration of its biological functions. Linking number (Lk) refers to the number of times the two helical strands are interwound. The Lk for a relaxed molecule, termed Lk_0, is equal to the number of base pairs divided by the period of the DNA helix. Most theoretical and computational modeling of DNA assumes it behaves as an isotropic elastic rod. Overwinding or underwinding of the helix changes twist (Tw), a property describing the number of times the individual strands coil around the helical axis. If DNA behaves as an isotropic elastic rod, then as the value of Tw increases, the associated torque should increase linearly. Thus, the free energy associated with Tw is a quadratic function of the elastic deformation (Hagerman, 1988). When Tw reaches a critical density, the molecule buckles to form plectonemic structures in which the helix coils about itself, a property known as writhe (Wr). This buckling acts to exchange twisting energy for bending energy. The sum of the real-valued geometric properties of Tw and Wr is equal to the linking number such that:

$$Lk = Tw + Wr \tag{1}$$

Any change in linking number must manifest as a change in the twist and/or writhe such that:

$$\Delta Lk = \Delta Tw + \Delta Wr \tag{2}$$

The introduction of writhe requires bending energy to overcome the intrinsic resistance of the DNA helix to bending. Therefore it is usually assumed that the writhe of a relaxed molecule is equal to zero, and hence $\Delta Wr = Wr$.

The coiling of the helix about itself is more commonly known to biologists as supercoiling. The discovery of DNA supercoiling in the 1960s by Vinograd and colleagues opened up a whole new branch of study: DNA topology (Vinograd *et al.*, 1965). Vinograd turned to mathematician F. Brock Fuller to provide a quantitative analysis of DNA supercoiling (Fuller, 1971; Fuller, 1978). Fuller's resultant elegant treatise led to the adoption of the terminology of linking number, twist and writhe to describe DNA topology, and these terms are now familiar to biologists studying DNA. The relationship between these properties had been proved previously without reference to DNA (White, 1969).

1.1. Rubber tubing model. It is often useful visually to model DNA supercoiling by using a piece of rubber tubing. As the rubber tubing is an isotropic elastic rod it should serve as a useful analogy for the idealized properties of DNA. The rubber tubing model is a simplified representation of supercoiled DNA; however, it serves well to illustrate the basic concepts of supercoiling. The principles hold true for a rope as shown in Figure 1. DNA is a right-handed helix, *i.e.* the helix spirals in a clockwise direction. Therefore, to simulate the effects of overwinding (positive supercoiling), one can introduce clockwise twist into the tubing (in the direction looking away from the observer along the axis of the tubing). This is most easily achieved by grasping one end of the tubing in each hand. The end held in the right hand should be facing towards the reader and twisted in a counter clockwise direction. When sufficient twist is added, the tubing coils about itself analogously to DNA writhe (Figure 1). Imagine that the rubber tubing represents a DNA helix. Looking down the superhelical axis, positively supercoiled DNA coils about itself in a left-handed or counter clockwise direction (Figure 1, left side). Negatively supercoiled DNA assumes a right-handed superhelical structure (Figure 1, right side). It is now almost universally accepted that the conformation associated with negative supercoiling is a right-handed plectonemic (interwound) superhelix, similar to the conformation illustrated by the rubber tubing model. This structure has been observed by electron microscopy (Adrian *et al.*, 1990; Bednar *et al.*, 1994; Boles *et al.*, 1990; Cherny and Jovin, 2001) and is supported by theoretical simulations of DNA supercoiling (Vologodskii *et al.*, 1992). Although limited data exist on the structure of positively supercoiled DNA, it is assumed to adopt a similar plectonemic conformation.

1.2. Importance of negative supercoiled DNA in biology. The DNA of all species examined, from bacteria to humans, is maintained in a homeostatically underwound, *i.e.* negatively supercoiled, state (reviewed in Cozzarelli *et al.*, 1990; Kramer and Sinden, 1997; Schvartzman and Stasiak, 2004; Travers and Muskhelishvili, 2005). Mathematically, this means that the value of Lk in cells is lower than that of Lk_0. The negative linking difference, ΔLk, is defined by:

$$\Delta Lk - Lk - Lk_0 \tag{3}$$

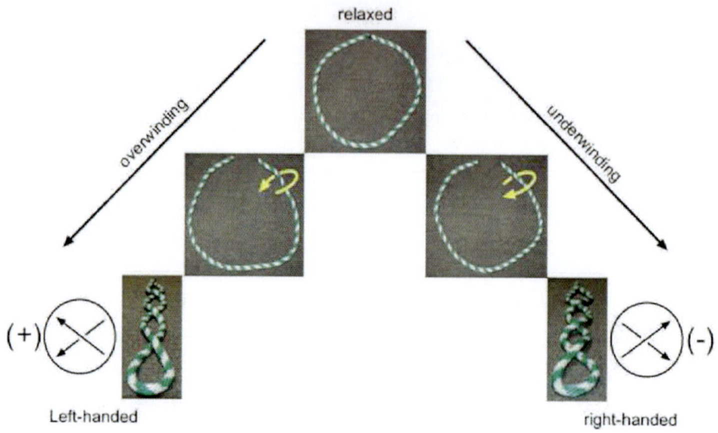

FIG. 1. *Modeling DNA supercoiling: DNA supercoiling can be modeled using a rubber tube or a rope (depicted). Even though the pictures might appear to be mirror images, DNA is a right-handed helix, and thus, there are fundamental differences between overwinding and underwinding. Holding the two ends of the rope separately, overwinding is modeled by twisting the rope in your right hand in the direction as shown by the arrow (left side). With overwinding, the rope will form a writhed polymer with left-handed crossovers. Underwinding (right side) has the opposite effect on writhe; twisting the rope in your right hand in the direction of the arrow will create a writhed polymer with right-handed crossovers.*

This underwinding, $\Delta Lk < 0$, has a profound effect on the structural and functional properties of DNA; it becomes "as it were alive, turning from a stable double helix into a diversely fluctuating entity" (Vologodskii *et al.*, 1979). For DNA to fulfill its role as a store of genetic information, it must exist in the cell on the edge of helix stability. Negative supercoiling of the DNA facilitates localized, controlled melting of the helix, thus facilitating the processes that require access to the DNA bases. Replication, the accurate copying of the genome, and transcription, the process converting the genetic code to an RNA message, require the helix to be denatured, which can only happen when the helix is slightly underwound. For many biological processes there is a distinct threshold level of negative supercoiling at which efficiency abruptly increases (Nollmann *et al.*, 2007; Travers and Muskhelishvili, 2005; Zechiedrich *et al.*, 1997; Zechiedrich *et al.*, 2000). Surely it is not accidental that DNA supercoiling is approximately this same threshold value when DNA is purified from cells. Almost every protein that interacts with DNA is influenced, to varying extents, by the degree of negative supercoiling. In spite of the critical importance for DNA function, relatively little is known about the structural transitions that accompany such underwinding. To date, the only crystal structures available in the Protein Data Bank (Berman *et al.*, 2000) have been determined using short duplex or oligomeric single stranded DNA, neither of which represent the native state of DNA inside cells.

1.3. Evidence that overwound DNA is also important in biology. Because it is the form obtained out of cells, negatively supercoiled DNA is much easier to generate than positively supercoiled DNA. As a consequence, we know much more about negative supercoiling than positive supercoiling. Thus, whereas the importance of negative supercoiling in DNA function has been well established, the role of positive supercoiling is not as well known. It is fairly well accepted that both DNA and RNA polymerase complexes generate transient positive supercoiling ahead of the replicating and transcribing processes, respectively (Crooke *et al.*, 1991; Dunaway and Dröge, 1989; Kreuzer and Alberts, 1984; Liu *et al.*, 1979a; Liu *et al.*, 1979b; Liu and Wang, 1987; Lockshon and Morris, 1983; Wahle and Kornberg, 1988). In a test tube, purified, reconstituted DNA replication machinery is only active on negatively supercoiled DNA. The polymerase stalls before replication is complete, and this is attributed to the build-up of positive supercoiling ahead of the replication machinery. The addition of enzymes, topoisomerases (see below), that relax positive supercoiling restores replication (Hiasa and Marians, 1996; Hiasa *et al.*, 1996, and references therein).

Reverse gyrase is an enzyme that introduces positive supercoils into DNA (Kikuchi and Asai, 1984). Reverse gyrase is found uniquely in hyperthermophilic organisms, which grow in environments where temperatures can exceed 80°C. Proteins from many hyperthermophilic archaea wrap DNA in nucleosome-like structures that constrain DNA in a positively supercoiled conformation (Musgrave *et al.*, 1991). The existence of positively supercoiled DNA in hyperthermophilic archaea is an elegant solution to the problem of preventing thermal denaturation in these organisms, but this concept is controversial (see excellent reviews by Musgrave *et al.*, 2002; Reeve *et al.*, 1997).

1.4. DNA topoisomerases. Topoisomerases are ubiquitous essential enzymes that unknot, decatenate, supercoil and relax DNA (Table 1 and Figure 2) (reviewed in Champoux, 2001; Wang, 2002; Schoeffler and Berger, 2008). These enzymes appear to undergo extraordinary gymnastics to pass DNA helices though each other. The topoisomerases break either one (type I) or two (type II) strands to subsequently change Lk by one or two, respectively. This DNA break is accompanied by a covalent phosphotyrosine bond between a tyrosine hydroxyl group on the enzyme and the DNA phosphate at the break site. A second single or double-stranded DNA segment is then passed through the break, which is subsequently resealed. Cells are equipped with both type I and type II topoisomerases. These enzymes differ in their molecular mechanisms and, consequently, in the reactions they carry out (Table 1).

One of the first hints that overwound DNA is not equal and opposite to underwound DNA was the finding that the bacterial type II enzyme, topoisomerase IV, is approximately 20-fold more efficient at removing pos-

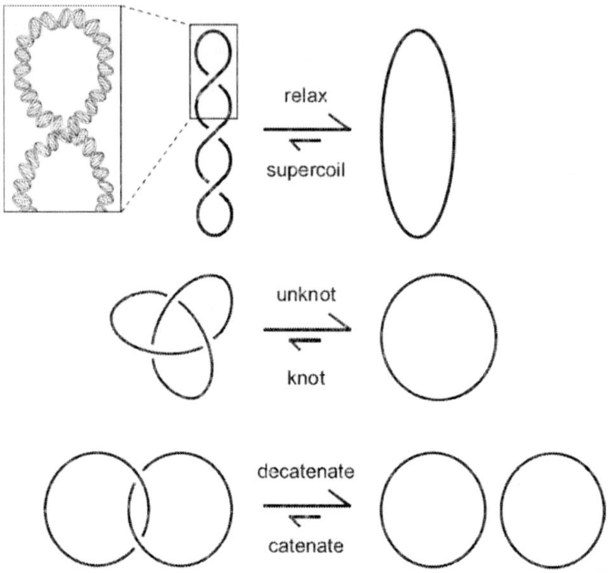

FIG. 2. *Substrates of topoisomerases: Depicted are schematics of topoisomerase substrates. For simplicity, each line represents a double-stranded DNA helix, as shown by the upper left inset. As drawn, most topoisomerases simplify DNA topology with the equilibrium shifted towards relaxing, unknotting, and decatenating. Bacterial DNA gyrase and archaeal reverse gyrase are unique enzymes because they introduce supercoils.*

itive than negative supercoils, in both ensemble and single-molecule assays (Crisona *et al.*, 2000). There must be something different between positive and negative supercoils that this type II enzyme distinguishes. Biologically, a preference of topoisomerase IV for positive over negative supercoiling makes sense because negative supercoiling is essential in bacterial cells and is very tightly regulated (Zechiedrich *et al.*, 2000). Topoisomerase IV is relatively inefficient at relaxing negative supercoils both *in vitro* (Ullsperger and Cozzarelli, 1996) and *in vivo* (Zechiedrich *et al.*, 2000). Thus, evolution appears to have provided a solution (a preference of positive over negative supercoils) to the need of removing potentially problematic positive supercoils that would prevent polymerase progression while simultaneously leaving the beneficial negative supercoils unaltered.

Preferential relaxation of positively supercoiled DNA is not a universal feature of type II topoisomerases. Yeast topoisomerase II, fruit fly topoisomerase II, and viral type II topoisomerases do not exhibit this preference (Charvin *et al.*, 2003; McClendon *et al.*, 2005; McClendon *et al.*, 2006; Strick *et al.*, 2000b). Humans express two closely related isoforms of topoisomerase II, α and β (Drake *et al.*, 1989) and these differ in their specificity for positively supercoiled DNA. Human topoisomerase IIα relaxes positively supercoiled DNA more than 10-times faster than negatively

TABLE 1
Subfamilies of topoisomerases and their activities.

Subfamily	Members	adds (+)	adds (-)	relaxes (+)	relaxes (-)	prefers (+)	decatenates	unknots
Type IA	Bacterial topoisomerase I	X	X	X	✓[1]	X	✓[2,3]	✓[3]
	Bacterial topoisomerase III	X	X	X	✓[2]	X	✓[3]	?
	Archaeal reverse gyrase	✓	X	X	X	X	?	?
Type IB	Vaccinia virus topoisomerase I	X	X	✓	✓	?	X	X
	Pox virus topoisomerase I	X	X	✓	✓	?	X	X
	Calf thymus topoisomerase I	X	X	✓	✓	?	X	X
	Human topoisomerase I	X	X	✓	✓	?	X	X
Type IC	Archaeal topoisomerase V	X	X	✓	✓	?	X	X
Type IIA	T4 phage topoisomerase II	X	X	✓	✓	X	✓	✓
	Bacterial DNA gyrase	X	✓	X	X	X	✓[2]	✓[2]
	Bacterial topoisomerase IV	X	X	✓	✓	✓	✓	✓
	Yeast topoisomerase II	X	X	✓	✓	X	✓	?
	Fruit fly topoisomerase II	X	X	✓	✓	X	✓	?
	Human topoisomerase IIα	X	X	✓	✓	✓	✓	✓
	Human topoisomerase IIβ	X	X	✓	✓	X	✓	?
Type IIB	Archaeal topoisomerase VI	X	X	✓	✓	?	✓	?

[1]Does not relax to completion
[2]Weak activity
[3]If single-stranded DNA region is present

supercoiled DNA; human topoisomerase IIβ has little or no preference (McClendon *et al.*, 2005). Interestingly, when an organism contains a type II topoisomerase that preferentially relaxes positively supercoiled DNA, there is always an additional type II topoisomerase in that organism that does not exhibit a preference (McClendon *et al.*, 2005). Each enzyme has distinct physiological roles (Grue *et al.*, 1998). Human topoisomerase IIα is expressed primarily in rapidly proliferating tissues (Nitiss, 1998), topoisomerase IIβ is found in all tissues, but predominantly in non-proliferating cells, such as neurons (Kondapi *et al.*, 2004). The preferential relaxation of positively supercoiled DNA by human topoisomerase IIα is clearly advantageous ahead of the highly active replication machinery in rapidly proliferating tissues.

1.5. Medical relevance of topoisomerases and supercoiled DNA. Topoisomerases are a major class of pharmaceutical targets. Billions of dollars each year are spent on drugs that target these enzymes. Fluoroquinolones target the bacterial enzymes, DNA gyrase and topoisomerase IV, and are among the most widely prescribed antibiotics in the United States (Wolters Kluwer Health, 2006). Fluoroquinolones are used to treat urinary and respiratory tract infections, prostatitis, and infections of the skin and sinuses. Camptothecin and its derivatives, which target human topoisomerase I (Pommier, 2006), are used to treat ovarian, colorectal, pancreatic, breast and prostate cancers. The epipodophyllotoxins (etoposide), anthracyclines, and anthracenediones, which target human topoisomerase IIα, are used to treat breast, lung, ovarian and bladder cancers as well as several forms of leukemia (reviewed in Martincic and Hande, 2005). These anti-topoisomerase drugs stabilize the normally fleeting catalytic intermediate where the enzyme is covalently attached to the cleaved DNA, converting the topoisomerase into a potent cellular toxin (Kreuzer and Cozzarelli, 1979). It is thought that the cellular replication and transcription machinery collide with these intermediates, ultimately leading to death of the bacterial or cancer cell (reviewed in Anderson and Osheroff, 2001; McClendon and Osheroff, 2007). Additional mechanisms of cell death, mediated by the anti-topoisomerase drugs, have also been proposed (reviewed in Drlica *et al.*, 2008). The anti-topoisomerase drugs are potent and specific, but their use has consequences: drug resistance, toxicity, and secondary malignancies caused by the drug. Therefore, understanding where and how these enzymes act and the effects of supercoiling on the drug activity should facilitate the development of more efficacious, potentially less toxic drugs.

For a better understanding of how the anti-topoisomerase drugs act it is imperative to consider where the topoisomerases act in the cell. Topoisomerases, with the notable exception of DNA gyrase, show at least 5- to 10-fold preference for binding and cleaving negatively supercoiled over relaxed (Lk_0) or linear DNA (Camilloni *et al.*, 1988; Camilloni *et al.*, 1989; Muller, 1985; Osheroff *et al.*, 1983; Osheroff, 1986; Osheroff and Zechiedrich, 1987; Zechiedrich and Osheroff, 1990). Although many studies of the mechanisms of anti-topoisomerase drugs have been performed with relaxed or linear DNA, these are not substrates for topoisomerases and, hence, the results may not be germane to our understanding of anti-topoisomerase drug activity. The finding that positively supercoiled DNA is the preferred substrate, over negatively supercoiled DNA, for the fluoroquinolones targets, gyrase and topoisomerase IV, and the anticancer drug target, human topoisomerase IIα, has important pharmacological implications. Preferential activity on positively supercoiled DNA leads to a higher probability of enzyme-DNA cleavage intermediates directly in the path of advancing polymerases. Whereas positively supercoiled DNA is, arguably, the more relevant substrate, few studies have addressed the drug action on positively supercoiled substrates. One study that addressed the drug response with positively and negatively supercoiled DNA uncovered a dramatic increased toxicity for positively supercoiled DNA (McClendon *et al.*, 2006). Another study found that anticancer drugs impede the relaxation of positive, but not negative, supercoils by human topoisomerase I (Koster *et al.*, 2007). This revealed a potential new mechanism for drug-induced cell death, previously unconsidered, and only revealed when experiments were performed with positively supercoiled DNA.

Toward a better understanding of the biological function of DNA, the DNA topoisomerases, and the drugs that target the DNA topoisomerases, here we review the properties of positively and negatively supercoiled DNA and the potential mechanisms that topoisomerases may use to distinguish between them.

2. Differences between positively and negatively supercoiled DNA.

2.1. Biological consequences of DNA twist and writhe. One of the most important things to consider when comparing the properties of underwound and overwound DNA is how supercoiling is partitioned into ΔTw and Wr. The consequences of changes in twist differ from those of writhe. One of the consequences of negative ΔTw is a higher propensity for the individual strands of the helix to separate. Thus, processes that require strand separation, such as DNA replication and transcription, are facilitated by increased negative ΔTw. Additionally, the formation of non-canonical alternative secondary structures, described below, requires strand separation, and therefore is also dependent on negative ΔTw. The consequence of positive ΔTw is resistance to strand separation and a concomitant increase in stability at higher temperatures. Other consequences of positive ΔTw include an increase in the length and a decrease in the diameter of the DNA helix upon overwinding (Gore *et al.*, 2006; Lionnet *et al.*, 2006). Because DNA is a highly charged molecule, changing twist alters the charge density and, as a consequence, also changes the location and density of solvent counterions (Randall G.L., Zechiedrich L. and Pettitt B.M., *submitted*).

Writhe is a property describing the coiling of the helical axis in three-dimensional space. The property Wr has a rigorous mathematical definition corresponding to the Gauss double integral over a closed curve in three-dimensional space (Levitt, 1983; Klenin and Langowski, 2000). One of the most important biological consequences of writhe is the resultant compaction of DNA, which facilitates its packaging into cells. In addition, because two sites are generally closer together in a writhed molecule (Vologodskii *et al.*, 1992), both the transfer of a protein from a random site to a specific site, as well as the synapsis of two sites, should be more efficient in writhed DNA (Embleton *et al.*, 2004; Gowers and Halford, 2003).

2.2. Partition between twist and writhe. In the absence of external factors such as bound proteins, the relative proportioning of ΔLk into ΔTw and Wr is governed by an exquisite equilibrium between twisting and bending energies, electrostatic potential energies within and between highly charged DNA helices, and the extent of electrostatic screening by solvent ions. As described above, partitioning of ΔLk into Wr reduces the torsional strain and thus the twisting energy. Writhing, on the other hand, requires the DNA bending rigidity to be overcome. Local environment also has a profound effect on the partitioning between ΔTw and Wr. Because writhe requires two highly charged helices to juxtapose, it is strongly dependent on electrostatic effects. We examine the current understanding of this partitioning and how it may differ for positively supercoiled DNA.

2.3. How to study DNA supercoiling. Innovative methods have been developed to generate supercoiled DNA (Figure 3). The most direct way to generate negatively supercoiled DNA is to incubate relaxed DNA with gyrase (Figure 3, panel A). To vary the Lk, the method shown in Figure 3 (panel B) shows how ethidium bromide or chloroquine can be used with topoisomerase I. Increasing intercalator concentrations results in increasingly supercoiled DNA. Incubation of negatively supercoiled or relaxed DNA with reverse gyrase (Figure 3, panel A) results in positively supercoiled DNA (Rodriguez, 2002). Other ways involve shifting Lk by binding chemicals (*e.g.*, netropsin) or proteins (*e.g.*, HmfB) and trapping the resultant ΔLk with enzymes to ligate nicked DNA or topoisomerases that remove compensatory ΔLk in closed circular molecules (LaMarr *et al.*, 1997; Snounou and Malcom, 1983). Strick *et al.* devised a way to introduce positive or negative supercoiling into a single DNA molecule in a controllable and reversible fashion (Strick *et al.*, 1996). Single molecule micromanipulation has provided an unparalleled insight into the elastic properties of DNA and also of enzymes that act on DNA (reviewed in Bustamante *et al.*, 2003; Charvin *et al.*, 2005b).

Analyzing DNA supercoiling experimentally is difficult, time-consuming, and requires specialized and often expensive equipment. The most direct methods are by electron microscopy or atomic force microscopy. Griffith and co-workers showed helix-helix juxtapositions, or crossovers that are characteristic of writhe are readily observed by electron microscopy of supercoiled DNA (Sperrazza *et al.*, 1984). The benefit of these methods is that, because the DNA is directly visualized, the shape information is much more easily extracted than it is from other indirect methods. Caution must be taken, however, in interpreting the results of transmission electron microscopy because in sample preparation the DNA is adsorbed onto a surface, air-dried and shadowed with metal. It is unclear as to what extent these manipulations might distort DNA. Similar caveats apply for atomic force microscopy (reviewed in Lyubchenko, 2004). Electron cryomicroscopy (cryoEM) should avoid some of these issues, because the sample is frozen in solution. As mentioned above, however, a downside to these powerful methods is that electron and atomic force microscopy requires specialized and expensive equipment. CryoEM facilities, in particular, are not readily available to most researchers and require considerable expertise and training to use. This has limited the application of these powerful techniques.

A much more common method of analysis of DNA topology is gel electrophoresis, a technique that is relatively simple to perform, inexpen-

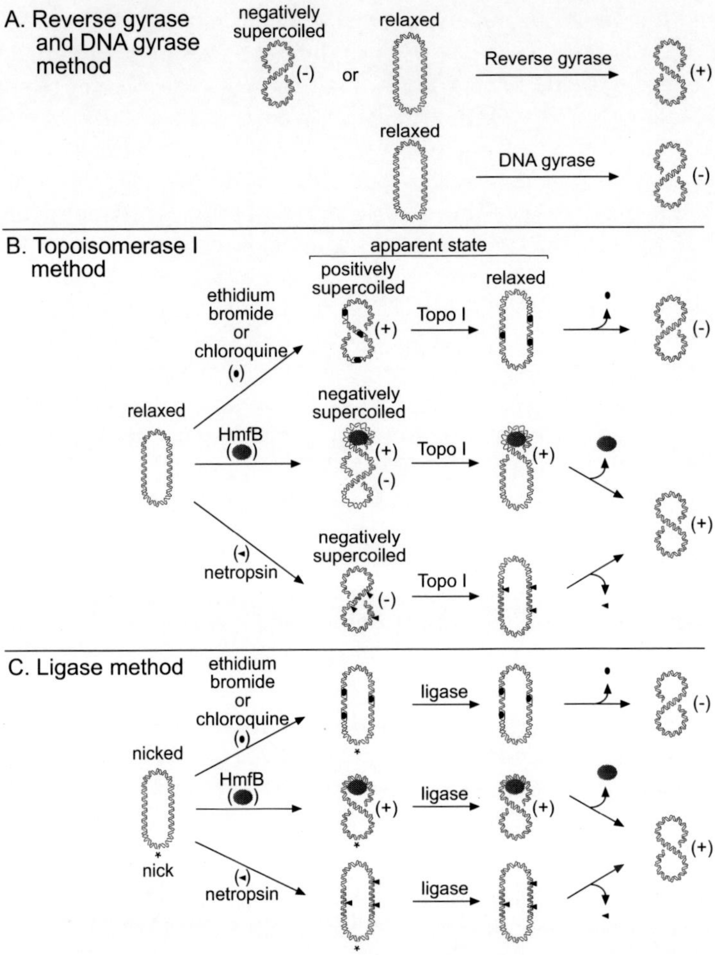

FIG. 3. *Methods to generate negatively and positively supercoiled DNA: A) Reverse gyrase and DNA gyrase method: Reverse gyrase directly introduces positive supercoils into DNA from either negatively supercoiled or relaxed DNA substrates. Conversely, DNA gyrase directly introduces negative supercoils into DNA from relaxed DNA substrates. B) Topoisomerase I (Topo I) method: Introduction of chemicals or proteins alters the relaxed DNA structure to appear supercoiled; however, the Lk does not change. Topo I relaxes the apparent supercoils, shifting the Lk. The Topo I reaction is then quenched, trapping the new topological state. Once the chemical or protein is removed, the new ΔLk repartitions between ΔTw and Wr, thus generating supercoiled DNA molecules. In the examples shown, negatively supercoiled DNA is generated by chemicals that unwind DNA (e.g., ethidium bromide or chloroquine, filled circles). Positively supercoiled DNA can be generated using chemicals that overwind DNA by introducing positive Tw (e.g., netropsin, filled triangles) or proteins that stabilize positive Wr. The most commonly used example is the protein HmfB. C) Ligase method: Beginning with nicked (*) DNA, the chemicals or proteins described in panel B shift Lk upon binding. Ligase is used to seal the nick and trap the supercoiling that has been introduced by the chemical or protein. Subsequent removal of the chemical or protein allows the supercoiling to repartition between ΔTw and Wr.*

sive and for which the apparatus can be found in almost every laboratory. Gel electrophoresis separates DNA of different sizes and conformations by their ability to snake their way through the pores of the agarose or polyacrylamide gel matrix (Deutsch, 1988; Weber *et al.*, 2006). DNA that is writhed migrates more rapidly through the gel matrix because of its compact shape (Keller and Wendel, 1975). In this way, electrophoresis provides a semi-quantitative measure of writhe. However, extracting quantitative values of writhe from electrophoresis is difficult and indirect because of the assumptions that must be made regarding how a given value of writhe affects the overall shape of the DNA molecule and, concomitantly, how this affects the speed at which the DNA can weave its way through the pores of the gel matrix. Gel electrophoresis is particularly robust for comparing relative writhe in DNAs of equal length. Faster mobility indicates a more compact structure and almost certainly a greater partition into writhe. We have observed that positively supercoiled 339 bp minicircles have a faster electrophoretic mobility than negatively supercoiled minicircles of comparable ΔLk (Fogg J.M., Catanese, D.J. and Zechiedrich L., *manuscript in preparation*). This faster mobility indicates a more compact structure and, therefore, we can conclude that these positively supercoiled minicircles have increased writhe.

The development of single-molecule experiments has provided an effective means to test models of DNA elasticity. The set-up consists of a single DNA molecule tethered at one end to a surface and at the other end to a small magnetic bead or micropipette (Smith *et al.*, 1992; Strick *et al.*, 1996; Strick *et al.*, 1998; reviewed in Bustamante *et al.*, 2003; Charvin *et al.*, 2005b; Strick *et al.*, 2000a). A rotating magnetic field induces rotation of the bead (Strick *et al.*, 1996), thereby underwinding or overwinding the DNA molecules in a controllable and reversible fashion. The change in the topology of the DNA is monitored in real time from the extension of the DNA. As the bead is rotated, torsional strain in the molecule accumulates until a critical point at which the DNA buckles and forms writhe. Writhe formation is observed by measuring changes in the extension of the DNA molecule. After the buckling transition is reached, the extension decreases in a linear fashion as the magnetic bead is turned. It should be noted that predicting the extension of a supercoiled molecule is non-trivial and requires numerous model-dependent assumptions (Strick *et al.*, 1998). Consequently, the conclusions from the single-molecule data are only as reliable as the model used to interpret them. Nevertheless, these single-molecule experiments have provided an unprecedented insight into the properties of DNA.

2.4. Results from electron and atomic force microscopy. In an experimental tour de force, Cozzarelli and co-workers (Boles *et al.*, 1990) measured writhe in negatively supercoiled DNA using transmission electron microscopy, and by analyzing the products of integrase recombination that traps writhe as catenane or knot nodes. They concluded that Wr changes linearly, proportional to decreasing ΔLk. The ratio of Wr to ΔTw remained at a constant ratio of $\sim$ 2.6:1 across the wide range of negative supercoiling tested. Dubochet and co-workers (Adrian *et al.*, 1990) used cryoEM to visualize supercoiled DNA in solution without the problems of sample preparation inherent to transmission electron microscopy. They found, similar to Cozzarelli and co-workers, that negatively supercoiled DNA forms a writhed, interwound superhelix in solution, with a ratio of Wr to ΔTw of $\sim$ 3-4:1. No analogous study has been performed for positively supercoiled DNA. Cherny and Jovin probed the structure of positively as well as negatively supercoiled DNA by transmission electron microscopy and atomic force microscopy (Cherny and Jovin, 2001). The positively supercoiled DNA tested, however, was only of low to moderate superhelix density because of limitations in the protocol used to supercoil the DNA. They reported that negatively or positively supercoiled DNA with comparable low values of ΔLk had a similar writhed structure when the samples were prepared and mounted under identical solution conditions. The configuration of more positively supercoiled and potentially more biologically relevant DNA remains to be determined.

2.5. Results from single molecule experiments. A useful feature of the single-molecule set-up is that the magnetic field can be made to pull on the bead and therefore apply tension to the DNA. This tension resists the contraction of the DNA to inhibit writhing. Under low tension (0.2 pN) the elastic behavior of DNA is symmetrical for positive and negative supercoiling. At higher forces, however, the chiral nature of DNA becomes apparent (Strick *et al.*, 1998). With the DNA held under higher, but still moderate tension (1 pN), overwinding causes the DNA to contract, and underwinding does not change the extension. At higher forces ($>$ 3 pN), writhing is completely inhibited for both overwinding and underwinding, consequently, the extension of the molecule changes very little for both negatively and positively supercoiled DNA. When writhing is suppressed, changes in Lk must be manifested as ΔTw. There is ample evidence, however, that negatively supercoiled DNA can only accommodate small negative changes in Tw before it denatures at susceptible sequences (Benham, 1979; Benham, 1992). This denaturation allows the DNA to maintain the remainder of the DNA in a near completely torsionally relaxed conformation (Randall G.R., Zechiedrich L. and Pettitt B.M., *submitted*). The phenomenon of localized denaturation of DNA was recognized early in the study of DNA supercoiling (Vinograd and Lebowitz, 1966), yet classical mechanical models of DNA have only very recently included such DNA denaturation (Liverpool *et al.*, 2008).

For positive supercoiling, denaturation to single strands is not applicable. Thc DNA is, therefore, obligated to writhe to relieve torsional stress. In this way, positively supercoiled DNA exhibits classical elastic rod behavior over a wider range of torsion and tension than does negatively supercoiled DNA. Negatively supercoiled DNA deviates from this idealized behavior under relatively low levels of tension and torsion, due to denaturation, whereas positively supercoiled DNA remains base paired. It is only at much higher forces, where writhe in positively supercoiled DNA is suppressed, that the elastic behavior of positively supercoiled differs from that predicted by the worm-like chain model. It appears that analogously to underwinding, the torsional stress of overwinding is not uniformly distributed throughout the molecule. Instead, under these conditions the excess twist is concentrated in regions of extremely overwound DNA with a helical repeat estimated from modeling to be $\sim$ 2.6 base pairs per turn compared with $\sim$ 10.5 bp per turn for B-DNA helix (Allemand *et al.*, 1998). This novel DNA conformation, named Pauling-DNA (P-DNA) will be discussed in more detail in a later section. For both negatively and positively supercoiled DNA it appears that changes in twist are concentrated in short regions of the helix, preserving the remainder of the DNA as a canonical B-form helix (Randall G.R., Zechiedrich L., and Pettitt B.M., *submitted*). There is distinct asymmetry, however, between positively and negatively supercoiled DNA regarding the amount of torsion required for these sharp structural transitions to occur. The assumption that twist is distributed uniformly throughout the molecule therefore only holds true for relatively small torsional deformations, especially for negatively supercoiled DNA.

2.6. Computational simulations of DNA supercoiling. To fully understand the experimental results requires knowledge of the underlying principles that govern the conformational changes. The study of DNA supercoiling has been enriched by collaborations between theoreticians and experimentalists. Computational simulations of DNA conformations have revealed much about DNA supercoiling. These simulations use well-established theories from polymer physics, which long pre-date the discovery of the structure of DNA. Because of the complexity of DNA, early computational models had to make coarse simplifying assumptions. Until very recently, the most common theoretical and computational models of large DNA systems have modeled DNA as an isotropic elastic polymer (Hagerman, 1988). These models have been tested and refined extensively to attempt to account for the unique characteristics of DNA (Benham and Mielke, 2005; Olson, 1996; Vologodskii *et al.*, 1992). As described above, measurements of the mechanical properties of single molecules of DNA provided a means to rigorously test these models over a wide range of negative and positive supercoiling (Smith *et al.*, 1992; Strick *et al.*, 1996). The data were best described by the Kratky-Porod worm-like chain model, which models the DNA as a continuous elastic medium (Bustamante *et al.*, 1994; Strick *et al.*, 1998). In contrast to the freely-jointed chain model, in which the molecule is divided into uncorrelated segments, the worm-like chain model treats the molecule as a continuous curve for which bending and twisting is distributed uniformly throughout the molecule.

Agreement between the structure computationally predicted and that observed experimentally does not necessarily mean that the model used is correct. Although the worm-like chain models accurately predict the behavior of the long DNA chains used in these experiments they fail to account for other experimental observations. For example, Cloutier and Widom reported that the bending and torsional flexibilities of certain short DNA sequences, $\sim$ 100 bp in length, exceeded the values predicted from a continuous elastic material by several orders of magnitude (Cloutier and Widom, 2004; Cloutier and Widom, 2005). The authors argued that conventional theories of DNA elasticity, determined for gently bent regimes, do not extrapolate to the sharply bent regimes found in, for example, small DNA circles. As discussed in more detail below, however, this conclusion ignores the unique structural properties of the DNA sequences studied by Cloutier and Widom. When the sequence-dependent structural properties are included in the computational model, the unusual DNA flexibility observed by Cloutier and Widom can be accounted for, within the framework of the existing theories (Czapla *et al.*, 2006).

A major shortcoming of theoretical studies of DNA supercoiling is their inability to accurately determine the relative contributions of enthalpic and entropic components to the free energy of supercoiling. Interactions between DNA and the surrounding solvent that are not explicitly considered in classical models almost certainly make a major contribution to the thermodynamics of supercoiling, especially entropy. Models of DNA elasticity consider the elastic free energy of supercoiling as determined from measured torsional and bending rigidities of DNA. These properties are determined experimentally in solution, and therefore should include interactions between the DNA and the solvent, including entropic effects. Although the entropic term is notoriously difficult to calculate, it is an important determinant of the mechanical properties of DNA (Harris *et al.*, 2005). Whenever models have been used in an attempt to determine the entropy of supercoiling, they incorrectly predict it to be negative, whereas experimental data overwhelmingly observe it to be positive (Benham and Mielke, 2005). In their 2005 review, Benham and Mielke eloquently describe reasons for the failure of models to predict the entropy and also argue that the elastic free energy may not be the primary determinant of the thermodynamics of supercoiling. The entropic term also becomes important when we consider the propensity of negatively supercoiled DNA to denature. Strand separation should increase the entropy of the system (Harris *et al.*, 2005; Harris *et al.*, 2008) and probably contributes to the large positive entropy observed experimentally. Therefore the entropy associated with positively supercoiled DNA, which has a much lower propensity to denature, is likely to significantly differ from that observed with negatively supercoiled DNA.

Accurately predicting conformations of positively supercoiled DNA, for which there is very little corroborating experimental data, will require a more detailed and quantitative understanding of the thermodynamics of supercoiling before we can be confident of the predictions. As simulations approach the all-atom level, we should gain a better understanding of these important thermodynamic terms.

In the classical models, the inability to allow DNA denaturation and the assumption that overwound DNA is equal and opposite to underwound DNA have obfuscated important general features of DNA. These two problems have led directly to the oversight of important differences between over- and underwound DNA. Extensions to the classical models have been made to attempt to address these differences, but the utility of such coarse-grained models has declined as atomic-scale modeling has become increasingly viable. Because of past limitations in computing power, all-atom molecular dynamics studies of the effects of supercoiling on DNA structure are few and far between. The computational time required for all atom molecular dynamics is approximately proportional to the number of atoms (including those from solvent) squared; therefore simulations of larger DNA systems are computationally expensive. Recent advances in supercomputing power, however, have made simulations of large DNA systems feasible. Unlike worm-like chain models, all-atom models explicitly take into account bond lengths, bond angles, dihedral angles, Van der Waals forces, and electrostatic forces, all allowing more realistic modeling of DNA dynamics. These models allow the simulation of DNA structural deformations such as kinking, twisting, and denaturing that are beyond the capabilities of worm-like chain models. In a recent study, Harris and co-workers (Harris *et al.*, 2008), performed all-atom molecular dynamics simulations on DNA minicircles ranging from 90 bp up to 178 bp in size. Writhing, because it requires a bending of the DNA helix, is suppressed in DNA minicircles because of the limited length of DNA over which the bending can be distributed. Harris and co-workers found that the simulated properties of positively and negatively supercoiled minicircles were highly dissimilar. The positively supercoiled minicircle simulations had a much higher propensity to writhe, even for minicircles as small as 90 bp. In contrast, the negatively supercoiled minicircles had a tendency to denature regions of the DNA instead of writhing. Writhing of negatively supercoiled minicircles was only observed for 118 bp and larger circles under conditions of high counterion screening, which favor writhing. We observe similar behavior experimentally with 339 bp DNA minicircles (Fogg, J.M., Catanese, D.J. and Zechiedrich L., *manuscript in preparation*). In the absence of added divalent metal ions to shield the negative charge on the DNA, writhing of the negatively supercoiled minicircles was suppressed. The ΔLk must be accommodated largely by denaturation in this case. The positively supercoiled minicircles writhe, however, even under the conditions of low counterion screening.

2.7. Torsional rigidity. The partition of supercoiling into twist and writhe is governed at least in part by the torsional and bending rigidities of the DNA molecule. These properties have been studied extensively using a number of methods including fluorescence anisotropy of intercalated ethidium bromide (Fujimoto and Schurr, 1990; Heath *et al.*, 1996), rates of cyclization of short DNA fragments (Cloutier and Widom, 2004; Cloutier and Widom, 2005; Crothers *et al.*, 1992; Shore and Baldwin, 1983a; Shore *et al.*, 1981), the equilibrium topoisomer distribution of ligated DNA circles (Horowitz and Wang, 1984; Shore and Baldwin, 1983b), and from direct mechanical measurements on single DNA molecules (Bryant *et al.*, 2003; Strick *et al.*, 1999). In order to accurately model DNA as an isotropic elastic rod, it is important to have a precise value for these properties. The values obtained for the torsional rigidity depend significantly upon which method is used. Common to all these methods is a requirement for sophisticated theoretical analyses of the data to estimate the relative contributions of twist and writhe to the results. Writhing acts to reduce the torsional strain and, unless writhe is accurately accounted for, will lead to underestimates of the torsional rigidity. For small DNA circles, it is generally assumed that supercoiling is partitioned mostly into twist. Writhing requires a bending motion of the DNA helix, and most of this bending is concentrated in the superhelical apices. It follows that as DNA circles get smaller there should be increased resistance to the introduction of writhe. For this reason, studies that measure the cyclization of short DNA fragments, either from rates of cyclization (Shore *et al.*, 1981) or from the resulting topoisomer distributions (Horowitz and Wang, 1984) are generally considered to yield the most reliable estimations of torsional and bending rigidity because the contribution of writhe is small in these assays. Nevertheless, Levene and Crothers (1986) predicted that the contribution of writhe, although small, is not negligible and assuming absence of writhe leads to an underestimate of the torsional rigidity.

Many of these studies, in particular ones that measure the cyclization of short DNA molecules, sample thermally accessible states close to the most relaxed topoisomer. Consequently, the dependence of the torsional and bending rigidities on superhelicity is unknown. It is often assumed that the torsional rigidity is that of an isotropic elastic rod, *i.e.* it is does not change as the DNA is underwound or overwound. Models of DNA elasticity assume a constant value for the torsional rigidity that is independent of supercoiling density. This assumption ignores the fact that DNA is a right-handed helix, thus the torque required to introduce positive ΔTw probably differs from the torque required to introduce negative ΔTw. A limited number of studies have sought to address the dependence of the torsional rigidity on supercoiling. From fluorescence anisotropy measurements of intercalated ethidium bromide, Selvin *et al.* (1992) concluded that positively supercoiled plasmid DNA is significantly more torsionally flexible than negatively supercoiled plasmid DNA. Under conditions of high ionic strength ($\sim$ 175 mM), the authors observed a linear dependence of the torsional rigidity on supercoiling density. Under low-salt conditions (ionic strength $\sim$ 7.5 mM), the observed relationship was more complex, which the authors attributed to the formation of alternative structures in negatively supercoiled DNA. Another source of inaccuracy in the reported values (at both high-salt and low-salt) was the assumption that the fraction of ΔLk that is partitioned into ΔTw is a constant value. This assumption is almost certainly incorrect and raises serious doubts about the validity of some of these findings.

The development of techniques to micro-manipulate single molecules of DNA provided a means to directly measure torque (Bryant *et al.*, 2003). The application of tension to the DNA in these measurements should act to suppress writhing, presumably leading to a more accurate determination of the torsional rigidity. The authors measured torque as a function of ΔTw over a wide range of negative and positive supercoiling. Torque was observed to increase almost linearly with ΔTw as the DNA was underwound, until a critical torque (−9.6 pN·nm) was reached, beyond which the torque remained constant even as the DNA was further underwound. This plateau probably reflects the denaturation of the helix to single strands beyond this critical torque. For overwinding, there was a similar near-linear increase in torque as ΔTw was increased. The near-linear region of the data extended much beyond that observed for underwinding, however, and the plateau was not reached until a critical torque of 34 pN·nm. Beyond this critical torque, no further increase in torque was observed even as the DNA was further overwound. The constant torque region was postulated to reflect the conversion of canonical B-DNA into the extremely overwound helical form called Pauling-DNA (Allemand *et al.*, 1998). For the quasi-linear region of the data, the data were fit best by the model proposed by Selvin *et al.* (1992), which includes a term to account for changes in torsional rigidity as the DNA is overwound. The observed deviation from linearity was small, however, and corresponds to a less than 10% decrease in the torsional rigidity for the most overwound DNA.

The presence of tension in the single-molecule experiments described above should act to completely suppress writhing. This should make analysis of the data more straightforward and consequently more reliable. The direct measurement of the torsional rigidity of single DNA molecules yielded a torsional modulus of 440 pN·nm^2 (Bryant *et al.*, 2003), that although higher than the widely used value of $\sim$ 300 pN·nm^2 (Hagerman, 1988), agrees with the highest estimates predicted from earlier cyclization and topoisomer distribution studies (Levene and Crothers 1986, Crothers *et al.*, 1992). This value is significantly higher than the torsional modulus $\sim$ 200 pN·nm^2, calculated by Selvin *et al.* (1992), using fluorescence anisotropy measurements of intercalated ethidium bromide. A potential explanation for this difference is that previous studies have not sufficiently accounted for writhing and therefore underestimate the torsional rigidity. It should be noted that the moduli extracted from all these experiments, including the single-molecule experiments, are apparent twisting rigidities and reflect many variables that are subsumed in the analysis.

It may be argued that the single-molecule experiments do not deform the DNA in a way that is comparable to the deformations that occur as a consequence of supercoiling. Benham and Mielke argued that because the DNA in the single molecule experiments is conformationally constrained, the measurements obtained from these studies are not necessarily applicable to free, unconstrained DNA in solution (Benham and Mielke, 2005). The torsional rigidity of the linear DNA molecule in these assays may not be the same as that of a writhed DNA molecule. It is often assumed that bending and twisting are independent of each other; however, this ignores potential effects of DNA bending on torsional rigidity and conversely potential effects of overwinding and underwinding on bending rigidity. By considering DNA as a right-handed helix with major and minor grooves, instead of an isotropic elastic rod, Marko and Siggia postulated a coupling between twist and bend (Marko and Siggia, 1994). They argued that DNA bending should lead to an unwinding of the helix. Analogously, underwinding should make the molecule easier to bend and overwinding should increase the resistance to bending.

The coupling between twist and bend was observed prior to the studies of Marko and Siggia. The relative rotation and displacement of neighboring base pairs can be described by six parameters: twist, tilt, roll, shift, slide and rise. The coupling of these parameters has long been recognized, the most notable example being the coupled changes in twist, roll and slide that accompany transitions between the B and A forms of DNA (Calladine and Drew, 1984). One of the most important correlations is between twist and roll (Gorin *et al.*, 1995). Unwinding of the helix is usually mediated by a decrease in twist between base pairs. This is generally accompanied by positive roll (a bend into the major groove). Analogously, overwinding of the helix, mediated by an increase in twist between base pairs is usually accompanied by negative roll (a bend into the minor groove).

Schurr and co-workers (Heath *et al.*, 1996) compared the torsional rigidity of small DNA circles to that of linear DNA by measurements of fluorescence anisotropy of intercalated ethidium bromide. Interestingly, they observed that the torsional rigidity was approximately 42% higher for the circular species, which they attributed to the bending strains in the small circles. The 181 bp circular species assayed in these experiments were significantly overwound (reported $\Delta Lk \sim +0.6$), which would suggest increased torsional rigidity for positively supercoiled DNA, in contradiction to the findings of Selvin *et al.* (1992) and Bryant *et al.* (2003).

Clearly, the dependence of torsional rigidities on supercoiling is far from being resolved. The single molecule results suggest that the assumption that DNA behaves as an isotropic elastic rod holds true more so for positively supercoiled DNA than it does for negatively supercoiled DNA. Given the potential question of how accurately these experiments reproduce the conformational changes induced by supercoiling, independent verification of these properties would help to confirm these findings.

2.8. DNA crossovers. The conformation associated with negative supercoiling is a right-handed interwound superhelix, similar to the conformation illustrated by the rubber tubing model (Figure 1). This plectonemic structure has been observed by transmission and cryo- electron microscopy (Adrian *et al.*, 1990; Bednar *et al.*, 1994; Boles *et al.*, 1990; Cherny and Jovin, 2001) and in theoretical simulations of DNA supercoiling (Vologodskii *et al.*, 1992). The rubber tubing model is, of course, a vastly oversimplified representation of supercoiled DNA. Real DNA is not a rubber tube, but a right-handed helix with major and minor grooves and varying sequence composition. The actual conformation of the superhelix is affected by a number of factors, including allosteric interactions between grooves, sequence effects and the charge of the DNA helix. Electrostatic repulsion between the interwound helices inhibits DNA segments from coming into direct contact. Under conditions of low counterion screening, the helical axis of the DNA can be readily traced in electron micrographs (Boles *et al.*, 1990). In these electron micrographs one can clearly see the points at which the two segments of the helical cross each other in projection. These DNA juxtapositions (as shown in Figure 4) are referred to as "DNA crossovers" or "superhelical nodes". The crossovers in positively supercoiled DNA are left-handed, whereas the ones in negatively supercoiled DNA are right-handed (Figure 4).

Vologodskii and Cozzarelli predicted the angle juxtaposed between the two helices from Monte Carlo simulations of negatively supercoiled DNA (Vologodskii and Cozzarelli, 1996). A broad distribution of juxtaposition angles was computed, and the most frequently found angle was $\sim 60°$. A similar distribution of juxtaposition angles was later predicted for positively supercoiled DNA (Stone *et al.*, 2003). The similarity in the predicted juxtaposition angles for positively and negatively supercoiled DNA is not surprising as the DNA was modeled as an isotropic elastic rod and, thus, ignored how one helix may fit into the grooves of another helix. It is indisputable that the geometry of the helix, its sequence, and its negative charge will all affect the way DNA juxtaposes at DNA crossovers. Timsit and co-workers have shown that these factors influence the angle between the juxtaposed helices (Timsit and Moras, 1994; Timsit *et al.*, 1998; Timsit *et al.*, 1999). The interaction between the grooves and backbones of the two juxtaposed helices differs greatly between left- and right-handed crossovers such as occur in positively and negatively supercoiled DNA, respectively. To illustrate the findings of Timsit and co-workers, we modeled the juxtaposition of two helices in a left-handed and right-handed superhelical crossover, respectively (Figure 5). The right-handed nature of the helix causes the two helices to fit differently in left-handed and right-handed orientations.

Under conditions of high counterion screening, supercoiled DNA forms a tightly interwound superhelix (Bednar *et al.*, 1994; Fogg *et al.*, 2006). In this conformation, the helices are in close proximity and, therefore, the

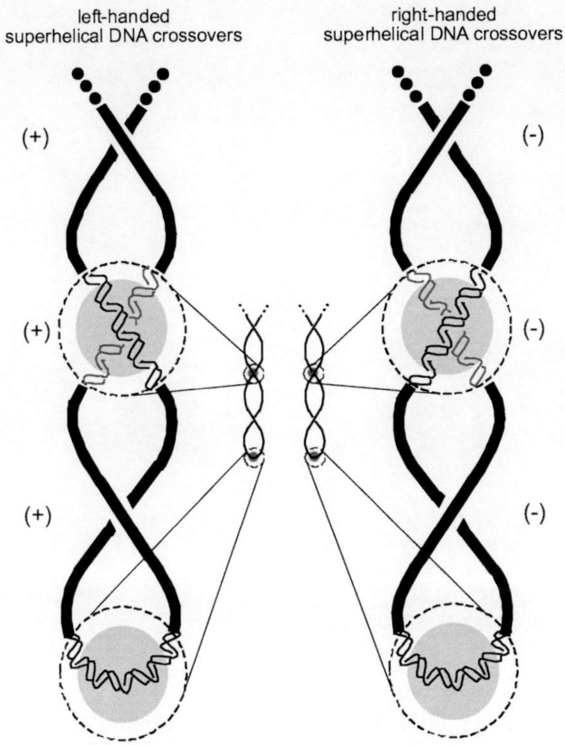

FIG. 4. *Left- and right-handed DNA-DNA juxtapositions: Positively supercoiled DNA contains left-handed crossovers whereas negatively supercoiled DNA contains right-handed crossovers. These DNA-DNA juxtapositions are longer than shown (denoted by ellipses). Many topoisomerases, as outlined later in the text, have been shown to bind superhelical crossovers and apices, both of which are shown magnified (dashed circles).*

geometric constraints imposed by the helix grooves are likely important. The angle juxtaposed between the two helices in positively supercoiled DNA therefore probably differs from the angle in negatively supercoiled DNA.

2.9. Electrostatics. Because DNA is a highly charged polymer, no discussion of DNA dynamics is complete without discussing the importance of electrostatics in mediating the structure and geometry of DNA helices. Every base pair has a net charge of −2 arising from the unshielded phosphate groups aligned along the surface of the backbone. Thus, over the length of a genome, a negatively charged electric field accumulates. The surrounding solvent and positively charged counterions play an important role in neutralizing this negative charge. Water screens electrostatic attractions because of its polarity—the electric field orients the positive poles of the water molecules towards the negatively charged nucleic acids. The

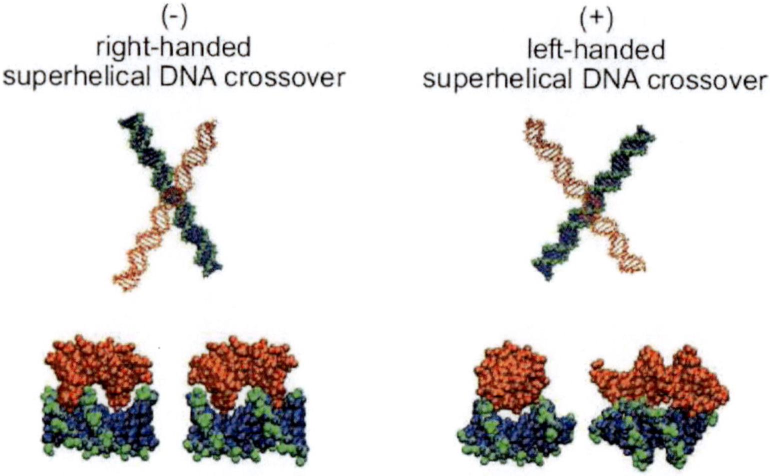

FIG. 5. *Differences between left- and right-handed juxtapositons: Left-handed juxtapositions are found in positively supercoiled DNA, whereas right-handed crossovers are found in negatively supercoiled DNA. Pictured are two DNA helices (one in red, the other in blue and green, space-filled) modeled in a left- and right-handed juxtaposition. The right-handed DNA helix causes the major and minor grooves of two helices to fit together differently in left- and right-handed orientations. Below each model are two alternative views of the crossover showing how the grooves and backbones fit together. There is a steric clash between the helices in a left-handed juxtaposition, but the helices fit well in a right-handed juxtaposition. The best fit angles differ. The relative orientation and interaction of DNA sequence differs. Even the way the DNA helices would interact during motion differs. This latter idea is best illustrated using two pieces of uncooked fusilli noodles and rolling them back and forth between your fingers. In one direction, the grooves fit nicely and slide with rolling. In the other direction, rolling is blocked. Movies showing panoramic views of each crossover are available in Supplemental Data.*

ease with which water orients itself in an electric field is the basis for its relatively high dielectric constant. Counterions are attracted to the negative charge density of DNA. Manning's counterion condensation theory predicts that positively charged ions, attracted to this plenitude of negative charge, condense on the DNA backbone and in its major and minor grooves (Manning, 1969a; Manning, 1969b). This creates a gradient distribution of counterions in the bulk solvent that decreases with distance from the DNA backbone, as DNA, in effect, absorbs counterions to neutralize its charge. The valence of the counterions affects the magnitude of counterion condensation. Multivalent cations are more strongly attracted to DNA and therefore more effective at screening the electrostatic forces (Pack *et al.*, 1999).

The condensation of counterions around the DNA molecule also affects its structure. Ignoring solution screening effects, it is easy to imagine that the electrostatic repulsion between the phosphate groups on the DNA backbone stretches the helical length and unwinds the helix. Therefore, it is not surprising that the helical repeat of DNA depends on the concentration of counterions, in particular multivalent cations, to neutralize the electrostatic repulsion between the backbone phosphate groups (Rybenkov *et al.*, 1997b; Xu and Bremer, 1997). This phenomenon was demonstrated early on from fiber-diffraction experiments of DNA. Wilkins, Hamilton and co-workers (Marvin *et al.*, 1958) observed a reversible transition of the diffraction pattern of DNA, in the presence of lithium and in conditions of reduced humidity, to a pattern similar but distinct from that observed at higher humidity. This pattern was consistent with a structure similar to B-DNA, but with a significantly reduced helical pitch that they predicted to be approximately 9.3 base pairs per turn, instead of the 10 base pairs per turn they predicted for B-DNA. They named this overtwisted DNA helix, C-form. Anderson and Bauer (1978) later systematically quantified the overtwisting of the DNA helix in the presence of monovalent cations, and determined that in each case as the salt concentration increased, the DNA helix became progressively overwound. Polyvalent cations bind DNA with higher affinity than monovalent ions, and consequently induce an even greater winding effect (Xu and Bremer, 1997).

Whereas the presence of counterions can affect the helical repeat of DNA, overtwisting and undertwisting could affect the concentration of counterions. In static atomic models, overtwisting increases the charge density of DNA by reducing the width of the minor groove, in particular, and bringing the phosphate groups closer together. This increased charge density should attract higher concentrations of cations. Conversely, undertwisting decreases the charge density by widening the grooves. As a result, we expect the linking number of DNA to have an effect on local counterion concentrations. Molecular dynamics simulations showed an increase in sodium ion concentrations in the grooves as DNA was overtwisted (Randall G.L., Zechiedrich L. and Pettitt B.M., *submitted*).

In addition to its effect on interhelical structure, the composition of the solvent is also a critical component in determining the global geometry of DNA. The solvent and its counterions not only screen interactions within the helix, they also screen intrahelical electrostatic interactions. Writhe is only possible because counterions in the solvent screen the electrostatic repulsions in the juxtaposition of two helical segments. The exchange between twist and writhe as a function of bulk solvent concentrations has been observed by cryoEM and atomic force microscopy (Adrian *et al.*, 1990; Bednar *et al.*, 1994; Cherny and Jovin, 2001). These studies observed that the number of crossovers in supercoiled plasmids increased with corresponding increases in the bulk concentrations of counterions. A study using Poisson-Boltzmann theory showed that the minimum concentration of monovalent counterions required to effectively screen two helices in a crossover separated by 1 nm is only 50 mM (Randall *et al.*, 2006), less than the estimated concentration of counterions in cells.

2.10. Deviations from canonical B-DNA. The stabilization of alternative, non-B DNA structures by negative supercoiling is well-documented. These include a cornucopia of unusual DNA conformations including cruciforms, left-handed Z-DNA, slipped DNA, DNA triplexes, and tetraplexes (reviewed in Bacolla and Wells, 2004; Wang and Vasquez, 2006; Wells *et al.*, 2005; Wells, 2007; Wang and Vasquez, 2007). The majority of these are induced by localized unwinding of a region of DNA, subsequent formation of the alternative DNA structure, accompanied by a concomitant decrease in the overall level of supercoiling. The free energy liberated from the reduction in supercoiling therefore compensates for the energy required to form the alternative structure. As a result, the structures mentioned above are exclusively found in negatively supercoiled DNA. These structures are also associated with certain DNA sequences, and consequently appear sporadically throughout the genome.

As we have described above, positively supercoiled DNA has a much lower propensity to denature. Therefore, it is difficult to envision how the alternative structures mentioned above may form in positively supercoiled DNA. There is, however, evidence for the existence of unusual non-B DNA conformations in positively supercoiled DNA. The first indication of the presence of non-canonical DNA structures in overwound, positively supercoiled DNA was its reactivity to osmium tetroxide, a chemical used to probe for exposed DNA bases (McClellan and Lilley, 1991). This sensitivity suggested, somewhat surprisingly, that positively supercoiled DNA may have unpaired bases exposed to solvent. The development of techniques to micromanipulate single molecules of DNA allowed researchers to study this structural transition in more detail. As described above, whereas negatively supercoiled DNA was observed to denature at relatively low levels of torsion and tension, much higher forces were required to denature positively supercoiled DNA (Allemand *et al.*, 1998; Strick *et al.*, 1998). Only when the tension was above a critical value, at which writhe was completely suppressed, was denaturation detected for positively supercoiled DNA (Allemand *et al.*, 1998). Under these conditions, supercoiling is manifested wholly as twist. When a critical amount of twist was added, torsional strain was relieved by a localized transition of the DNA into an extended and highly twisted structure. To demonstrate that the bases in this novel structure are exposed to the solvent, the authors used glyoxal, a reagent that reacts with unpaired DNA bases. Lavery used molecular modeling to investigate possible structures for this novel DNA form (Allemand *et al.*, 1998). When a critical threshold of twist was reached, this could no longer be accommodated as B-DNA and the bases were expelled from the double helix. This enabled the DNA backbones to move towards the helical axis, resulting in an inside-out double helix. This novel DNA conformation, with the two backbones wrapped about each on the inside and unpaired bases on the outside, is reminiscent of the DNA structure originally proposed by Linus Pauling (Pauling and Corey, 1953) and was thus named Pauling DNA (P-DNA). With a helical repeat of $\sim$ 2.6 base pairs per turn, as estimated from molecular modeling, and an extension $\sim$ 75% longer than B-DNA, this is an extremely overwound and extended DNA conformation.

It is pertinent to ask the question; does P-DNA exist under physiological conditions? An analogous DNA structure has been reported for the DNA genome of the PfI virus (Liu and Day, 1994). The Pf1 virus genome is, however, single-stranded. The P-DNA like structure adopted involves interwinding of the single strands and is stabilized by interactions with viral coat proteins. There is as yet no clear evidence for the existence of P-DNA under biologically relevant conditions. The forces required in the single molecule experiments to inhibit writhing ($\sim$ 3 pN), as required for P-DNA generation, are similar in magnitude to the forces exerted by polymerases during transcription and replication ($\sim$ 14 pN, (Yin *et al.*, 1995)). Therefore, it is not inconceivable for P-DNA to form transiently during these processes. Because formation of P-DNA involves a complete loss of base pairing, if it does exist *in vivo* it will most likely be associated with certain DNA sequences that are more susceptible to localized melting of the DNA strands. Indeed, McClellan and Lilley (1991), observed that sensitivity to osmium tetroxide in positively supercoiled DNA was associated with alternating adenine-thymine sequences. These sequences are susceptible to localized melting and form cruciform structures in negatively supercoiled DNA. Consequently, certain DNA sequences may be susceptible to deformation both by overwinding or underwinding.

2.11. Effect of supercoiling on protein-DNA interactions. Relatively little is known about how proteins interact with supercoiled DNA. All crystal structures of protein-DNA complexes and almost all data for proteins binding to DNA come from experiments using the linear form of DNA. In many of these structures, individual base-pair steps are overtwisted or undertwisted (Olson *et al.*, 1998). The ΔTw either arises from, or is stabilized by, interactions with the protein and is not detected in protein-free linear DNA. The likelihood of these changes occurring should be strongly influenced by DNA supercoiling. In the absence of a crystal structure with a supercoiled DNA molecule, the details of how a protein interacts with a physiologically relevant supercoiled substrate remains speculative. We will discuss in more detail below how type IA topoisomerases recognize helical twist. Here we describe general concepts that should apply to all protein-DNA interactions.

One way that DNA supercoiling may facilitate the formation of protein-DNA complexes is by facilitating DNA bending. Dramatically bent or protein-wrapped DNA is observed in the crystal structures of many protein-DNA complexes (Dickerson, 1998; Olson *et al.*, 1998). Examples include the Lac repressor (Lewis *et al.*, 1996), TATA binding protein (Kim *et al.*, 1993a; Kim *et al.*, 1993b), nucleosomes (Luger *et al.*, 1997) and yeast topoisomerase II (Dong and Berger, 2007). Considerable free energy must be expended in order to overcome the intrinsic resistance of a short DNA helix to bending. Despite this energetic cost, DNA bending is a common feature of many site-specific protein-DNA interactions (Maher, 1998). Conventional wisdom states that the necessary DNA bending energy is provided by the favorable protein-DNA interactions that result from formation of the protein-DNA complex (Maher, 1998). This mechanism assumes an almost passive role for the DNA with the protein providing the driving force for DNA bending. Harnessing the free energy of supercoiling should reduce the additional energy that must be provided by a protein. Protein-DNA interactions may also be modulated by the intrinsic deformability of particular DNA sequences *e.g.*, the kinking of the central TA step in the sequence recognized by the restriction enzyme EcoRV (Hiller *et al.*, 2005), or the sequence-dependent deformations of nucleosomal DNA (Tolstorukov *et al.*, 2007). Olson and co-workers (Olson *et al.*, 1998) used structural information extracted from the Protein Data Bank (Berman *et al.*, 2000) to determine the contribution of DNA sequence to DNA flexibility. Most theories of DNA flexibility do not consider the unique features of a particular sequence and therefore may misestimate the bending energy required. Most, if not all, protein-DNA interactions should also be influenced by DNA supercoiling. Indeed, in rare cases where binding to supercoiled DNA has been assayed, the interaction is generally observed to be promoted by DNA supercoiling, including for example, Lac repressor (Whitson *et al.*, 1987). How does supercoiling promote protein-DNA interactions?

A fundamental way in which protein-DNA interactions may be modulated is through pre-existing curvature in the DNA, such as that observed for the TATA binding protein. When the binding site on the DNA was "pre-bent" (by virtue of being constrained in a 156 bp minicircle), the protein bound with 100-fold higher affinity than to linear DNA of identical sequence (Parvin *et al.*, 1995). The presence of a pre-existing bend in the DNA should reduce the required bending energy and therefore promote complex formation (Maher, 1998). Supercoiled DNA has considerable curvature, both at the superhelical apices (Pavlicek *et al.*, 2004), and at the helix-helix juxtapositions in writhed DNA (Figure 4). Lyubchenko and co-workers postulated that, because supercoiled DNA is very dynamic, protein binding sites can migrate to the superhelical apices and become bent (Pavlicek *et al.*, 2004). Superhelical apices are common to both positively and negatively supercoiled DNA (Figure 4), therefore this facilitated binding is thought to be equal for supercoils of either handedness.

Isotropic elastic rod models of DNA elasticity predict that smooth, continuous planar bending should be the most energetically favorable. From analysis of 86 published structures of protein-DNA complexes, Dickerson found that for 79 of the 86 structures, the DNA in the complex was bent (Dickerson, 1998). Dickerson classified the DNA bending in these 79 structures into one of three categories: (i) smooth, continuous bending in a single plane, (ii) nonplanar bending (writhe), or (iii) kinks, in which the helix makes a sharp turn at one or two discrete loci. Each type of bend has distinct characteristics and consequently supercoiling would be expected to affect each type differently. Only 11 of the 79 structures were classified with smooth, continuous, planar DNA bends. The reasons for this became evident when the coupling of the parameters (twist, tilt, roll, shift, slide and rise) that describe the relative rotation and displacement of neighboring base pairs were examined (Dickerson, 1998; Olson *et al.*, 1998). When these parameters were considered, planar bending was much more difficult to accommodate than either writhing or kinking (Dickerson, 1998). Consequently, writhing and kinking are much more common. Indeed, 39 of the 79 complexes analyzed had writhe, 23 had kinks, and a combination of kinking and writhing was seen in 6 of the complexes.

Many of the bends observed in protein-DNA complexes do not lie in a single two-dimensional plane, but instead exhibit three-dimensional writhe. The non-planar nature has important consequences, as illustrated by the example of the recombination protein, integrase host factor (IHF), which is observed bound to DNA with a 160° bend (Rice *et al.*, 1996). An important feature of integrase-mediated recombination is its strong dependence upon negative (Mizuuchi *et al.*, 1978), but not positive (Richet *et al.*, 1986) supercoiling (indeed this requirement led to the discovery of DNA gyrase, the topoisomerase responsible for introducing negative supercoils into bacterial DNA (Gellert *et al.*, 1976)). DNA supercoiling, and more specifically, DNA writhe, promotes formation of a productive integration complex, which includes IHF and λ integrase, by promoting and stabilizing DNA bending at the recombination site (Richet *et al.*, 1986). Landy and co-workers proposed that IHF is localized to superhelical apices where the DNA is sharply bent and this facilitates the correct positioning of the DNA sequence at the bend (Nunes-Duby *et al.*, 1995). Thus, if the only purpose of supercoiling is to reduce the energetic cost of bending, we would expect negatively and positively supercoiled DNA to be acted upon equally by IHF. The crystal structure of the IHF-DNA complex (Rice *et al.*, 1996) revealed that the DNA is not bent in a single plane, but instead makes a dihedral angle of $\sim 15°$, creating a small, but significant amount of negative writhe. The handedness of this writhe may explain why positively supercoiled DNA is an insufficient substrate for integrase-mediated recombination.

The frequency of kinks in the structures of protein-DNA complexes (Dickerson, 1998) is difficult to rationalize with traditional isotropic elastic rod models of DNA flexibility. These isotropic elastic rods, by definition, do not kink spontaneously (Maher, 2006), which, perhaps erroneously, led to the common assumption that the kinks must be induced entirely by mechanical forces generated by protein binding. Indeed, there are a number of mechanisms a protein may employ to induce or stabilize a kink in the DNA. One example comes from the recent crystal structure of yeast topoisomerase II bound to DNA. In this complex, an amino acid sidechain from each subunit of the protein intercalates between base pairs, trapping or generating two sharp kinks in the bound DNA (Dong and Berger, 2007). These two kinks result in an overall global bend of $\sim 150°$.

Although long predicted (Crick and Klug, 1975), the existence of kinks in protein-free DNA is controversial. Recent findings, however, suggest that spontaneous kinking of protein-free DNA may occur, and is facilitated by the bending and torsional stress of DNA supercoiling. Interest in the phenomenon of localized sharp DNA bending and kinking was stimulated by the somewhat surprising results from Cloutier and Widom (2004). As discussed above, the authors reported that short DNA fragments (< 100 bp) formed DNA circles several orders of magnitude more efficiently than current DNA bending theory predicts computationally. Vologodskii and co-workers found that by modifying traditional DNA bending theory to allow for "rare" transient appearance of sharp DNA kinks (Du *et al.*, 2005), the high cyclization efficiencies reported by Cloutier and Widom could be explained. Missing from either analysis, however, was a consideration of sequence-dependent DNA deformability. Olson and co-workers (Czapla *et al.*, 2006) have developed a new coarse-grained model of DNA flexibility to account for sequence-dependent variability of each base pair step. Using this new sequence-dependent representation of DNA, Olson and co-workers could account for the high intrinsic flexibility of the nucleosome-positioning sequences studied by Cloutier and Widom without needing to invoke large distortions, *e.g.*, kinks, of the DNA.

In molecular dynamics simulations of relaxed and positively supercoiled 94 bp minicircles, Lankas *et al.* observed that even a relaxed minicircle kinked at a single location as a consequence of extreme curvature, but two kinks relieved the bending and torsional strain for the positively supercoiled minicircle (Lankas *et al.*, 2006). Because of the scale of these simulations, they were groundbreaking at the time of publication, but the results have since been questioned as further studies (Svozil *et al.*, 2008, Perez *et al.*, 2008) have revealed flaws, such as underestimating the helical twist and falsely predicting α/γ backbone flips, in the AMBER ff94 force field. These flaws have since been rectified in more recent versions of the AMBER force field. In a more recent study, Harris and co-workers also observed sharp kinking during molecular dynamics simulations of positively supercoiled 90 bp minicircles (Harris *et al.*, 2008). In contrast, the same study found that negatively supercoiled minicircles were observed to relieve torsional strain by localized melting of the DNA.

In experiments just published at the time of writing this review, Vologodskii and co-workers sought to determine if the sharp DNA kinks predicted from theory and simulations could be detected experimentally (Du *et al.*, 2008). Using enzymes specific for distorted structures in DNA the researchers probed for disruption of the DNA helix caused by the bending strain in 64 and 65 bp minicircles. They observed cleavage with Bal31 nuclease, an enzyme that cleaves at distorted structures in DNA, but not with

S1 nuclease, an enzyme that specifically cleaves single-stranded DNA. The authors postulated that this was evidence that the distortions observed do not disrupt base pairing, but, instead, represent sharp kinks of the helix. If the conclusions are correct, these results represent the first experimental evidence of spontaneous kinking in protein-free DNA, driven by bending strain. However, cleavage by Bal31 nuclease may not necessarily represent kinking of the helix. Although the conclusions are highly speculative, sharp DNA kinking may be more common than previously thought. The torsional strain in negatively supercoiled DNA may be partially relieved by localized denaturation of the helix to single strands. Positively supercoiled DNA is, however, resistant to strand separation and therefore the torsional strain must be relieved by other means, potentially including spontaneous kinking. The molecular dynamics simulations of minicircle DNA suggest that kinking is an inherent response of DNA to the torsional and bending strain associated with DNA supercoiling. Protein-DNA interactions that invoke DNA bending and kinking should therefore be facilitated by DNA supercoiling.

3. How topoisomerases recognize their DNA substrate and distinguish positive from negative supercoils. Just like for any enzyme that acts on DNA, before a topoisomerase can act, it must first find its DNA substrate and decide where and when to act. Without a doubt, the initial interaction with DNA must be non-specific because DNA is very long and the enzyme, relative to DNA, is very small. At the same time, there are many topoisomerases in cells. Precisely determining the number of enzymes in cells is technically difficult, but chicken cells, for example, have been estimated to have $\sim 7.5 \times 10^4$ topoisomerase II copies per cell. With the chicken genome being 2.4 billion base pairs, there is about one functional enzyme dimer per 16,000 base pairs (Heck and Earnshaw, 1986). This means that there are plenty of enzymes available to handle what seemed early on to be a mathematically insurmountable problem associated with replicating and segregating the double helix structure of DNA. In cancer cells, human topoisomerase IIα has been estimated to be as high as 1×10^6 copies per cell, 100-fold more than normal human cells (Heck and Earnshaw, 1986). This topoisomerase overexpression may be a consequence of the highly active replication machinery in rapidly proliferating cancer cells.

If topoisomerases were to act indiscriminately, cleaving and passing DNA strands through each other at random, they would simply generate an equilibrium distribution of knots, catenanes and supercoils (Vologodskii *et al.*, 2001). However, topoisomerases do not treat DNA molecules as so-called "phantom chains", where indiscriminate strand passage freely occurs (Sikorav and Jannink, 1994). If they did, then because DNA is compacted into a small volume in the cell, the knotting probability would be very high (Arsuaga *et al.*, 2002). Such DNA knotting can lead to disastrous consequences (Arai *et al.*, 1999; Deibler *et al.*, 2007; Portugal and

Rodriguez-Campos, 1996; Rodriguez-Campos, 1996; Saitta *et al.*, 1999). A testament to the efficient and precise action of the topoisomerases is the low level of knotting observed in cells, thus preserving the integrity of the genomic material.

Topoisomerases likely employ a number of strategies to find their DNA substrates. The enzymes all contain positively-charged regions that are required to approach closely to the negatively-charged backbone of DNA. As will be discussed below, there are classes and subclasses of topoisomerases, each with preferred DNA substrates. Through a series of hops (three-dimensional) and slides (one-dimensional) (Terry *et al.*, 1985; reviewed in von Hippel, 2007), all topoisomerases search through non-specific DNA for features that are characteristic of its substrate. Such features may include for negatively supercoiled DNA, single-stranded regions, and for both positively and negatively supercoiled DNA, helix juxtaposition geometry or DNA curvature. Most topoisomerases probably do not impose specific geometry, but rather recognize pre-existing local information. Some topoisomerases also are able to distinguish positively from negatively supercoiled DNA. Binding to the correct substrate DNA may trigger conformational changes in the topoisomerases (Crisona and Cozzarelli, 2006) so that they act only at the correct target site.

How do topoisomerases recognize the global topology of DNA and how do some topoisomerases preferentially relax positively over negatively supercoiled DNA? What are the unique features of positively supercoiled DNA that these enzymes recognize? Supercoil chiral sensing cannot be a prerequisite of topoisomerase function or it would be exhibited by all topoisomerases. As described above, one key difference between positively and negatively supercoiled DNA is twist. Positively supercoiled DNA has increased twist, resulting in a decreased helical repeat and is resistant to denaturation, whereas negatively supercoiled DNA has decreased twist, resulting in an increased helical repeat and a high propensity for denaturation. Could topoisomerases recognize a difference in twist? This is almost certainly the case for a number of type I topoisomerases.

3.1. Mechanism of type I topoisomerases. Type I topoisomerases can be subdivided further into three categories (types IA, IB and IC) depending upon differences in their structure and mechanism (Table 1) (Champoux, 2001; Corbett and Berger, 2004; Wang, 1996; Wang, 2002; Forterre, 2006; Forterre *et al.*, 2007; Schoeffler and Berger, 2008). Type IA topoisomerases, which include the *E. coli* enzymes topoisomerase I and topoisomerase III, can relax negative, but not positive supercoils (Wang, 1971). This is a consequence of the way these enzymes recognize supercoiled DNA. *E. coli* topoisomerase I preferentially binds to DNA that contains a short single-stranded loop (Kirkegaard and Wang, 1985). If DNA is constructed to contain an unpaired single-stranded region of DNA and the experiment is performed in the presence of ethidium bromide, *E. coli* topoisomerase I is now able to relax the apparent positively supercoiled

DNA (Kirkegaard and Wang, 1985). Crystal structures of *E. coli* topoisomerase I (Lima *et al.*, 1994) and *E. coli* topoisomerase III (Mondragon and DiGate, 1999) revealed single-stranded DNA binding grooves in these enzymes that were later confirmed to bind single-stranded oligonucleotides (Changela *et al.*, 2001; Perry and Mondragon, 2003). These structures revealed a "decatenation loop" present in topoisomerase III, but not in topoisomerase I, which influences the preference of topoisomerase III to decatenate much more readily than it relaxes DNA. Topoisomerase I, which lacks the decatenation loop, is a poor decatenase, but excellent relaxase. To relax negatively supercoiled DNA, the type IA enzymes unpair a short region of DNA, catalyze a transient break in the single-stranded region, pass a single strand through the transient, protein bound break, then reseal the break (Wang, 1996). The free energy of enzyme binding, alone, is insufficient to unpair the helix without the assistance of a certain degree of negative supercoiling. Because of its natural propensity to denature, negatively supercoiled DNA requires much less energy to be provided by the enzyme to bind a single-stranded region. As the negative supercoiling is relaxed, type IA enzymes can no longer overcome the energy, accounting for the inability of the type IA enzymes to relax the DNA to completion either *in vitro* (Wang, 1971) or *in vivo* (Zechiedrich *et al.*, 2000). Type IA enzymes relax DNA strictly one Lk at a time (Dekker *et al.*, 2002, and also see commentary by Champoux, 2002). As will be discussed below, other type I enzymes do not have this feature.

Twist recognition is patently an efficient mechanism for recognizing negative supercoiling. However, it falls short as a means to recognize positive supercoiling for at least two reasons. Positive supercoiling is more likely to manifest as writhe (Fogg, J.M., Catanese, D.J., and Zechiedrich, L., *manuscript in preparation*), rather than through changes in twist. In addition, the structural characteristics of overtwisting are highly dissimilar to those of undertwisting, making it impossible for one enzyme to recognize both undertwisting and overtwisting through a conserved mechanism.

In contrast to their type IA counterparts, the type IB topoisomerases, which include human topoisomerase I and vaccinia virus topoisomerase, are adept at relaxing both positively and negatively supercoiled DNA. Indeed, one of the probable cellular roles of human topoisomerase I is relaxation of positive supercoils ahead of the replication machinery (Leppard and Champoux, 2005). Therefore, we can hypothesize that the type IB enzymes must recognize features that are common to both positive and negative supercoils. An attractive mechanism for sensing supercoiling for the type IB topoisomerases (and also some type II topoisomerases as discussed below) is that they may recognize an obvious common structural feature present in both positively and negatively supercoiled DNA, the crossovers where two DNA helices juxtapose. Zechiedrich and Osheroff examined the interaction of topoisomerases with supercoiled DNA using transmission electron microscopy (Zechiedrich and Osheroff, 1990). The vast majority of calf thymus topoisomerase I, a type IB topoisomerase, and fruit fly topoisomerase II, a type II topoisomerase, were seen bound to crossovers in negatively supercoiled DNA. In contrast, the type IA enzyme, *E. coli* topoisomerase I, did not bind DNA crossovers. Later work confirmed the findings for type IB topoisomerases (Madden *et al.*, 1995).

The type IB topoisomerases have been proposed to relax DNA through a "controlled rotation" mechanism (Stewart *et al.*, 1998), based primarily on evidence obtained from multiple crystal structures of the type IB enzyme human topoisomerase I in covalent and noncovalent complexes with short, double-stranded, linear DNA duplexes (Redinbo *et al.*, 1998; Redinbo *et al.*, 1999; Redinbo *et al.*, 2000). Evidence for this model was also provided from earlier ensemble supercoil relaxation experiments with vaccinia virus topoisomerase (Stivers *et al.*, 1997). Subsequent single-molecule assays, also performed with the vaccinia virus topoisomerase, provided further support for this mechanism (Koster *et al.*, 2005). The topoisomerase binds to a single DNA duplex forming a clamp around the double-stranded helix. The enzyme then cleaves one of the strands, forming a transient covalent enzyme-DNA intermediate. Once cleavage occurs, the clamp is thought to open partially to facilitate rotation of the helix downstream of the cleavage site. This rotation, driven by the torque normally present in supercoiled DNA, allows relaxation. In this model, multiple rotation events may occur during each cleavage/religation cycle (Koster *et al.*, 2005; Stivers *et al.*, 1997). The preferential binding of a type IB topoisomerase to DNA crossovers is, however, perplexing because the controlled rotation model never requires the enzyme to interact with two DNA helices. Perhaps the enzyme binding DNA crossovers serves as a regulatory function; one that cannot be seen in crystal structures. In this respect, the enzyme may recognize DNA crossovers to discriminate supercoiled DNA from relaxed DNA.

Human topoisomerase I exhibited a strong preference for binding supercoiled over relaxed DNA, independent of whether the DNA was positively or negatively supercoiled (Madden *et al.*, 1995). DNA with increased superhelical density better competed relaxed DNA for human topoisomerase I binding (Madden *et al.*, 1995), perhaps as a consequence of an increasing number of crossovers in the supercoiled DNA (Boles *et al.*, 1990). If the enzyme bound all DNA crossovers equally, it might be expected that a simple linear inverse relationship exists between the amount of DNA required to compete for enzyme binding and the number of crossovers in the competitor, *e.g.*, DNA with 28 crossovers would be expected to be $\sim$ 3.5-fold more effective than DNA with 8 crossovers. However, a 7 kilobase plasmid DNA, which, at least from electron microscopy studies should have $\sim$ 28 crossovers, was approximately thirty-times more effective as a competitor for binding human topoisomerase I than a more relaxed 7 kilobase plasmid DNA, which should have $\sim$ 8 crossovers (Boles *et al.*, 1990). Clearly, it is not understood how these enzymes find and act on their DNA substrates and these are important questions to address.

Type IC topoisomerases, such as topoisomerase V from the hyperthermophilic bacterium *Methanopyrus kandleri*, are a distinct class of topoisomerases with structures unrelated to type IA and IB (reviewed in Forterre, 2006; Forterre *et al.*, 2007). Although the physiological role of topoisomerase V is unknown, single molecule experiments have revealed that the enzyme relaxes both positively and negatively supercoiled DNA using a "controlled rotation" mechanism similar to type IB enzymes (Taneja *et al.*, 2007). In fact, for several years topoisomerase V was incorrectly designated a type IB enzyme (Belova *et al.*, 2001; Belova *et al.*, 2002). The amino acid sequence of topoisomerase V is, however, highly dissimilar to the sequences of the classical type IB topoisomerases (Belova *et al.*, 2001). The enzyme structure, solved in 2006, revealed topoisomerase V had a different domain architecture and contained extra domains not previously found in other type I enzymes (Taneja *et al.*, 2006). These protein sequence, structural and activity differences warranted the new class distinction, type IC. Nevertheless, conclusions from type IB experiments should be broadly applicable to type IC enzymes as a consequence of the similar mechanisms of relaxation by both classes. Only time will tell if type IC topoisomerases are solely found in archaea or if they will be discovered in other kingdoms.

3.2. Type IB topoisomerases and DNA bending. As discussed above, extensive data show proteins bound to bent DNA. Thus supercoiling may influence the interaction of proteins with DNA by providing bending potential. We propose that a subtle interplay of both pre-existing DNA curvature attracting or stabilizing protein interactions with DNA, combined with protein interacting with DNA and therefore influencing DNA bending exists.

Although DNA juxtapositions are a recognition site for topoisomerases, the simple intersection of two DNA strands alone may not be sufficient for substrate recognition. Other structural elements, such as juxtaposition angle and DNA curvature, are likely important. To describe the observed binding of type IB topoisomerases to supercoiled DNA, Madden *et al.* postulated that only a small subset of crossovers in DNA with low levels of supercoiling may fulfill the structural requirements for a high affinity enzyme binding site (Madden *et al.*, 1995). Additionally, in the electron microscopy experiments by Zechiedrich and Osheroff, the type IB enzyme was never found at crossovers generated by two separate fragments of linear DNA (Zechiedrich and Osheroff, 1990), suggesting that the preferential binding observed of type IB topoisomerases for DNA juxtapositions may require the DNA also to be under torsional stress. These data argue that DNA crossover geometry may not be independent of superhelical stress.

As DNA becomes more supercoiled, it forms a superhelical more tightly interwound with the helices in close contact where they juxtapose (Adrian *et al.*, 1990; Bednar *et al.*, 1994; Fogg *et al.*, 2006). As the DNA becomes more writhed, it must have a higher number of bends with decreasing bend angles at or close to the helical juxtapositions to accommodate the supercoiling. This DNA bending could be what type IB topoisomerases recognize to sense torsional stress in positively and negatively supercoiled DNA. In support of this hypothesis, it has been shown that type IB topoisomerase binding is markedly stimulated by the inclusion of sequence-dependent bends in linear duplex DNA (Krogh *et al.*, 1991; Wang, 1996).

3.3. Mechanism of type II topoisomerases. Type II topoisomerases generate transient double-stranded breaks in one DNA helix (the so-called "gate" segment), pass a second DNA helix (the so-called "transfer" segment) through the gate, and then close the gate (Champoux, 2001; Schoeffler and Berger, 2005; Wang, 1996; Schoeffler and Berger, 2008). This $\Delta Lk = 2$ stepsize mechanism allows type II topoisomerases to act on knots and catenanes in closed circular double-stranded DNA, as well as to relax DNA supercoils (Table 1 and Figure 2). Because the type II topoisomerases pass intact helices through each other, the consequences of acting at the wrong place or at the wrong time are potentially disastrous (Liu *et al.*, in press). Therefore, the enzymes must have precise mechanisms for recognizing their DNA substrates. Presumably, the type II topoisomerases use the same DNA strand passage mechanism to unknot and decatenate as they do either to relax or add supercoils, which would suggest that the enzymes recognize structural features that are common to knots, catenanes and DNA supercoils.

One complication in building a unifying model for type II topoisomerase action is that results from different enzymes are used to test different models, and these enzymes exhibit seemingly subtle, but ultimately important, differences in the way they recognize DNA. Indeed, in the study that first pointed out that type II enzymes reduce the steady-state levels of knots and catenanes to below equilibrium (more on this below), whereas all type II enzymes tested did so, the magnitude of the effect varied widely, ranging from $\sim$ 5- to 80-fold (Rybenkov *et al.*, 1997a). It may be futile, therefore, to try to build one unifying model that fits all type II topoisomerases or to assume they all use identical mechanisms.

DNA gyrase was the first type II topoisomerase to be identified (Gellert *et al.*, 1976). Although it is, arguably, the most thoroughly studied, and the best understood of all the type II topoisomerases (reviewed in Nollmann *et al.*, 2007), it is actually an atypical topoisomerase. Gyrase is unique in its ability to introduce negative supercoils into DNA. Liu and Wang showed that this enzyme imposes the directionality of introducing negative supercoils by wrapping DNA about itself to generate a left-handed crossover in the DNA (Liu and Wang, 1978). Strand passage, driven by ATP, converts this to a right-handed crossover, thereby converting a positive supercoil into a negative one. It is the wrap, dictated by how the DNA interacts with its binding pockets on gyrase that determines both the recognition of substrate and of its essentially unidirectional action. No other topoisomerases are known to wrap DNA; therefore, it is unknown how the other type II topoisomerases know where to act.

3.4. Models for type II topoisomerase action. Type II topoisomerases use the free energy from ATP hydrolysis to drive the necessary large scale protein conformation changes used to cleave, strand-pass, and religate DNA (reviewed in Bates and Maxwell, 2007). This allows the type II enzymes to catalyze otherwise energetically unfavorable reactions. This is in contrast to the type I topoisomerases, which, with the exception of reverse gyrase, do not strictly require ATP, and must use the free energy of supercoiling to drive the reaction.

Rybenkov *et al.* used long linear DNA with sticky ends to measure the steady-state (thermodynamic[1]) equilibrium of ring closure and the steady-state probability that the annealed sticky ends would either trap a knot or would close around a separate intact, circular DNA plasmid (Rybenkov *et al.*, 1997a). This relaxed knot and catenane probability was found not to change with time. The addition of type II topoisomerases to the steady-state equilibrium shifted it to less likely to be knotted or catenated. This finding led to the conclusion that topoisomerases lower the probability of knots and catenanes. To explain their findings, Rybenkov *et al.* proposed that a type II topoisomerase binds to two DNA helices simultaneously, one of them the gate segment, and then tracks along the second helix to capture a third segment, which becomes the transfer segment. Roca and co-workers also proposed a "three site" binding model, but differing from the Rybenkov model in that the enzyme would bind both a gate and a transfer helix, and then bind a second transfer segment (Trigueros *et al.*, 2004). The enzyme uses the energy of ATP hydrolysis to drive the enzyme in one direction towards the second transfer segment, and in this movement, is able to determine whether action is necessary. Marko and co-workers (Yan *et al.*, 1999, 2001) proposed a "kinetic proofreading mechanism" (Hopfield, 1974; Ninio, 1975) for type II topoisomerases. They argued that the type II topoisomerases may initially bind the transfer segment and then use ATP hydrolysis to change enzyme conformation. This new conformation would then allow for the passage of the transfer segment if it is the correct substrate or the loss of the transfer segment if it is not supposed to be acted upon.

3.5. Type II topoisomerases and DNA bending. Buck and Zechiedrich argued that the models described above ignore extensive local information already present on the DNA substrate (Buck and Zechiedrich, 2004). They proposed a more DNA-centric model in which a type II enzyme

[1]We prefer the term "steady-state" because the system used by Rybenkov *et al.*, 1997 is largely undefined thermodynamically. The energetics of sticky single-strand annealing and additional complications of potential topoisomerase interaction with DNA ends, double-stranded DNA and single-stranded DNA, are unknown and have not been quantified.

acts preferentially at "hooked" juxtapositions, where the DNA helices curve toward each other, and not at "free" juxtapositions where the DNA helices curve away from each other. Such hooked juxtapositions are much more prevalent in knotted, catenated, or extremely supercoiled DNA than unknotted, unlinked, less supercoiled or relaxed DNA (Buck and Zechiedrich, 2004). Subsequent numerical analyses using lattice (Liu *et al.*, 2006a; Liu *et al.*, 2006b) and continuum random-flight (Burnier *et al.*, 2007) models demonstrated that preferential segment passages at juxtapositions with a pre-existing hooked rather than a free local geometry can lead to steady-states with substantial reductions in catenane (Liu *et al.*, 2006b) and knot (Burnier *et al.*, 2007; Liu *et al.*, 2006a) probabilities. More generally and notably, an extensive consideration of the catenane and knot reduction potentials of $\sim$ 700 different juxtaposition geometries on a lattice (Liu *et al.*, 2006b) indicated a robust scaling relation, which provided the first rationalization of a strikingly similar and otherwise puzzling trend observed experimentally for several type II topoisomerases from different organisms (Rybenkov *et al.*, 1997a). The preponderance of agreement between experiment and theoretical results lend credence to the Buck and Zechiedrich hypothesis.

Another model that involves DNA bending, the so-called "hairpin model," puts forth that instead of recognizing existing hooked DNA juxtapositions, type II topoisomerases impart hairpins to provide unidirectionality to the strand-passage event. Vologodskii and co-workers found that putting a hairpin into modeled DNA lowered the steady-state probability of knotting and linking (Vologodskii *et al.*, 2001). They proposed that type II topoisomerases induce DNA hairpins of varying degrees, and the probabilities of juxtaposing a straight transfer segment with the bent gate segment at varying bend angles was simulated (Klenin *et al.*, 2002). The results suggested that when the gate segment is more sharply bent, the probability of juxtaposing the gate and transfer segments is higher in positively supercoiled DNA than negatively supercoiled DNA. The magnitude of this difference depends strongly on the angle of the DNA bend, approaching 1,000-fold for a 240° bend and diminishing to $\sim$ 4-fold for a bend angle of 180°.

The recent crystal structure of the DNA binding and cleavage core of yeast topoisomerase II in a complex with a putative gate segment revealed that the DNA was indeed bent $\sim$ 150° within the enzyme (Dong and Berger, 2007). A 150° bend, according to the Vologodskii correlation described above, could explain the inability of the yeast enzyme to distinguish positive and negative supercoiling. Whether the enzymes bend DNA (Vologodskii *et al.*, 2001), recognize pre-existing curvature (Buck and Zechiedrich, 2004), or do both, is an open question.

3.6. The decatenation paradox. Stone *et al.* studied the relaxation of DNA braids (two helices with ends fixed to surfaces and wrapped about each other) to dissect the activity of *E. coli* topoisomerase IV (Stone *et al.*, 2003). Unlike supercoiled DNA, braids lack superhelical apices. In addition, the braided molecules contained a single-stranded break, or nick, in one of the DNA strands. The nicks prevent the braids from having any torsional strain, either when overtwisted or undertwisted. Left-handed braids were relaxed preferentially by topoisomerase IV, suggesting that the enzyme recognizes the left-handed crossovers in positively supercoiled DNA. In a separate study, fruit fly topoisomerase II, an enzyme that does not distinguish positive from negative supercoiled DNA, relaxed left-handed and right-handed braids with equal efficacy (Charvin *et al.*, 2003). If *E. coli* topoisomerase IV does preferentially act at left-handed DNA crossings, it raises the question: how is it that the enzyme so efficiently unlinks right-handed catenanes *in vivo* (Zechiedrich and Cozzarelli, 1995; Zechiedrich *et al.*, 1997)? Charvin *et al.* addressed this "decatenation paradox" and showed that, in fact, right-handed DNA braids could be efficiently relaxed by topoisomerase IV if the braids were allowed to form a left-handed plectonemic superhelix (Charvin *et al.*, 2003). The left-handed second order DNA crossings were proposed to provide an efficient substrate for topoisomerase IV. This may explain why decatenation by topoisomerase IV *in vivo* is strongly influenced by DNA supercoiling (Ullsperger and Cozzarelli, 1996; Zechiedrich and Cozzarelli, 1995; Zechiedrich *et al.*, 1997).

3.7. At what step of catalysis do type II topoisomerases distinguish positive from negative supercoiling? How type II topoisomerases preferentially act on left-handed DNA crossings has become nearly assumed in spite of existing evidence to the contrary. It is widely held that initial binding to DNA by type II topoisomerases allows the handedness discrimination. Measuring enzyme binding requires the experiments to be performed in the absence of ATP to prevent the additional steps of catalysis and turnover. While it is certainly true that all type II topoisomerases (with the exception of DNA gyrase (Higgins and Cozzarelli, 1982)), bind preferentially to supercoiled DNA over relaxed DNA in the absence of ATP (Osheroff *et al.*, 1983; Osheroff, 1986; Osheroff, 1987; Peng and Marians, 1995), *E. coli* topoisomerase IV and human topoisomerase IIα bind equally well to positively and negatively supercoiled DNA (Charvin *et al.*, 2005a; Crisona *et al.*, 2000; McClendon *et al.*, 2005; Stone *et al.*, 2003). The positive and negative distinction by *E. coli* topoisomerase IV and human topoisomerase IIα must, therefore, take place at a step subsequent to the initial binding step. Crisona *et al.* found that *E. coli* topoisomerase IV cleaved positively supercoiled substrates preferentially over negatively supercoiled substrates (Crisona *et al.*, 2000). This suggests that the preferential action of topoisomerase IV on positively supercoiled DNA may be manifested, at least in part, prior to passage of the transfer segment. In a later study, Crisona and Cozzarelli found that even though topoisomerase IV binds with equal affinity to positively and negatively supercoiled DNA, it adopts a different conformation upon binding to positively supercoiled

DNA (Crisona and Cozzarelli, 2006). The authors postulated that when topoisomerase IV is bound to a gate segment on positively supercoiled DNA, the putative transfer segments are in a suitable orientation for capture by the enzyme. Binding of the transfer segment is proposed to result in an additional altered conformation of the enzyme, which may facilitate cleavage of the gate segment and passage of the transfer segment. An additional possibility is that it is faster or easier to recruit the transfer segment in positively supercoiled DNA.

3.8. Potential role of the topoisomerase C-terminus in distinguishing positive and negative supercoiling. Comparisons of protein sequence combined with biochemical and structural data for the type II topoisomerases agree that the carboxyl terminus (C-terminus) may be responsible for distinguishing positive from negative supercoiling. Although DNA gyrase and topoisomerase IV share a high degree of structural similarity, as stated above and shown in Table 1, the enzymes have distinct mechanistic characteristics. Topoisomerase IV, because it only protects $\sim$ 34 bp, is unable to wrap DNA and, therefore, introduces a negligible amount of writhe upon binding in comparison to DNA gyrase, which protects $\sim$ 110–160 bp (Peng and Marians, 1995). Fruit fly topoisomerase II (Lee *et al.*, 1989) and calf thymus topoisomerase II (Thomsen *et al.*, 1990) protect DNA lengths of 25 bp and 28 bp, respectively. Deleting the C-terminus of *E. coli* topoisomerase IV dramatically reduced the relaxation activity of the enzyme of both positively and negatively supercoiled DNA (Corbett *et al.*, 2005). In this C-terminus truncated form, however, the enzyme is no longer able to distinguish positive from negative supercoiling. Generally, the structure of the C-terminal domains of topoisomerase IV and DNA gyrase are similar, but there are critical structural and functional differences (Corbett *et al.*, 2005). The gyrase C-terminal domain is capable of bending DNA by $> 180°$ on its own (Corbett *et al.*, 2004). The topoisomerase IV C-terminal domain appears to have lost the ability to bend DNA, except at very high protein:DNA ratios (Corbett *et al.*, 2004). Topoisomerase IV C-terminal domain contains a positively charged outer surface, which may comprise a DNA-binding surface. The best way to fit two DNA helices into the topoisomerase IV structure is for them to curve towards each other, much like hooked juxtapositions (Corbett *et al.*, 2005).

As stated above, human topoisomerase IIα preferentially relaxes positive supercoils whereas human topoisomerase IIβ exhibits no preference (McClendon *et al.*, 2005). The most dissimilar region by sequence between human topoisomerase IIα and IIβ is the C-terminal domain, suggesting that the C-terminal domain of human topoisomerase IIα acts as a sensor of DNA topology. Viral topoisomerase II, which naturally lacks the C-terminal domain segment present in all other type II topoisomerases, shows no preference for relaxation of positive supercoils (McClendon *et al.*, 2006b). These data point to a possible role for the C-terminus in distinguishing positively and negatively supercoiled DNA. By secondary structure analyzing algorithms, the C-terminus is predicted to have minimal organized structure, making crystallization difficult. As a result, this region of type II topoisomerase remains structurally and functionally obscure.

4. Open questions regarding DNA supercoiling. Throughout this review, we have mentioned some of the many open questions that remain regarding the study of DNA supercoiling, the proteins that act on DNA, DNA topoisomerases, and the chemotherapeutic agents that target the topoisomerases. Below, we list the big questions comprehensively:

Regarding DNA:

1. How do twist and writhe partition as a function of positive and negative supercoiling?
2. How do positive and negative supercoiling influence DNA structure, geometry and chemistry?
3. How does DNA sequence influence DNA structure and topology?
4. How do electrostatics and counterions vary with positive and negative supercoiling?
5. Are positive and negative knot nodes the same (geometrically) as positive and negative supercoil nodes?

Regarding proteins that act on DNA:

6. How do proteins such as topoisomerases search or interact with non-specific DNA?
7. How do proteins such as topoisomerases 'decide' where to act?
8. How do proteins such as topoisomerases distinguish positive and negative supercoils?
9. What are the thermodynamic parameters that describe sticky single-strand annealing and probability of knotting and linking with and without type II topoisomerases?
10. What is the role of ATP binding and hydrolysis for type II topoisomerase function?

Regarding anti-topoisomerase drugs that interact with DNA:

11. How do anti-topoisomerase drugs such as the fluoroquinolone antibiotics and the epipodophyllotoxin and camptothecin anticancer drugs act?

Acknowledgements. We would like to thank: Professor Hue Sun Chan of University of Toronto, for thoughtful discussions, critical review of the manuscript and for generously communicating unpublished data; Professor Gregory Buck of St. Anselm College for numerous helpful discussions; Professor Youri Timsit for sending manuscripts otherwise unavailable to us; Ms. Allison Kriel, Baylor College of Medicine, for critical discussions and technical assistance; Dr. Lauren Becnel Boyd, Baylor College of Medicine, for invaluable revisions and proofreading.

REFERENCES

Adrian M., ten Heggeler-Bordier B., Wahli W., Stasiak A.Z., Stasiak A., and Dubochet J. (1990). *Direct visualization of supercoiled DNA molecules in solution.* EMBO J. **9**, 4551–4554.

Allemand J.F., Bensimon D., Lavery R., and Croquette V. (1998). *Stretched and overwound DNA forms a Pauling-like structure with exposed bases.* Proc. Natl. Acad. Sci. USA **95**, 14152–14157.

Anderson P. and Bauer W. (1978) *Supercoiling in closed circular DNA: dependence upon ion type and concentration.* Biochemistry. **17**, 594–601.

Anderson V.E. and Osheroff N. (2001). *Type II topoisomerases as targets for quinolone antibacterials: turning Dr. Jekyll into Mr. Hyde.* Curr. Pharm. Des. **7**, 337–353.

Arai Y., Yasuda R., Akashi K., Harada Y., Miyata H., Kinosita K.J., and Itoh H. (1999). *Tying a molecular knot with optical tweezers.* Nature **399**, 446–448.

Arsuaga J., Vazquez M., Trigueros S., Sumners de W., and Roca J. (2002). *Knotting probability of DNA molecules confined in restricted volumes: DNA knotting in phage capsids.* Proc. Natl. Acad. Sci. USA **99**, 5373–5377.

Bacolla A. and Wells R.D. (2004). *Non-B DNA conformations, genomic rearrangements, and human disease.* J. Biol. Chem. **279**, 47411–47414.

Baldwin G.S., Brooks N.J., Robson R.E., Wynveen A., Goldar A., Leikin S., Seddon J.M., and Kornyshev A.A.(2008) *DNA double helices recognize mutual sequence homology in a protein free environment.* J. Phys. Chem. B. **112**, 1060–1064.

Bates A.D. and Maxwell A. (2007). *Energy coupling in type II topoisomerases: why do they hydrolyze ATP?* Biochemistry **46**, 7929–7941.

Bednar J., Furrer P., Stasiak A., Dubochet J., Egelman E.H., and Bates A.D. (1994). *The twist, writhe and overall shape of supercoiled DNA change during counterion-induced transition from a loosely to a tightly interwound superhelix.* J. Mol. Biol. **235**, 825–847.

Belova G.I., Prasad R., Kozyavkin S.A., Lake J.A., Wilson S.H., and Slesarev A.I. (2001). *A type IB topoisomerase with DNA repair activities.* Proc. Natl. Acad. Sci. USA. **98**, 6015–6020.

Belova G.I., Prasad R., Nazimov I.V., Wilson S.H., and Slesarev A.I. (2002). *The domain organization and properties of individual domains of DNA topoisomerase V, a type 1B topoisomerase with DNA repair activities.* J. Biol. Chem. **277**, 4959–4965.

Benham C.J. (1979). *Torsional stress and local denaturation in supercoiled DNA.* Proc. Natl. Acad. Sci. USA **76**, 3870–3874.

Benham C.J. (1992). *Energetics of the strand separation transition in superhelical DNA.* J. Mol. Biol. **225**, 835–847.

Benham C.J. and Mielke S.P. (2005). *DNA mechanics.* Ann. Rev. Biomed. Eng. **7**, 21–53.

Berman H.M., Westbrook J., Feng Z., Gilliland G., Bhat T.N., Weissig H., Shindyalov I.N., and Bourne P.E. (2000). *The Protein Data Bank.* Nucleic Acids Res. **28**, 235–242.

Boles T.C., White J.H., and Cozzarelli N.R. (1990). *Structure of plectonemically supercoiled DNA.* J. Mol. Biol. **213**, 931–951.

Bryant Z., Stone M.D., Gore J., Smith S.B., Cozzarelli N.R., and Bustamante C. (2003). *Structural transitions and elasticity from torque measurements on DNA.* Nature **424**, 338–341.

Buck G.R. and Zechiedrich E.L. (2004). *DNA disentangling by type-2 topoisomerases.* J. Mol. Biol. **340**, 933–939.

Burnier Y., Weber C., Flammini A., and Stasiak A. (2007). *Local selection rules that can determine specific pathways of DNA unknotting by type II DNA topoisomerases.* Nucleic Acids Res. **35**, 5223–5231.

BUSTAMANTE C., BRYANT Z., AND SMITH S.B. (2003). *Ten years of tension: single-molecule DNA mechanics.* Nature **421**, 423–427.

BUSTAMANTE C., MARKO J.F., SIGGIA E.D., AND SMITH S. (1994). *Entropic elasticity of lambda-phage DNA.* Science **265**, 1599–1600.

CALLADINE C.R. AND DREW H.R. (1984) *A base-centred explanation of the B-to-A transition in DNA.* J. Mol. Biol. **178**, 773–782.

CAMILLONI G., DI MARTINO E., CASERTA M., AND DI MAURO E. (1988). *Eukaryotic DNA topoisomerase I reaction is topology dependent.* Nucleic Acids Res. **14**, 7071–7085.

CAMILLONI G., MARTINO E., DI MAURO E., AND CASERTA M. (1989). *Regulation of the function of eukaryotic DNA topoisomerase I: topological conditions for inactivity.* Proc. Natl. Acad. Sci. USA **86**, 3080–3084.

CHAMPOUX J.J. (2001). *DNA topoisomerases: structure, function, and mechanism.* Annu. Rev. Biochem. **70**, 369–413.

CHAMPOUX J.J. (2002). *Type IA DNA topoisomerases: strictly one step at a time.* Proc. Natl. Acad. Sci. USA. **99**, 11998–12000.

CHANGELA A., DIGATE R.J., AND MONDRAGON A. (2001). *Crystal structure of a complex of a type IA DNA topoisomerase with a single-stranded DNA molecule.* Nature **411**, 1077–1081.

CHARVIN G., BENSIMON D., AND CROQUETTE V. (2003). *Single-molecule study of DNA unlinking by eukaryotic and prokaryotic type-II topoisomerases.* Proc. Natl. Acad. Sci. USA **100**, 9820–9825.

CHARVIN G., STRICK T.R., BENSIMON D., AND CROQUETTE V. (2005a). *Topoisomerase IV bends and overtwists DNA upon binding.* Biophys. J. **89**, 384–392.

CHARVIN G., STRICK T.R., BENSIMON D., AND CROQUETTE V. (2005b). *Tracking topoisomerase activity at the single-molecule level.* Ann. Rev. Biophys. Biomol. Struct. **34**, 201–219.

CHERNY D.I. AND JOVIN T.M. (2001). *Electron and scanning force microscopy studies of alterations in supercoiled DNA tertiary structure.* J. Mol. Biol. **313**, 295–307.

CLOUTIER T.E. AND WIDOM J. (2004). *Spontaneous sharp bending of double-stranded DNA.* Mol. Cell **14**, 355–362.

CLOUTIER T.E. AND WIDOM J. (2005). *DNA twisting flexibility and the formation of sharply looped protein-DNA complexes.* Proc. Natl. Acad. Sci. USA **102**, 3645–3650.

CORBETT K.D. AND BERGER J.M. (2004). *Structure, molecular mechanisms, and evolutionary relationships in DNA topoisomerases.* Ann. Rev. Biophys. Biomol. Struct. **33**, 95–118.

CORBETT K.D., SCHOEFFLER A.J., THOMSEN N.D., AND BERGER J.M. (2005). *The structural basis for substrate specificity in DNA topoisomerase IV.* J. Mol. Biol. **351**, 545–561.

CORBETT K.D., SHULTZABERGER R.K., AND BERGER J.M. (2004). *The C-terminal domain of DNA gyrase A adopts a DNA-bending beta-pinwheel fold.* Proc. Natl. Acad. Sci. USA **101**, 7293–7298.

COZZARELLI N.R., BOLES T.C., AND WHITE J.H. (1990). *Primer on the topology and geometry of DNA supercoiling. In DNA topology and its biological effects.* Cozzarelli N.R. and Wang, J.C. (eds.) Cold Spring Harbor Laboratory Press.

CRICK F.H. AND KLUG A. (1975). *Kinky helix.* Nature **255**, 530–533.

CRISONA N.J. AND COZZARELLI N.R. (2006). *Alteration of* Escherichia coli *topoisomerase IV conformation upon enzyme binding to positively supercoiled DNA.* J. Biol. Chem. **281**, 18927–18932.

CRISONA N.J., STRICK T.R., BENSIMON D., CROQUETTE V., AND COZZARELLI N.R. (2000). *Preferential relaxation of positively supercoiled DNA by* E. coli *topoisomerase IV in single-molecule and ensemble measurements.* Genes Dev. **14**, 2881–2892.

CROOKE E., HWANG D.S., SKARSTAD K., THONY B., AND KORNBERG A. (1991). E. coli *minichromosome replication: regulation of initiation at oriC.* Res. Microbiol. **142**, 127–130.

Crothers D.M., Drak J., Kahn J.D., and Levene S.D. (1992). *DNA bending, flexibility, and helical repeat by cyclization kinetics.* Methods Enzymol. **212**, 3–29.

Czapla L., Swigon D. and Olson W.K. (2006) *Sequence-Dependent Effects in the Cyclization of Short DNA.* J. Chem. Theory Comput. **2**, 685–695.

Deibler R.W., Mann J.K., Sumners D.W.L., and Zechiedrich L. (2007). *Hin-mediated DNA knotting and recombination promote replicon dysfunction and mutation.* BMC Mol. Biol. **8**, 44

Dekker N.H., Rybenkov V.V., Duguet M., Crisona N.J., Cozzarelli N.R., Bensimon D., and Croquette V. (2002). *The mechanism of type IA topoisomerases.* Proc. Natl. Acad. Sci. USA **99**, 12126–12131.

Deutsch J.M. (1988). *Theoretical studies of DNA during gel electrophoresis.* Science **240**, 922–924.

Dickerson R.E. (1998). *DNA bending: the prevalence of kinkiness and the virtues of normality.* Nucleic Acids Res. **26**, 1906–1926.

Dong K.C. and Berger J.M. (2007). *Structural basis for gate-DNA recognition and bending by type IIA topoisomerases.* Nature **450**, 1201–1205.

Drake F.H., Hofmann G.A., Bartus H.F., Mattern M.R., Crooke S.T., and Mirabelli C.K. (1989). *Biochemical and pharmacological properties of p170 and p180 forms of topoisomerase II.* Biochemistry **28**, 8154–8160.

Du Q., Kotlyar A., and Vologodskii A. (2008). *Kinking the double helix by bending deformation. Nucleic Acids Res.* **36**, 1120–1128.

Du Q., Smith C., Shiffeldrim N., Vologodskaia M., and Vologodskii A. (2005). *Cyclization of short DNA fragments and bending fluctuations of the double helix.* Proc. Natl. Acad. Sci. USA **102**, 5397–5402.

Dunaway M. and Dröge P. (1989). *Transactivation of the* Xenopus *rRNA gene promoter by its enhancer.* Nature **341**, 657–659.

Drlica K., Malik M., Kerns R.J., Zhao X. (2008) *Quinolone-mediated bacterial death.* Antimicrob. Agents Chemother. **52**, 385–392.

Embleton M.L., Vologodskii A.V., and Halford S.E. (2004). *Dynamics of DNA loop capture by the SfiI restriction endonuclease on supercoiled and relaxed DNA.* J. Mol. Biol. **339**, 53–66.

Fogg J.M., Kolmakova N., Rees I., Magonov S., Hansma H., Perona J.J., and Zechiedrich E.L. (2006). *Exploring writhe in supercoiled minicircle DNA.* J. Phys: Condens. Matter **18**, S145–S159.

Forterre P. (2006) *DNA topoisomerase V: a new fold of mysterious origin.*Trends Biotechnol. **24**, 245–247.

Forterre P., Gribaldo S., Gadelle D., and Serre M.C. (2007) *Origin and evolution of DNA topoisomerases.* Biochimie. **89**, 427–446.

Fujimoto B.S. and Schurr J.M. (1990). *Dependence of the torsional rigidity of DNA on base composition.* Nature **344**, 175–178.

Fuller F.B. (1971). *The writhing number of a space curve.* Proc. Natl. Acad. Sci. USA **68**, 815–819.

Fuller F.B. (1978). *Decomposition of the linking of a closed ribbon: A problem of molecular biology.* Proc. Natl. Acad. Sci. USA **75**, 3557–3561.

Gellert M., Mizuuchi K., O'Dea M.H., and Nash H.A. (1976). *DNA gyrase: An enzyme that introduces superhelical turns into DNA.* Proc. Natl. Acad. Sci. USA **73**, 3872–3876.

Gore J., Bryant Z., Nollmann M., Le M.U., Cozzarelli N.R., and Bustamante C. (2006). *DNA overwinds when stretched.* Nature **442**, 836–839.

Gorin A.A, Zhurkin V.B., and Olson W.K. (1995) *B-DNA twisting correlates with base-pair morphology.* J. Mol. Biol. **247**, 34–48.

Gowers D.M. and Halford S.E. (2003). *Protein motion from non-specific to specific DNA by three-dimensional routes aided by supercoiling.* EMBO J. **22**, 1410–1418.

Grue P., Grasser A., Sehested M., Jensen P.B., Uhse A., Straub T., Ness W., and Boege F. (1998). *Essential mitotic functions of DNA topoisomerase IIalpha are not adopted by topoisomerase IIbeta in human H69 cells.* J. Biol. Chem. **273**, 33660–33666.

Hagerman P.J. (1988). *Flexibility of DNA.* Ann. Rev. Biophys. Biophys. Chem. **17**, 265–286.

Harris S.A., Laughton C.A., and Liverpool T.B. (2008). *Mapping the phase diagram of the writhe of DNA nanocircles using atomistic molecular dynamics simulations.* Nucleic Acids Res. **36**, 21–29.

Harris S.A., Sands Z.A., and Laughton C.A. (2005). *Molecular dynamics simulations of duplex stretching reveal the importance of entropy in determining the biomechanical properties of DNA.* Biophys. J. **88**, 1684–1691.

Heath P.J., Clendenning J.B., Fujimoto B.S., and Schurr J.M. (1996). *Effect of bending strain on the torsion elastic constant of DNA.* J. Mol. Biol. **260**, 718–730.

Heck M.M. and Earnshaw W.C. (1986). *Topoisomerase II: A specific marker for cell proliferation.* J. Cell. Biol. **103**, 2569–2581.

Hiasa H. and Marians K.J. (1996). *Two distinct modes of topological processing during theta-type DNA replication.* J. Biol. Chem. **271**, 21529–21535.

Hiasa H., Yousef D.O., and Marians K.J. (1996). *DNA strand cleavage is required for replication fork arrest by a frozen topoisomerase-quinolone-DNA ternary complex.* J. Biol. Chem. **271**, 26424–26429.

Hiller D.A., Rodriguez A.M., Perona J.J. *Non-cognate enzyme-DNA complex: structural and kinetic analysis of EcoRV endonuclease bound to the EcoRI recognition site GAATTC.* J. Mol. Biol. **354**, 121–136.

Higgins N.P. and Cozzarelli N.R. (1982). *The binding of gyrase to DNA: analysis by retention by nitrocellulose filters.* Nucleic Acids Res. **10**, 6833–6847.

Hopfield J.J. (1974). *Kinetic proofreading: a new mechanism for reducing errors in biosynthetic processes requiring high specificity.* Proc. Natl. Acad. Sci. USA **71**, 4135–4139.

Horowitz D.S. and Wang J.C. (1984). *Torsional rigidity of DNA and length dependence of the free energy of DNA supercoiling.* J. Mol. Biol. **173**, 75–91.

Keller W. and Wendel I. (1975). *Stepwise relaxation of supercoiled SV40 DNA.* Cold Spring Harbor Symposia on Quantitative Biology **39** (Part 1), 199–208.

Kikuchi A. and Asai K. (1984). *Reverse gyrase–a topoisomerase which introduces positive superhelical turns into DNA.* Nature **309**, 677–681.

Kim J.L., Nikolov D.B., and Burley S.K. (1993a). *Co-crystal structure of TBP recognizing the minor groove of a TATA element.* Nature **365**, 520–527.

Kim Y., Geiger J.H., Hahn S., and Sigler P.B. (1993b). *Crystal structure of a yeast TBP/TATA-box complex.* Nature **365**, 512–520.

Kirkegaard K. and Wang J.C. (1985). *Bacterial DNA topoisomerase I can relax positively supercoiled DNA containing a single-stranded loop.* J. Mol. Biol. **185**, 625–637.

Klenin K. and Langowski J. (2000). *Computation of writhe in modeling of supercoiled DNA.* Biopolymers **54**, 307–317.

Klenin K., Langowski J., and Vologodskii A.V. (2002). *Computational analysis of the chiral action of type II DNA topoisomerases.* J. Mol. Biol. **320**, 359–367.

Kondapi A.K., Mulpuri N., Mandraju R.K., Sasikaran B., and Subba Rao K. (2004) *Analysis of age dependent changes of Topoisomerase II alpha and beta in rat brain.* Int. J. Dev. Neurosci. **22**, 19–30.

Koster D.A., Croquette V., Dekker C., Shuman S., and Dekker N.H. (2005). *Friction and torque govern the relaxation of DNA supercoils by eukaryotic topoisomerase IB.* Nature **434**, 671–674.

Koster D.A., Palle K., Bot E.S., Bjornsti M.A., Dekker N.H. (2007). *Antitumour drugs impede DNA uncoiling by topoisomerase I.* Nature. **448**, 213–217.

Kramer P.R. and Sinden, R.R. (1997). *Measurement of unrestrained negative supercoiling and topological domain size in living human cells.* Biochemistry **36**, 3151–3158.

Kreuzer K.N. and Alberts B.M. (1984). *Site-specific recognition of bacteriophage T4 DNA by T4 type II DNA topoisomerase and* Escherichia coli *DNA gyrase.* J. Biol. Chem. **259**, 5339–5346.

KREUZER K.N. AND COZZARELLI N.R. (1979). Escherichia coli *mutants thermosensitive for deoxyribonucleic acid gyrase subunit A: effects on deoxyribonucleic acid replication, transcription, and bacteriophage growth.* J. Bacteriol. **140**, 424–435.

KROGH S., MORTENSEN U.H., WESTERGAARD O., AND BONVEN B.J. (1991). *Eukaryotic topoisomerase I-DNA interaction is stabilized by helix curvature.* Nucleic Acids Res. **19**, 1235–1241.

LAMARR W.A., SANDMAN K.M., REEVE J.N., AND DEDON P.C. (1997). *Large scale preparation of positively supercoiled DNA using the archaeal histone HMf.* Nucleic Acids Res. **25**, 1660–1661.

LANKAS F., LAVERY R., AND MADDOCKS J.H. (2006). *Kinking occurs during molecular dynamics simulations of small DNA minicircles.* Structure **14**, 1527–1534.

LEE M.P., SANDER M., AND HSIEH T. (1989). *Nuclease protection by* Drosophila *DNA topoisomerase II. Enzyme/DNA contacts at the strong topoisomerase II cleavage sites.* J. Biol. Chem. **264**, 21779–21787.

LEPPARD J.B. AND CHAMPOUX J.J. (2005). *Human DNA topoisomerase I: relaxation, roles, and damage control.* Chromosoma **114**, 75–85.

LEVENE S.D. AND CROTHERS D.M. (1986). *Topological distributions and the torsional rigidity of DNA. A Monte Carlo study of DNA circles.* J. Mol. Biol. **189**, 73–83.

LEVITT M. (1983) *Protein folding by restrained energy minimization and molecular dynamics.* J. Mol. Biol. **170**, 723–764.

LEWIS M., CHANG G., HORTON N.C., KERCHER M.A., PACE H.C., SCHUMACHER M.A., BRENNAN R.G., AND LU P. (1996). *Crystal structure of the lactose operon repressor and its complexes with DNA and inducer.* Science **271**, 1247–1254.

LIMA C.D., WANG J.C., AND MONDRAGON A. (1994). *Three-dimensional structure of the 67K N-terminal fragment of* E. coli *DNA topoisomerase I.* Nature **367**, 138–146.

LIONNET T., JOUBAUD S., LAVERY R., BENSIMON D., AND CROQUETTE V. (2006). *Wringing out DNA.* Phys. Rev. Lett. **96**, 178102.

LIU C.C., BURKE R.L., HIBNER U., BARRY J., AND ALBERTS B. (1979a). *Probing DNA replication mechanisms with the T4 bacteriophage in vitro system.* Cold Spring Harbor Symposia on Quantitative Biology **43** (Part 1), 469–487.

LIU D.J. AND DAY L.A. (1994). *Pf1 virus structure: helical coat protein and DNA with paraxial phosphates.* Science **265**, 671–674.

LIU L.F., LIU C.C., AND ALBERTS B.M. (1979b). *T4 DNA topoisomerase: a new ATP-dependent enzyme essential for initiation of T4 bacteriophage DNA replication.* Nature **281**, 456–461.

LIU L.F. AND WANG J.C. (1978). *DNA-DNA gyrase complex: the wrapping of the DNA duplex outside the enzyme.* Cell **15**, 979–984.

LIU L.F. AND WANG J.C. (1987). *Supercoiling of the DNA template during transcription.* Proc. Natl. Acad. Sci. USA **84**, 7024–7027.

LIU Z., MANN J.K., ZECHIEDRICH E.L., AND CHAN H.S. (2006a). *Topological information embodied in local juxtaposition geometry provides a statistical mechanical basis for unknotting by type-2 DNA topoisomerases.* J. Mol. Biol. **361**, 268–285.

LIU Z., ZECHIEDRICH E.L., AND CHAN H.S. (2006b). *Inferring global topology from local juxtaposition geometry: interlinking polymer rings and ramifications for topoisomerase action.* Biophys. J. **90**, 2344–2355.

LIU Z., DEIBLER R.W., CHAN H.S., AND ZECHIEDRICH (2008). Hooked on DNA: the why and how of DNA untangling. Nucleic Acids Res. *in press.*

LIVERPOOL T.B., HARRIS S.A., AND LAUGHTON C.A. (2008). *Supercoiling and denaturation of DNA loops.* Phys. Rev. Lett. **100**, 238103.

LOCKSHON D. AND MORRIS D.R. (1983). *Positively supercoiled plasmid DNA is produced by treatment of* Escherichia coli *with DNA gyrase inhibitors.* Nucleic Acids Res. **11**, 2999–3017.

LUGER K., MADER A.W., RICHMOND R.K., SARGENT D.F., AND RICHMOND T.J. (1997). *Crystal structure of the nucleosome core particle at 2.8Å resolution.* Nature **389**, 251–260.

LYUBCHENKO Y.L. (2004). *DNA structure and dynamics: an atomic force microscopy study.* Cell Biochem. Biophys. **41**, 75–98.

MADDEN K.R., STEWART L., AND CHAMPOUX J.J. (1995). *Preferential binding of human topoisomerase I to superhelical DNA.* EMBO J. **14**, 5399–5409.

MAHER L.J., 3rd (1998). *Mechanisms of DNA bending.* Current Opin. Chem. Biol. **2**, 688–694.

MAHER L.J., 3rd (2006). *DNA kinks available...if needed.* Structure **14**, 1479–1480.

MANNING G.S. (1969a). *Limiting laws and counterion condensation in polyelectrolyte solutions. I. Colligative properties.* J. Chem. Phys. **51**, 924–933.

MANNING G.S. (1969b). *Limiting laws and counterion condensation in polyelectrolyte solutions. II. Self-diffusion of the small ions.* J. Chem. Phys. **51**, 934–938.

MARKO J.F. AND SIGGIA E.D. (1994). *Bending and Twisting Elasticity of DNA.* Macromolecules **27**, 981–988.

MARTINCIC D. AND HANDE K.R. (2005). *Topoisomerase II inhibitors.* Cancer Chemother. Biol. Response Modif. **22**, 101–121.

MARVIN D.A., SPENCER M., WILKINS M.H., AND HAMILTON L.D. (1958) *A new configuration of deoxyribonucleic acid.* Nature **182**, 387–388.

MCCLELLAN J.A. AND LILLEY D.M. (1991). *Structural alteration in alternating adenine-thymine sequences in positively supercoiled DNA.* J. Mol. Biol. **219**, 145–149.

MCCLENDON A.K., DICKEY J.S., AND OSHEROFF N. (2006a). *The geometry of DNA supercoils modulates topoisomerase-mediated DNA cleavage and enzyme response to anticancer drugs.* Biochemistry. **45**, 3040–3050.

MCCLENDON A.K., DICKEY J.S., AND OSHEROFF N. (2006b). *Ability of viral topoisomerase II to discern the handedness of supercoiled DNA: bimodal recognition of DNA geometry by type II enzymes.* Biochemistry **45**, 11674–11680.

MCCLENDON A.K. AND OSHEROFF N. (2007). *DNA topoisomerase II, genotoxicity, and cancer.* Mut. Res. **623**, 83–97.

MCCLENDON A.K., RODRIGUEZ A.C., AND OSHEROFF N. (2005). *Human topoisomerase IIα rapidly relaxes positively supercoiled DNA: implications for enzyme action ahead of replication forks.* J. Biol. Chem. **280**, 39337–39345.

MIZUUCHI K., GELLERT M., AND NASH H.A. (1978). *Involvement of supertwisted DNA in integrative recombination of bacteriophage lambda.* J. Mol. Biol. **121**, 375–392.

MONDRAGON A. AND DIGATE R. (1999). *The structure of* Escherichia *coli DNA topoisomerase III.* Structure **7**, 1373–1383.

MULLER M.T. (1985). *Quantitation of eukaryotic topoisomerase I reactivity with DNA. Preferential cleavage of supercoiled DNA.* Biochim. Biophys. Acta **824**, 263–267.

MUSGRAVE D., ZHANG X., AND DINGER M. (2002). *Archaeal genome organization and stress responses: implications for the origin and evolution of cellular life.* Astrobiology **2**, 241–253.

MUSGRAVE D.R., SANDMAN K.M., AND REEVE J.N. (1991). *DNA binding by the archaeal histone HMf results in positive supercoiling.* Proc. Natl. Acad. Sci. USA **88**, 10397–10401.

NINIO J. (1975). *Kinetic amplification of enzyme discrimination.* Biochimie **57**, 587–595.

NITISS J.L. (1998). *Investigating the biological functions of DNA topoisomerases in eukaryotic cells.* Biochim. Biophys. Acta. **1400**, 63–81.

NÖLLMANN M., CRISONA N.J., AND ARIMONDO P.B. (2007). *Thirty years of* Escherichia coli *DNA gyrase: from in vivo function to single-molecule mechanism.* Biochimie **89**, 490–499.

NUNES-DUBY S.E., SMITH-MUNGO L.I., AND LANDY A. (1995). *Single base-pair precision and structural rigidity in a small IHF-induced DNA loop.* J. Mol. Biol. **253**, 228–242.

OLSON W.K. (1996). *Simulating DNA at low resolution.* Current Opin. Struct. Biol. **6**, 242–256.

OLSON W.K., GORIN A.A., LU X.J., HOCK L.M., AND ZHURKIN V.B. (1998). *DNA sequence-dependent deformability deduced from protein-DNA crystal complexes.* Proc. Natl. Acad. Sci. USA **95**, 11163–11168.

Osheroff N. (1986). *Eukaryotic topoisomerase II. Characterization of enzyme turnover.* J. Biol. Chem. **261**, 9944–9950.

Osheroff N. (1987). *Role of the divalent cation in topoisomerase II mediated reactions.* Biochemistry **26**, 6402–6406.

Osheroff N., Shelton E.R., and Brutlag D.L. (1983). *DNA topoisomerase II from* Drosophila melanogaster. *Relaxation of supercoiled DNA.* J. Biol. Chem **258**, 9536–9543.

Osheroff N. and Zechiedrich E.L. (1987). *Calcium-promoted DNA cleavage by eukaryotic topoisomerase II: Trapping the covalent enzyme-DNA complex in an active form.* Biochemistry **26**, 4303–4309.

Pack G.R., Wong L., and Lamm G. (1999). *Divalent cations and the electrostatic potential around DNA: Monte Carlo and Poisson-Boltzmann calculations.* Biopolymers **49**, 575–590.

Parvin J.D., McCormick R.J., Sharp P.A., and Fisher D.E. (1995). *Pre-bending of a promoter sequence enhances affinity for the TATA-binding factor.* Nature **373**, 724–727.

Pauling L. and Corey R.B. (1953). *A proposed structure for the nucleic acids.* Proc. Natl. Acad. Sci. USA **39**, 84–97.

Pavlicek J.W., Oussatcheva E.A., Sinden R.R., Potaman V.N., Sankey O.F., and Lyubchenko Y.L. (2004). *Supercoiling-induced DNA bending.* Biochemistry **43**, 10664–10668.

Peng H. and Marians K. (1995). *The interaction of* Escherichia coli *topoisomerase IV with DNA.* J. Biol. Chem. **270**, 25286–25290.

Pérez A., Lankas F., Luque F.J., Orozco M. (2008) *Towards a molecular dynamics consensus view of B-DNA flexibility.* Nucleic Acids Res. **36**, 2379–2394.

Perry K. and Mondragon A. (2003). *Structure of a complex between* E. coli *DNA topoisomerase I and single-stranded DNA.* Structure **11**, 1349–1358.

Pohl W.F. and Roberts G.W. (1978). *Topological considerations in the theory of replication of DNA.* J. Math. Biol. **6**, 383–402.

Pommier Y. (2006). *Topoisomerase I inhibitors: camptothecins and beyond.* Nat. Rev. Cancer **6**, 789–802.

Portugal J. and Rodriguez-Campos A. (1996). *T7 RNA polymerase cannot transcribe through a highly knotted DNA template.* Nucleic Acids Res. **24**, 4890–4894.

Randall G.L., Pettitt B.M., Buck G., and Zechiedrich E.L. (2006). *Electrostatics of DNA-DNA juxtapositions: consequences for type II topoisomerase function.* J. Phys: Condens. Matter **18**, S173–S185.

Redinbo M.R., Champoux J.J., and Hol W.G. (2000). *Novel insights into catalytic mechanism from a crystal structure of human topoisomerase I in complex with DNA.* Biochemistry **39**, 6832–6840.

Redinbo M.R., Stewart L., Champoux J.J., and Hol W.G. (1999). *Structural flexibility in human topoisomerase I revealed in multiple non-isomorphous crystal structures.* J. Mol. Biol. **292**, 685–696.

Redinbo M.R., Stewart L., Kuhn P., Champoux J.J., and Hol W.G. (1998). *Crystal structures of human topoisomerase I in covalent and noncovalent complexes with DNA.* Science **279**, 1504–1513.

Reeve J.N., Sandman K., and Daniels C.J. (1997). *Archaeal histones, nucleosomes, and transcription initiation.* Cell **89**, 999–1002.

Rice P.A., Yang S., Mizuuchi K., and Nash H.A. (1996). *Crystal structure of an IHF-DNA complex: a protein-induced DNA U-turn.* Cell **87**, 1295–1306.

Richet E., Abcarian P., and Nash H.A. (1986). *The interaction of recombination proteins with supercoiled DNA: defining the role of supercoiling in lambda integrative recombination.* Cell **46**, 1011–1021.

Rodley G.A., Scobie R.S., Bates R.H.T., and Lewitt R.M. (1976) *A possible conformation for double-stranded polynucleotides.* Proc. Natl. Acad. Sci. USA. **73**, 2929–2963.

RODRIGUEZ-CAMPOS A. (1996). *DNA knotting abolishes in vitro chromatin assembly.* J. Biol. Chem. **271**, 14150–14155.

RODRIGUEZ A.C. (2002). *Studies of a positive supercoiling machine. Nucleotide hydrolysis and a multifunctional "latch" in the mechanism of reverse gyrase.* J. Biol. Chem. **277**, 29865–29873.

RYBENKOV V.V., ULLSPERGER C., VOLOGODSKII A.V., AND COZZARELLI N.R. (1997a). *Simplification of DNA topology below equilibrium values by type II topoisomerases.* Science **277**, 690–693.

RYBENKOV V.V., VOLOGODSKII A.V., AND COZZARELLI N.R. (1997b). *The effect of ionic conditions on DNA helical repeat, effective diameter, and free energy of supercoiling.* Nucleic Acids Res. **25**, 1412–1418.

SAITTA A.M., SOPER P.D., WASSERMAN E., AND KLEIN M.L. (1999). *Influence of a knot on the strength of a polymer strand.* Nature **399**, 46–48.

SCHOEFFLER A.J. AND BERGER J.M. (2005). *Recent advances in understanding structure-function relationships in the type II topoisomerase mechanism.* Biochem. Soc. Trans. **33**, 1465–1470.

SCHOEFFLER A.J. AND BERGER J.M. (2008). *DNA topoisomerases: harnessing and constraining energy to govern chromosome topology.* Quart. Rev. Biophys. **41**, 41–101.

SCHVARTZMAN J.B. AND STASIAK A. (2004). *A topological view of the replicon.* EMBO Rep. **5**, 256–261.

SELVIN P.R., COOK D.N., PON N.G., BAUER W.R., KLEIN M.P., AND HEARST J.E. (1992). *Torsional rigidity of positively and negatively supercoiled DNA.* Science **255**, 82–85.

SHORE D. AND BALDWIN R.L. (1983a). *Energetics of DNA twisting. I. Relation between twist and cyclization probability.* J. Mol. Biol. **170**, 957–981.

SHORE D. AND BALDWIN R.L. (1983b). *Energetics of DNA twisting. II. Topoisomer analysis.* J. Mol. Biol. **170**, 983–1007.

SHORE D., LANGOWSKI J., AND BALDWIN R.L. (1981). *DNA flexibility studied by covalent closure of short fragments into circles.* Proc. Natl. Acad. Sci USA **78**, 4833–4837.

SIKORAV J.L. AND JANNINK G. (1994). *Kinetics of chromosome condensation in the presence of topoisomerases: a phantom chain model.* Biophys. J. **66**, 827–837.

SMITH S.B., FINZI L., AND BUSTAMANTE C. (1992). *Direct mechanical measurements of the elasticity of single DNA molecules by using magnetic beads.* Science **258**, 1122–1126.

SNOUNOU G. AND MALCOM A.D. (1983). *Production of positively supercoiled DNA by netropsin.* J. Mol. Biol. **167**, 211–216.

SPERRAZZA J.M., REGISTER J.C. 3RD., AND GRIFFITH J. (1984). *Electron microscopy can be used to measure DNA supertwisting.* Gene **31**, 17–22.

STEWART L., REDINBO M.R., QIU X., HOL W.G., AND CHAMPOUX J.J. (1998). *A model for the mechanism of human topoisomerase I.* Science **279**, 1534–1541.

STIVERS J.T., HARRIS T.K., AND MILDVAN A.S. (1997). *Vaccinia DNA topoisomerase I: evidence supporting a free rotation mechanism for DNA supercoil relaxation.* Biochemistry **36**, 5212–5222.

STONE M.D., BRYANT Z., CRISONA N.J., SMITH S.B., VOLOGODSKII A., BUSTAMANTE C., AND COZZARELLI N.R. (2003). *Chirality sensing by* Escherichia coli *topoisomerase IV and the mechanism of type II topoisomerases.* Proc. Natl. Acad. Sci. USA **100**, 8654–8659.

STRICK T.R., ALLEMAND J.F., BENSIMON D., BENSIMON A., AND CROQUETTE V. (1996). *The elasticity of a single supercoiled DNA molecule.* Science **271**, 1835–1837.

STRICK T.R., ALLEMAND J.F., BENSIMON D., AND CROQUETTE V. (1998). *Behavior of supercoiled DNA.* Biophys. J. **74**, 2016–2028.

STRICK T.R., ALLEMAND J.F., BENSIMON D., AND CROQUETTE V. (2000a). *Stress-induced structural transitions in DNA and proteins.* Ann. Rev. Biophys. Biomol. Struct. **29**, 523–543.

Strick T.R., Bensimon D., and Croquette V. (1999). *Micro-mechanical measurement of the torsional modulus of DNA*. Genetica **106**, 57–62.

Strick T.R., Croquette V., and Bensimon D. (2000b). *Single-molecule analysis of DNA uncoiling by a type II topoisomerase*. Nature **404**, 901–904.

Svozil D., Sponer J.E., Marchan I., Pérez A., Cheatham T.E. 3rd, Forti. F., Luque F.J., Orozco M., Sponer J. (2008) *Geometrical and electronic structure variability of the sugar-phosphate backbone in nucleic acids*. J. Phys. Chem. B **112**, 8188–8197.

Taneja B., Patel A., Slesarev A., Mondragón A. (2006) *Structure of the N-terminal fragment of topoisomerase V reveals a new family of topoisomerases*. EMBO J. **25**, 398–408.

Taneja B., Schnurr B., Slesarev A., Marko J.F., Mondragón A. (2007) *Topoisomerase V relaxes supercoiled DNA by a constrained swiveling mechanism*. Proc. Natl. Acad. Sci. USA. **104**, 14670–14675.

Terry B.J., Jack W.E., and Modrich P. (1985). *Facilitated diffusion during catalysis by EcoRI endonuclease. Nonspecific interactions in EcoRI catalysis*. J. Biol. Chem. **260**, 13130–13137.

Thomsen B., Bendixen C., Lund K., Andersen A.H., Sorensen B.S., and Westergaard O. (1990). *Characterization of the interaction between topoisomerase II and DNA by transcriptional footprinting*. J. Mol. Biol. **215**, 237–244.

Timsit Y., Duplantier B., Jannink G., and Sikorav J.-L. (1998). *Symmetry and chirality in topoisomerases II-DNA crossover recognition*. J. Mol. Biol. **284**, 1289–1299.

Timsit Y. and Moras D. (1994). *DNA self-fitting: the double helix directs the geometry of its supramolecular assembly*. EMBO J. **13**, 2737–2746.

Timsit Y., Shatzky-Schwartz M., and Shakked Z. (1999). *Left-handed DNA crossovers. Implications for DNA-DNA recognition and structural alterations*. J. Biomol. Struct. Dyn. **16**, 775–785.

Tolstorukov M.Y., Colasanti A.V., McCandlish D.M., Olson W.K., Zhurkin V.B. (2007) *A novel roll-and-slide mechanism of DNA folding in chromatin: implications for nucleosome positioning*. J. Mol. Biol. **371**, 725–738.

Travers A. and Muskhelishvili G. (2005). *DNA supercoiling - a global transcriptional regulator for enterobacterial growth?* Nat. Rev. Microbiol. **3**, 157–169.

Trigueros S., Salceda J., Bermudez I., Fernandez X., and Roca J. (2004). *Asymmetric removal of supercoils suggests how topoisomerase II simplifies DNA topology*. J. Mol. Biol. **335**, 723–731.

Ullsperger C. and Cozzarelli N.R. (1996). *Contrasting enzymatic activities of topoisomerase IV and DNA gyrase from* Escherichia coli. J. Biol. Chem. **271**, 31549–31555.

Vinograd J. and Lebowitz J. (1966). *Physical and topological properties of circular DNA*. J. Gen. Physiol. **49**, 103–125.

Vinograd J., Lebowitz J., Radloff R., Watson R., and Laipis P. (1965). *The twisted circular form of polyoma viral DNA*. Proc. Natl. Acad. Sci. USA **53**, 1104–1111.

Vologodskii A.V. and Cozzarelli N.R. (1996). *Effect of supercoiling on the juxtaposition and relative orientation of DNA sites*. Biophys. J. **70**, 2548–2556.

Vologodskii A.V., Levene S.D., Klenin K.V., Frank-Kamenetskii M., and Cozzarelli N.R. (1992). *Conformational and thermodynamic properties of supercoiled DNA*. J. Mol. Biol. **227**, 1224–1243.

Vologodskii A.V., Lukashin A.V., Anshelevich V.V., and Frank-Kamenetskii M.D. (1979). *Fluctuations in superhelical DNA*. Nucleic Acids Res. **6**, 967–982.

Vologodskii A.V., Zhang W., Rybenkov V.V., Podtelezhnikov A.A., Subramanian D., Griffith J.D., and Cozzarelli N.R. (2001). *Mechanism of topology simplification by type II DNA topoisomerases*. Proc. Natl. Acad. Sci. USA **98**, 3045–3049.

VON HIPPEL P.H. (2007). *From "simple" DNA-protein interactions to the macromolecular machines of gene expression.* Ann. Rev. Biophys. Biomol. Struct. **36**, 79–105.

WAHLE E. AND KORNBERG A. (1988). *The partition locus of plasmid pSC101 is a specific binding site for DNA gyrase.* EMBO J. **7**, 1889–1895.

WANG G. AND VASQUEZ K.M. (2006). *Non-B DNA structure-induced genetic instability.* Mut. Res. **598**, 103–119.

WANG G. AND VASQUEZ K.M. (2007). *Z-DNA, an active element in the genome.* Front. Biosci. **12**, 4424–38.

WANG J.C. (1971). *Interaction between DNA and an* Escherichia coli protein ω. J. Mol. Biol. **55**, 523–533.

WANG J.C. (1996). *DNA topoisomerases.* Annu. Rev. Biochem. **65**, 635–692.

WANG J.C. (2002). *Cellular roles of DNA topoisomerases: A molecular perspective.* Nat. Rev. Mol. Cell Biol. **3**, 430–440.

WATSON J.D. AND CRICK F.H.C. (1953). *Molecular structure of nucleic acids: A structure for deoxyribose nucleic acid.* Nature **171**, 737–738.

WEBER C., STASIAK A., DE LOS RIOS P., AND DIETLER G. (2006). *Numerical simulation of gel electrophoresis of DNA knots in weak and strong electric fields.* Biophys. J. **90**, 3100–3105.

WELLS R.D. (2007). *Non-B DNA conformations, mutagenesis and disease.* Trends Biochem. Sci. **32**, 271–278.

WELLS R.D., DERE R., HEBERT M.L., NAPIERALA M., AND SON L.S. (2005). *Advances in mechanisms of genetic instability related to hereditary neurological diseases.* Nucleic Acids Res. **33**, 3785–3798.

WHITE J.H. (1969). *Self-linking and the Gauss integral in higher dimensions.* Am. J. Math.91, 693–728.

WHITSON P.A., HSIEH W.T., WELLS R.D., AND MATTHEWS K.S. (1987). *Supercoiling facilitates lac operator-repressor-pseudooperator interactions.* J. Biol. Chem. **262**, 4943–4946.

WOLTERS KLUWER HEALTH, *Pharmaceutical Audit Suite (PHAST)*, January to December 2006.

XU Y.C. AND BREMER H. (1997). *Winding of the DNA helix by divalent metal ions.* Nucleic Acids Res. **25**, 4067–4071.

YAN J., MAGNASCO M.O., AND MARKO J.F. (1999). *A kinetic proofreading mechanism for disentanglement of DNA by topoisomerases.* Nature **401**, 932–935.

YAN J., MAGNASCO M.O., AND MARKO J.F. (2001). *Kinetic proofreading can explain the supression of supercoiling of circular DNA molecules by type-II topoisomerases.* Phys. Rev. E Stat. Nonlin. Soft Matter Phys. **63**, 031909.

YIN H., WANG M.D., SVOBODA K., LANDICK R., BLOCK S.M., AND GELLES J. (1995). *Transcription against an applied force.* Science **270**, 1653–1657.

ZECHIEDRICH E.L. AND COZZARELLI N.R. (1995). *Roles of topoisomerase IV and DNA gyrase in DNA unlinking during replication in* Escherichia coli. Genes Dev. **9**, 2859–2869.

ZECHIEDRICH E.L., KHODURSKY A.B., BACHELLIER S., SCHNEIDER R., CHEN D., LILLEY D.M., AND COZZARELLI N.R. (2000). *Roles of topoisomerases in maintaining steady-state DNA supercoiling in* Escherichia coli. J. Biol. Chem. **275**, 8103–8113.

ZECHIEDRICH E.L., KHODURSKY A.B., AND COZZARELLI N.R. (1997). *Topoisomerase IV, not gyrase, decatenates products of site-specific recombination in* Escherichia coli. Genes Dev. **11**, 2580–2592.

ZECHIEDRICH E.L. AND OSHEROFF N. (1990). *Eukaryotic topoisomerases recognize nucleic acid topology by preferentially interacting with DNA crossovers.* EMBO J. **9**, 4555–4562.

CALIBRATION OF TETHERED PARTICLE MOTION EXPERIMENTS

LIN HAN*, BERTRAND H. LUI*†, SETH BLUMBERG*‡, JOHN F. BEAUSANG§, PHILIP C. NELSON§, AND ROB PHILLIPS*¶

Abstract. The Tethered Particle Motion (TPM) method has been used to observe and characterize a variety of protein-DNA interactions including DNA looping and transcription. TPM experiments exploit the Brownian motion of a DNA-tethered bead to probe biologically relevant conformational changes of the tether. In these experiments, a change in the extent of the bead's random motion is used as a reporter of the underlying macromolecular dynamics and is often deemed sufficient for TPM analysis. However, a complete understanding of how the motion depends on the physical properties of the tethered particle complex would permit more quantitative and accurate evaluation of TPM data. For instance, such understanding can help extract details about a looped complex geometry (or multiple coexisting geometries) from TPM data. To better characterize the measurement capabilities of TPM experiments involving DNA tethers, we have carried out a detailed calibration of TPM magnitude as a function of DNA length and particle size. We also explore how experimental parameters such as acquisition time and exposure time affect the apparent motion of the tethered particle. We vary the DNA length from 200 bp to 2.6 kbp and consider particle diameters of 200, 490 and 970 nm. We also present a systematic comparison between measured particle excursions and theoretical expectations, which helps clarify both the experiments and models of DNA conformation.

Key words. Tethered particle, DNA, Brownian motion, calibration, single molecule.

AMS(MOS) subject classifications. Primary 92C05; secondary 92C40, 92C37.

1. Introduction.

Single molecule studies are enriching our understanding of biological processes by providing a unique window on the micro-trajectories of individual molecules rather than their ensemble-averaged behavior. Many of these studies are devoted to exploring the intricacies of protein–DNA interactions that are central to gene regulation, DNA replication and DNA repair. The resolution of nanometer-scale distances involved in such interactions poses a significant challenge. The emergence of the tethered particle motion (TPM) method offers a practical and relatively simple solution. In this method, a biopolymer is tethered between

*Department of Applied Physics, California Institute of Technology, Pasadena, CA 91125. Partially supported by the Keck Foundation, National Science Foundation grants CMS-0301657 and CMS-0404031, and the National Institutes of Health Director's Pioneer Award grant DP1 OD000217.

†Current address: Department of Bioengineering, Stanford University, Stanford, CA.

‡Current address: University of Michigan Medical Scientist Training Program, Ann Arbor, MI 48109.

§Department of Physics and Astronomy, University of Pennsylvania, Philadelphia, PA 19104. Partially supported by NSF grants DGE-0221664, DMR04-25780, and DMR-0404674.

¶Corresponding author: phillips@pboc.caltech.edu.

C.J. Benham et al. (eds.), *Mathematics of DNA Structure, Function and Interactions*, The IMA Volumes in Mathematics and its Applications 150, DOI 10.1007/978-1-4419-0670-0_6, a stationary substrate and a micrometer-scale sphere (a "bead"), which is large enough to be imaged with conventional optical microscopy (Fig. 1). The constrained Brownian motion of the bead serves as a reporter of the underlying macromolecular dynamics, either by observing its blurred image in a long exposure [5], or by tracking its actual trajectory in time (e.g. as done in [11] and the present work). Changes in the extent of the motion (which we will call "excursion") reflect conformational transformations of the tethered molecule. Such changes may be caused by processive walking of RNA polymerase [12, 23], DNA looping [5, 17, 24, 25, 22, 19, 2], DNA hybridization [14], DNA bending [15], Holliday junction formation [11] or RNA translation [20].

Although TPM is simple in principle, there are a variety of technical challenges that must be addressed for successful implementation. For example, sample preparation can be compromised by multiply-tethered beads, non-specific adsorption, transient sticking events and dissociation of the tether joints [11, 19, 17, 3, 9]. In addition, image analysis of TPM data is complicated by instrumental drift and the stochastic nature of the tethered particle's motion. Several time scales must be considered, including the total observation time, exposure time, and the intrinsic diffusive time scale of the tethered particle. We will show that quantification of the spatial and temporal resolution of TPM measurements requires an understanding of how particle motion depends on tether length, particle size and other controllable parameters. We focus exclusively on TPM behavior in the absence of externally applied force (as might be applied via magnetic or optical tweezers).

The aims of this article are to: (1) review how data acquisition and data analysis affect TPM measurements; (2) explain a practical scheme of data selection and quantify the fractions of typical data that are rejected by each of our criteria; (3) calibrate particle motion, tether length and observation time so that subsequent TPM experiments can be quantitatively interpreted; and (4) discuss the physical processes that govern TPM. Calibration of the particle motion allows precise predictions of how a particular conformational change of the tether, such as Lac repressor induced looping of DNA, affects TPM.

Some of our experimental results were outlined in [8]. Theoretical work leading up to the present results on TPM motion appeared in [13, 9, 16]. For example, Segall et al. predicted effects of changing the size of the bead and tether length, which we document experimentally in the present work. Our results are preparatory to experimental [7] and theoretical [16] work on DNA looping in the *lac* operon system.

2. Results and discussion. Using differential interference contrast (DIC) microscopy, the projected position of several beads in a field of view are recorded using a CCD camera. Sub-pixel resolution position traces for each bead in the image is determined using a cross-correlation method [6].

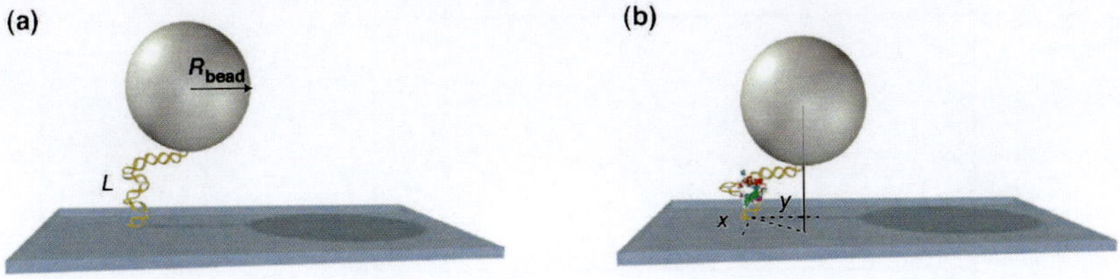

FIG. 1. *Idea of the tethered particle motion method. Cartoons showing the tethered bead in the (a) absence and (b) presence of a DNA-binding protein, which changes the effective tether length by looping and/or bending the DNA. For example, Lac repressor protein (LacR) has two binding sites, which recognize and bind to two specific sequences ("operators") on DNA.*

Standard microscopy systems such as ours are limited to two spatial dimensions; tracking of three dimensions has been accomplished using evanescent fields or diffraction rings [3], but this involves additional calibration and technical challenges. Two-dimensional tracking is sufficient for the applications we have in mind, such as DNA-looping studies. The tracked position of the bead is subject to slow drift, due to vibrations of the experimental apparatus, which we removed using a first order Butterworth filter at 0.05 Hz cutoff frequency [19]. To quantify bead excursion, we then used the square root of the sum of the variances of the drift corrected particle position (x, y) along two orthogonal image-plane axes:

$$\mathrm{RMS}_t = \sqrt{\langle (x-\bar{x})^2 + (y-\bar{y})^2 \rangle_t}. \tag{2.1}$$

Here t is the time interval over which the RMS motion is measured (typically 4 s); $\bar{x}$ and $\bar{y}$ represent the average of x and y over time t. Eq. (2.1) is evaluated as a sliding filter at each point along the trajectory, and permits us to capture the tether dynamics using a single scalar quantity, as illustrated in Fig. 7 below. The finite-sample means $\bar{x}, \bar{y}$ are subtracted as an additional method of eliminating instrumental drift not removed by the Butterworth filter; in practice, this subtraction has little effect. When simulating the motion numerically, we will compute the same quantity as Eq. (2.1), in order to make an appropriate comparison.

2.1. Data selection criteria. Although single-particle tracking data can reveal detailed features of the dynamics of protein-DNA interactions, care must be taken to minimize experimental artifacts such as non-specific binding of the bead and DNA to each other and the surface, as well as multiple DNA attachments on the same bead. To get acceptable calibration data, we implemented several selection criteria called "minimum motion," "motion symmetry," and "uniformity."

"Minimum motion" discards beads that cannot be differentiated from beads stuck to the glass substrate (or otherwise compromised in their mobility). $\mathrm{RMS}_{4\,\mathrm{s}}$ from a control experiment with beads but no DNA is shown

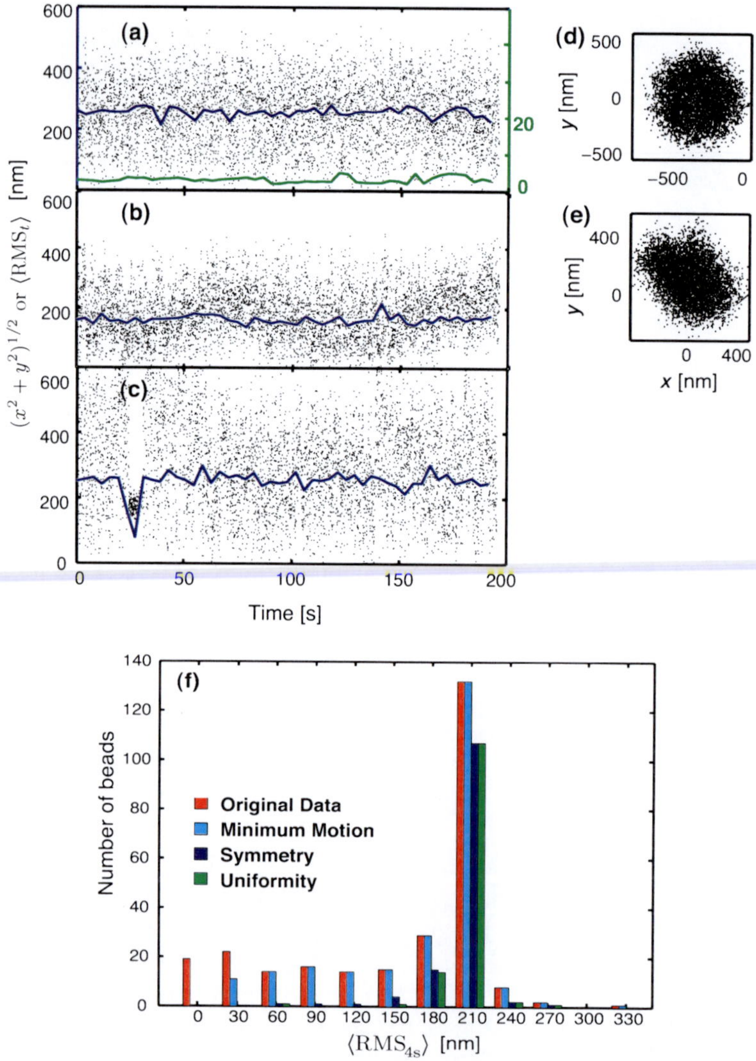

FIG. 2. *Selection of qualified tethers. In (a–e), the dots show instantaneous positions after drift subtraction; the lines show* $RMS_{4\,s}$. *(a) Trajectory associated with an accepted data set (blue) and stuck bead (green). (b) Trajectory for a bead that passed "minimum motion" but failed the "motion symmetry" test (see (e)). (c) Trajectory associated with nonuniform motion caused by transient, nonspecific binding, seen as a downward spike between 0 and 50 seconds. (d) xy scatter plot of the trajectory in (a) shows it to be symmetric. (e) Scatter plot of the motion in (b) shows it to be asymmetric. The DNA used in (a–e) are 1206 bp long and the bead size is 490 nm in diameter. (f) Distribution of bead excursions and the number of beads that pass successive application of the selection criteria (see text) [3]. Red: original data. Cyan: after application of minimal motion filter. Blue: after application of symmetry filter. Green: after application of uniformity filter. The DNA used in (f) are 901 bp long and bead size is 490 nm diameter.*

in Fig. 2 (green line), and is substantially smaller than that for a tethered bead (blue line). Data sets exhibiting average excursions, $\mathrm{RMS}_{4\,\mathrm{s}}$, lower than 30 nm cannot be differentiated from stuck beads and are therefore rejected.

"Motion symmetry" requires that a tethered particle should exhibit symmetric in-plane motion about its anchor point, and is calculated from the covariance matrix [3, 9]:

$$C = \begin{pmatrix} \sigma_{x_1x_1} & \sigma_{x_1x_2} \\ \sigma_{x_2x_1} & \sigma_{x_2x_2} \end{pmatrix}, \tag{2.2}$$

where

$$\sigma_{x_ix_j} = \frac{1}{N}\sum_{k=1}^{N} x_i^k x_j^k - \bar{x}_i\bar{x}_j \tag{2.3}$$

are the second moments of the bead's position. Here N is the number of video frames and x_1^k, x_2^k are the in-plane coordinates (i.e. the position x, y) of the microsphere for frame k as obtained from the drift-corrected data. The eigenvalues (λ_1, λ_2) of the covariance matrix indicate the squares of the major and minor axes corresponding to the in-plane displacement of the bead and are equal for a perfectly symmetric motion. We took $s = \sqrt{\lambda_{\max}/\lambda_{\min}} \leq 1.1$ as our acceptable threshold. Fig. 2(d,e) displays scatter plots for the in-plane motion of two beads to illustrate the distinction between symmetric and asymmetric tethers. The first plot passes the symmetry test and would serve as a qualified tether; the second would be rejected. Asymmetric bead trajectories may be caused by multiple DNA tethers [11].

"Uniformity" qualifies tethers on the basis of the consistency of their motion over time and eliminates beads showing non-specific binding events, such as binding of DNA to the bead or glass surface for short periods. To detect these events automatically, we refine a procedure used in Ref. [3]. We first divide the entire time series into 10 subsets labeled by $i = 1, \ldots 10$. In subset i, we calculate $\mathrm{RMS}_{4\,\mathrm{s}}$ over each 4 s window and then average these, defining $A_i \equiv \langle \mathrm{RMS}_{4\,\mathrm{s}} \rangle_i$. Then we define u as the standard deviation of $\{A_1, \ldots A_{10}\}$, normalized by the overall average $\mathrm{RMS}_{4\,\mathrm{s}}$. Only data sets with relative standard deviation $u < 0.2$ are accepted. For example, the bead shown in Fig. 2(c) meets the motion and symmetry criterions; however, it displays a non-specific binding event at 30s. In short, our third criterion removes tethers with temporal inconsistency in their Brownian motion.

The first two selection criteria discard tethers that are permanently defective, whereas the third eliminates time series with undesirable transient events. Note that if the purpose of the experiment is to identify interesting molecular binding events, such as those leading to DNA looping or bending, then the last criterion cannot be applied, because these transient events can appear similar to the sticking events rejected by the uniformity criterion. In the present work we aimed at characterizing uniform DNA tethers, so we enforced all three criteria. Prior to applying the selection criteria, Fig. 2(f) displays a broad distribution in the measured $\mathrm{RMS}_{4\,\mathrm{s}}$ (red). Afterwards, $\sim 50\%$ of the data are qualified and exhibit well-defined Brownian motion (green bars). This figure shows that the primary cause of bead rejection is asymmetric in-plane motion. Experimentally, beads with multiple tethers can be minimized by reducing the concentration of DNA.

2.2. Acquisition time. The drift-corrected (x, y) trajectories are noisy due to the stochastic Brownian motion of the particle, and are thus filtered using Eq. (2.1) over a particular time window t (usually four seconds). Although analysis methods exist that make no use of this windowing step [2, 1], nevertheless many experiments do use it, and so we investigated its effect on reported bead excursion. Too short a window will increase the noise, leading to broad peaks in the distribution of RMS_t that make signals from differently sized tethers too difficult to distinguish. Moreover, for short t the bead will not adequately explore its full range of accessible configurations, leading to an underestimate of RMS_t, as we document below. At the other extreme, however, too long a window will result in a loss of temporal resolution.

To determine the optimum TPM window size, we recorded data for 200 s, for several bead sizes and a wide range of tether lengths, then found the mean ($\langle \mathrm{RMS}_t \rangle$) and standard deviation ($\mathrm{std}_t = \sqrt{\langle \mathrm{RMS}_t^2 \rangle - \langle \mathrm{RMS}_t \rangle^2}$) of the RMS-filtered trajectory for various values of window size t (see Fig. 3). Here $\langle \ldots \rangle$ denotes two averages: (1) over the $(200\,\mathrm{s}/t)$ windows that make up each bead's time series, and (2) over nominally identical tethered particles with the same bead size and tether length. The DNA lengths varied from 199 bp to 2625 bp, and we tested beads with three different diameters: 200 nm, 490 nm and 970 nm.

Fig. 3 shows the trends as we vary t, R_{bead}, and tether length L. We first notice that for fixed R_{bead} and L, each curve levels off as $t \to \infty$, giving an asymptote that is the true RMS excursion. (For short times, the bead has not had a chance to explore its full range of motion in any given window, and so each $\mathrm{RMS}_t \to 0$, and hence so does $\langle \mathrm{RMS}_t \rangle$.) To make the tradeoff discussed earlier, we now ask: How long must we choose the window time t in order to get a reliable estimate of the true excursion?

Naively we might suppose that each video frame gives an independent draw from a distribution of bead positions whose RMS value we seek. In that case, we would expect that as soon as $t/(30\,\mathrm{msec})$ becomes large, we would have a good estimate of the true RMS excursion. But the top row of Fig. 3 shows that, on the contrary, the minimum required observation time increases both with increasing bead radius (moving between the three panels) and with increasing tether length (moving between the curves on a given panel). Physically, the point is that successive video frames are

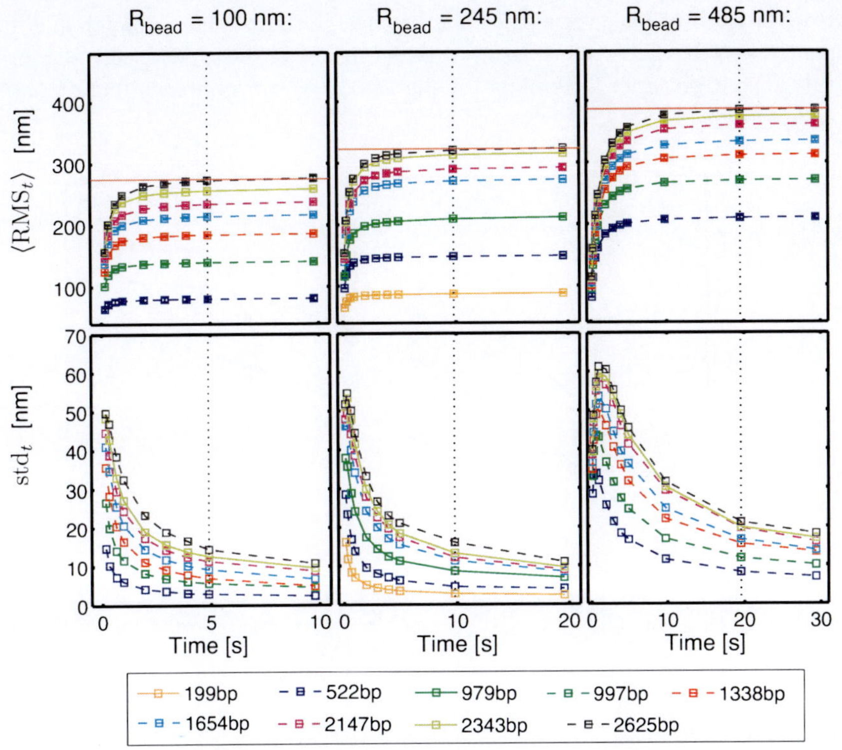

FIG. 3. *Top: Average RMS excursion and bottom: standard deviation of* RMS_t *as functions of window time* t *in Eq. (2.1) for different bead sizes (columns) and lengths* L *of the DNA tether (colored lines). As discussed in the text, black dotted lines indicate "large enough" choices of* t*.*

not independent draws from the distribution of particle positions, because the particle's motion is diffusive. The diffusion time $\tau_{\rm diff}$ of a particle in a trap increases with increasing trap radius and with increasing viscous drag constant for the particle, giving rise to the trends observed in the figure. (For a theoretical discussion see the Supplement to [16].)

Similarly, the second row of graphs in Fig. 3 shows that the scatter between successive determinations of RMS_t decreases with increasing t. This "sharpening" effect also explains how RMS filtering takes rather diffuse raw data (e.g. Fig. 2a–c) and transforms it into a fairly well-defined "state" (e.g. the individual states visible in filtered traces such as Fig. 7). In both rows of Fig. 3, we have drawn dotted lines to illustrate a value of t that is "safe" (long enough) for tether lengths up to 2600 bp.

2.3. Calibration of motion. In order for TPM experiments to detect discrete conformational changes of biopolymers such as in DNA looping, it is necessary to quantify how tether length affects particle motion.

Precise calibration data also indicates the minimum detectable change in tether length. Sensitive measurements may also allow detection of more subtle changes, such as kinking of the DNA upon protein binding or multiple loop topologies.

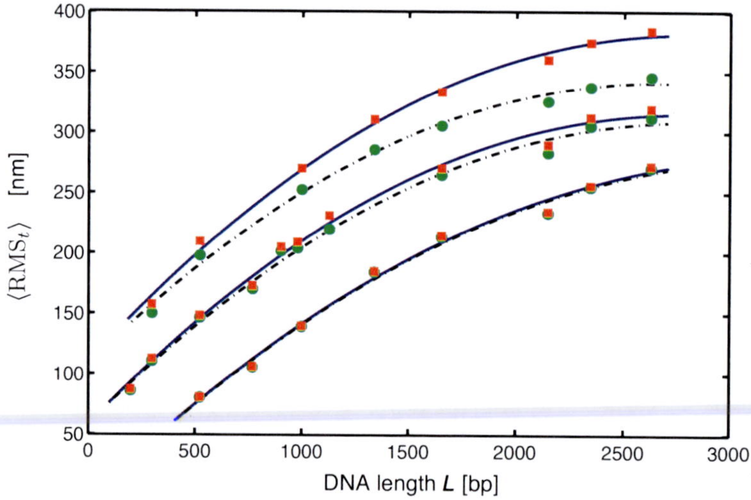

FIG. 4. *RMS excursion of bead as a function of the tether length for different sized microspheres, for random-sequence DNAs of various lengths. Each red square is the average of equilibrium amplitude of RMS motion over 20 to 200 qualified beads, which is calculated by using Eq. (2.1) with $t = 5$ s for $R = 100$ nm (bottom data set), $t = 10$ s for $R = 245$ nm (middle data set) and $t = 20$ s for $R = 485$ nm (top data set). Using $t = 4$ s for the same data systematically underestimates the motion of larger beads (green circles). The curves are empirical polynomial fits to the datasets (see Table 1).*

To find the empirical calibration curve, we created many DNA tethers of varying lengths, and attached beads of three different sizes. For each bead size, we estimated the RMS excursion by its finite-sample estimate RMS_t, taking t to be the lowest "safe" value as estimated in the previous subsection: $t = 5$ s, 10 s and 20 s for beads with diameters of 200 nm, 490 nm and 970 nm respectively, with results shown in Fig. 4. (For comparison, we also show corresponding results with t fixed to 4 s, which deviate significantly from the longer observations for the larger beads.) We summarized all these data with polynomial fits shown in the figure and given explicitly in Table 1.

2.4. Theoretical predictions. We also compared the experimental data in Fig. 4 to a mathematical simulation of the bead-tether-wall system (Fig. 5). The excursion of the bead away from its attachment point on the microscope slide is affected by the length and stiffness of the DNA tether, the size of the bead, and the various interactions between the bead/wall, bead/tether, and wall/tether. To account for all these effects, we modified

TABLE 1
Parameters of quadratic function $ax^2 + bx + c$ obtained for fitting both the equilibrium motion data (red squares in Fig. 4) and 4 s interval data (green circles in Fig. 4).

Time [s]	Diameter	$a \times 10^{-5}$	b	c
5	200 nm	-2.58 ± 0.68	0.17 ± 0.02	-4.5±14.8
10	490 nm	-3.37 ± 0.47	0.19 ± 0.01	57.3± 7.2
20	970 nm	–3.49 ±0.46	0.20 ± 0.01	109.5 ± 8.7
4	200 nm	-2.60 ±0.69	0.17 ±0.02	-4.75±14.7
4	490 nm	-3.17 ± 0.41	0.18± 0.01	58.05± 6.6
4	970 nm	–3.31 ±0.48	0.18 ± 0.01	107.7 ± 8.7

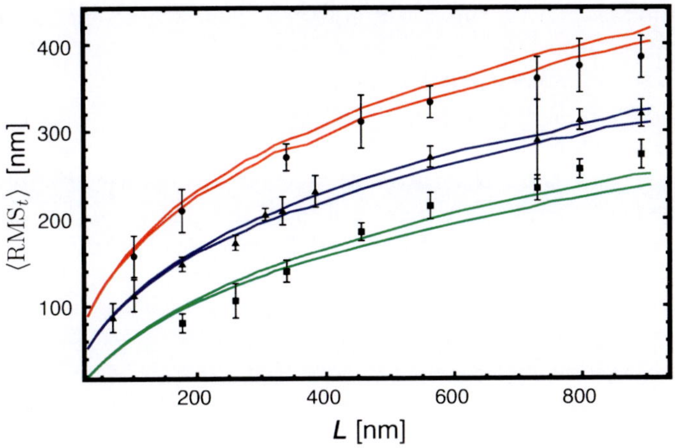

FIG. 5. *Theoretical prediction of equilibrium bead excursion, following a method introduced in [13, 9]. Dots: Experimental values (same bead sizes as red squares in Fig. 4). Each dot represents 20–200 different observed beads, with the given tether length. Each such bead was observed for about 200 s, yielding (200 s)/t measurements of the RMS motion, which were averaged; here $t = 20$, 10, and 5 s as in Fig. 4. Each data point shown is the average of these averages; error bars represent the variation (standard deviation) among the beads. Curves: Theoretically predicted RMS motion, corrected for the blurring effect of finite shutter time. For each of the three bead sizes studied, two curves are shown. From top to bottom, each pair of curves assumes persistence length values 47 and 39 nm, respectively, a range appropriate for the solution conditions we used [21]. There are no fit parameters; the theoretical model uses values for bead diameter given by the manufacturer's specification. The bumpiness in the curves reflects the statistical character of the Monte Carlo algorithm that generated them.*

the Gaussian sampling Monte Carlo technique previously used in [13, 9, 4, 8] (see [16] for details).

Suppose first that a semiflexible polymer chain is anchored at one point in space, but is otherwise unconstrained. At the anchored point we suppose we are given a probability distribution of different possible initial orientations for the first chain segment. The distribution of positions and orientations of the other end is then a convolution of this initial distribution with a kernel representing a particular diffusion process (random walk) on the group manifold of the three-dimensional Euclidean group.

We can numerically compute moments of this final distribution, or its various marginal distributions, by a Monte Carlo procedure. Idealizing the polymer as a chain of finite elements, each is related to its predecessor by a shift along the latter's 3-axis, a twist about the same axis, and some random bend and twist. Rather than represent the random part using Euler angles, a more invariant formulation is to draw a 3×3 generator matrix from a Gaussian distribution on the Lie algebra so (3), then exponentiate it. The Gaussian distribution is determined by a covariance matrix, which represents the bend and twist elasticity of the DNA, together with bend-twist couplings. We estimated it up to an overall rescaling factor from structural data on DNA, then chose the overall factor to yield a desired value of the persistence length of DNA.

Turning from the idealized problem above to TPM, we see that we must implement steric constraints: One end of the DNA tether is attached to a wall, which the DNA may not penetrate. Moreover, the other end is attached to the sphere, which itself must not penetrate the wall. Nevertheless, each segment of the intervening DNA is otherwise free to bend, independently of its neighbors. Thus the same Monte Carlo generation just described continues to be valid, except that some sterically forbidden chains must be discarded. Thus our computer code generated many simulated DNA chains and bead orientations in a Boltzmann distribution, applied the steric constraints [13], and tabulated the resulting values of the distance from the projected bead center to the attachment point. The necessary calculations were coded in *Mathematica* and ran conveniently on a laptop computer.

We chose to compare to experimental data with "safe" values of the window time t, so we simply had the code evaluate the RMS value of this distance. (For a procedure valid for any t, see the Supplement to [16].) We also applied a correction to this theoretical result, to account for the bead's motion during the rather long shutter time (see the following subsection). Fig. 5 shows that an *a priori* calculation of the expected motion matches the data fairly well, with a value of persistence length consistent with others' experiments; there were no other fitting parameters.

2.5. Blurring effect. In our experiments the camera had a long shutter time ($\delta t = 31\,\mathrm{msec}$). During each exposure, the bead moved, creating a blurred image whose center is not quite the same as the instantaneous center. This blurring effect reduces the apparent bead excursion. Suppose for example that the bead has a momentary excursion to a large value of x. Subsequently, its stretched tether will pull it inward, so that the average position during the video frame has a smaller value of x. We quantified this

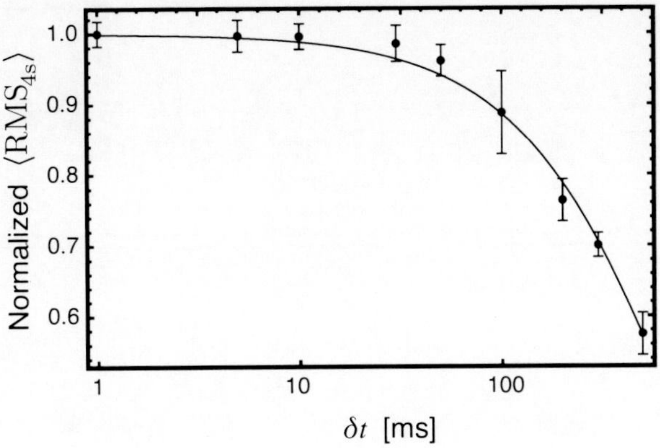

FIG. 6. *RMS bead excursion as a function of camera shutter time in milliseconds. Dots: Experimental data. Each dot represents about 20 different observed beads, with a tether of length* $L = 901$ *bp and a 490* nm *diameter bead. Error bars were drawn using the same method as in Fig. 5. Each point has been normalized to the data at 1 ms to give a dimensionless quantity on the vertical axis. Curve: Expected correction due to finite shutter speed, calculated by the method in the text (Eq. (2.4)), with shutter time given on the horizontal axis (see also the Supplement to [16]).*

effect using a 901 bp DNA and a 490 nm diameter bead at 1, 5, 10, 30, 50, 100, 200, 300 and 500 msec exposures (Fig. 6). Longer exposures indeed reduce the apparent RMS motion of the bead. The effect is minimal for exposure times smaller than 30 ms, but decreases sharply above this value.

These effects can be considered from a theoretical perspective (see the Supplement to [16]). The effect of the tether on the bead may be approximated as a harmonic restoring force. If the bead starts at a distance ρ_0 from the center, then its average position drifts inward under the influence of this force. Averaging that trajectory over the video frame gives a blurred trajectory with center at $S(\rho_0)\rho_0$, where the blur factor is

$$S(\rho_0) = \frac{T_s}{\delta t}\left[1 - \mathrm{e}^{-\delta t/T_s}\right]. \tag{2.4}$$

The time constant T_s can in principle be estimated from first principles, but in practice we fit it to data such as those in Fig. 6. For very small δt we get $S \to 1$. For large δt, we have $S \to 0$.

To predict the experimental data we should thus take the theoretical prediction and correct it by a factor of S. This correction is trivial to apply (comes out of the statistical averaging), because S is independent of ρ_0. The curves in Fig. 6 show that a correction of the form of Eq. (2.4) fits the data well; this correction was applied when drawing the curves in Fig. 5.

3. Applications to DNA looping. One of the key applications of the tethered particle method has been its use in studying DNA looping.

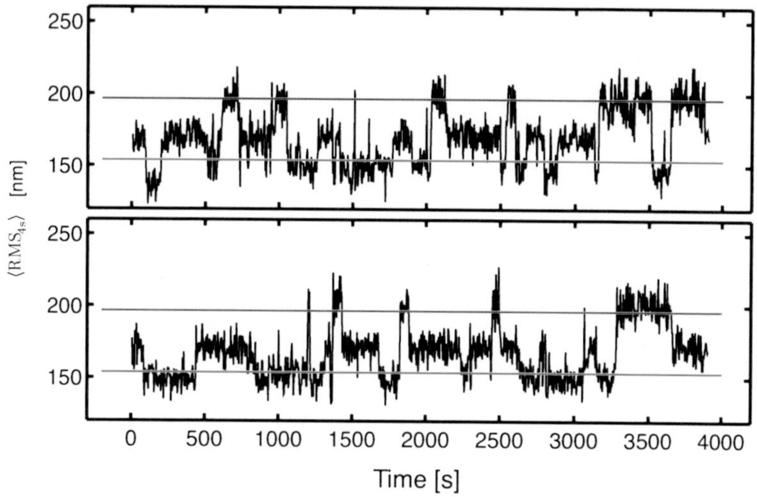

FIG. 7. *Two typical $RMS_{4\,\mathrm{s}}$ trajectories in the presence of Lac repressor, showing events of loop formation and breakdown. Total length of the DNA tether is $L = 901$ bp; bead diameter is 490 nm. Operator center to center distance is 325.5 bp. The upper horizontal line is the expected excursion from the calibration curve for the full tether; the lower horizontal line is the expected excursion for a tether of length 901-325 bp.*

Many transcriptional regulatory motifs involve the binding of transcription factors that bind at more than one site simultaneously, forming a loop of the intervening DNA (Fig. 1). The TPM technique has been used to explore these problems. The calibration analysis performed here can serve as the basis of a more careful evaluation of DNA looping and bending by DNA-binding proteins, and a guide to optimize the design of subsequent DNA looping experiments. For example, one may ask, what is the optimal total DNA length and bead size needed to reliably detect a particular type of loop? To answer such questions, first note that Fig. 3 quantifies how smaller beads and shorter tethers both allow us to work with small window size t, while still giving the narrow peak widths necessary to resolve substructure in the distribution of RMS_t. Fig. 4 reinforces this point and also quantifies how smaller beads also optimize the resolution of TPM by maintaining a high slope to the calibration curve over a wide range of L. There are limits to what can be achieved in this way, of course: Small beads are hard to observe, and short DNA tethers tend to collapse (due to surface absorption). Our work helps the experimenter to make appropriate tradeoffs when designing experiments.

Another benefit derived from the calibration curve is a better understanding of the geometry of the conformational changes we have studied. For example, Fig. 7 clearly shows the existence of a *third* state, not coinciding with either of the horizontal lines naively predicted from the calibration curve [22, 10, 7]. More detailed simulations can then shed light on the geometries of the two distinct looped species disclosed by TPM assays [16, 7].

4. Conclusions. The tethered particle motion method is one of the simplest tools for performing single-molecule experiments on DNA-protein complexes. In contrast to other methods involving fluorescence, TPM never bleaches, allowing very long observations. The central idea is to use the Brownian motion of a small particle tethered to a DNA molecule as a reporter of the underlying macromolecular dynamics of the DNA in its complexes with DNA-binding proteins. The point of this paper has been to examine the challenges that are inherent in making useful quantitative measurements using this method. One of the main outcomes of that effort has been the development of calibration curves that illustrate how tethered-particle excursions depend upon both bead size and tether length.

5. Materials and methods.

5.1. Sample preparation. The first step in any TPM experiment is construction of the relevant DNA tethers with their associated reporter beads. Polymerase Chain Reaction (PCR) was used to amplify labeled DNA with two modified primers. The primers were either biotin or digoxigenin labeled at the 5' ends (MWG Biotech AG, Ebersberg, Germany). The labels permit specific linkage of the DNA to a polystyrene microsphere or glass coverslip, respectively. The PCR templates were taken from lambda phage or modified pUC19 plasmid (sequences available upon request). The PCR products were purified by gel extraction (QIAquick Gel Extraction Kit, QIAGEN).

Streptavidin (Bangs lab) or neutravidin (Molecular Probes) coated microspheres of diameter 200, 490 and 970 nm served as our tethered particles. In contrast to the 490 and 970 nm microspheres, the 200 nm microspheres were fluorescent. Prior to incubation with DNA, a buffer exchange on the beads was performed by three cycles of centrifugation and resuspension in TPB buffer (20mM Tris-acetate, pH=8.0, 130mM KCl, 4mM $MgCl_2$, 0.1mM DTT, 0.1mM EDTA, 20 μg/ml acetylated BSA (Sigma-Aldrich), 80 μg/ml heparin(Sigma-Aldrich) and 3 mg/ml casein (Sigma), filtered with 300kD MWCO polysulfone membrane (Millipore)). This combination of reagents was chosen in an attempt to maximize sample yield and longevity, while minimizing non-specific adsorption of DNA and microspheres onto the coverslip.

The second step is DNA tether assembly. Tethered particle samples were created inside a 20-30 μl flow cell made out of a glass slide, glass coverslip, double-sided tape and tygon tubing. The coverslip and slide were cleaned with 4N HCl for 24 hours and then the flow cell was constructed in the same manner as described by van Oijen *et al.* [18]. Next, the flow chamber was incubated with 20 μg/mg anti-digoxigenin (Sigma) in PBS buffer for 30 minutes, and then rinsed with 400 μl wash buffer (TPB buffer with no casein) followed by 400 μl of TPB buffer. Microsphere-DNA complexes were created by incubating approximately 100 pM microspheres with 10 pM labeled DNA in TPB buffer for at least an hour. The DNA concentration was estimated via gel band strength. The 10:1 ratio of beads to DNA was designed to minimize the occurrence of multiple DNA strands attached to a single microsphere. The tethering procedure was completed by introducing 50 μl of the microsphere-DNA complexes into the flow cell for four to ten minutes. Additional tethering yield could be accomplished by another round of incubation with fresh microsphere-DNA complexes. Finally, unbound microspheres were removed by flushing the chamber with 1 mL TPB buffer. Once microspheres were introduced into the flow cell, tether integrity was improved by taking care to minimize flow rates within the sample chamber.

5.2. Data acquisition and analysis. The sample is imaged on an inverted microscope using Differential Interference Contrast (DIC) optics and a 1.3 NA 100x oil-objective (Olympus). The tethered particle's motion was captured using an Andor Ixon camera. Each pixel dimension corresponds to 102 nm in the sample plane. Image transfer and storage was either controlled through Ixon software (Andor Technology) or custom Matlab code (all of our Matlab acquisition and analysis code is available upon request). The former recorded 8-14 bits per pixel, while the latter captured 14 bits per pixel. However, a comparison of the capture methods showed insignificant differences (data not shown). Care was taken to ensure that the image intensity exhibited broad dynamic range without saturation. Some data was obtained using a Matlab-based autofocus routine that interfaced with a Prior controller. However, for acquisition times shorter than five minutes, the paraxial drift was small and autofocus was not needed.

The first step in analyzing TPM data is to compute trajectories for every tethered particle. The particle's X and Y displacement as a function of time was extracted from the raw data using a cross-correlation tracking algorithm [6]. Such raw positional data are subject to a slow drift due to vibrations of the experimental apparatus. A drift correction is then applied using high pass first-order Butterworth filter at cutoff frequency 0.05 Hz [19].

Acknowledgements. The senior authors gratefully acknowledge Nick Cozarelli's direct and indirect influence on our work. We especially recall Nick's tactful, wise counsel at a time when one of us was an embryonic biological physicist with an interesting, but poorly presented, idea. Multiplied manyfold, such attentions have shaped a generation of researchers.

We are grateful to generous colleagues who have advised us on many aspects of this work, including: Meredith Betterton, David Dunlap, Laura Finzi, Arivalagan Gajraj, Jeff Gelles, Jané Kondev, Chris Meiners, Keir Neuman, Matthew Pennington, Tom Perkins, Bob Schleif, Kevin Towles.

REFERENCES

[1] J.F. Beausang and P.C. Nelson, *Diffusive hidden Markov model characterization of DNA looping dynamics in tethered particle experiments*, Phys. Biol., **4** (2007), pp. 205–219.

[2] J.F. Beausang, C. Zurla, C. Manzo, D. Dunlap, L. Finzi, and P.C. Nelson, *DNA looping kinetics analyzed using diffusive hidden Markov model*, Biophys. J., **92** (2007), pp. L64–6.

[3] S. Blumberg, A. Gajraj, M.W. Pennington, and J. Meiners, *Three-dimensional characterization of tethered microspheres by total internal reflection fluorescence microscopy*, Biophys. J. (2005), pp. 1272–1281.

[4] L. Czapla, D. Swigon, and W.K. Olson, *Sequence-dependent effects in the cyclization of short DNA*, Journal of Chemical Theory and Computation, **2** (2006), pp. 685–695.

[5] L. Finzi and J. Gelles, *Measurement of lactose repressor-mediated loop formation and breakdown in single DNA molecules*, Science, **267** (1995), pp. 378–80.

[6] J. Gelles, B. Schnapp, and M. Sheetz, *Tracking kinesin-driven movements with nanometre-scale precision*, Nature, **331** (1988), pp. 450–453.

[7] L. Han, H.G. Garcia, S. Blumberg, K.B. Towles, J.F. Beausang, P.C. Nelson, and R. Phillips, *Concentration and length dependence of DNA looping in transcriptional regulation.* Submitted; available at `http://arxiv.org/abs/0806.1860`, 2008.

[8] P.C. Nelson, *Colloidal particle motion as a diagnostic of DNA conformational transitions*, Curr. Op. Colloid Intef. Sci., **12** (2007), pp. 307–313.

[9] P.C. Nelson, C. Zurla, D. Brogioli, J.F. Beausang, L. Finzi, and D. Dunlap, *Tethered particle motion as a diagnostic of DNA tether length*, J. Phys. Chem. B, **110** (2006), pp. 17260–17267.

[10] D. Normanno, F. Vanzi, and F. Pavone, *Single-molecule manipulation reveals supercoiling-dependent modulation of lac repressor-mediated DNA looping*, Nucl. Acids Res., **36** (2008), pp. 2505–2513.

[11] N. Pouget, C. Dennis, C. Turlan, M. Grigoriev, M. Chandler, and L. Salome, *Single-particle tracking for DNA tether length monitoring*, Nucl. Acids Res., **32** (2004), pp. e73–(1–7).

[12] D.A. Schafer, J. Gelles, M.P. Sheetz, and R. Landick, *Transcription by single molecules of RNA polymerase observed by light microscopy*, Nature, **352** (1991), pp. 444–8.

[13] D.E. Segall, P.C. Nelson, and R. Phillips, *Volume-exclusion effects in tethered-particle experiments: Bead size matters*, Phys. Rev. Lett., **96** (2006), pp. 088306–(1–4).

[14] M. Singh-Zocchi, S. Dixit, V. Ivanov, and G. Zocchi, *Single-molecule detection of DNA hybridization*, Proc. Natl. Acad. Sci. USA, **100** (2003), pp. 7605–10.

[15] S. Tolic-Norrelykke, M. Rasmussen, F. Pavone, K. Berg-Sorensen, and L. Oddershede, *Stepwise bending of DNA by a single TATA-box binding protein*, Biophys. J., **90** (2006), pp. 3694–703.

[16] K. Towles, J.F. Beausang, H.G. Garcia, R. Phillips, and P.C. Nelson, *First-principles calculation of DNA looping in tethered particle experiments.* Submitted; available at `http://arxiv.org/abs/0806.1551`, 2008.

[17] B. van den Broek, F. Vanzi, D. Normanno, F.S. Pavone, and G.J. Wuite, *Real-time observation of DNA looping dynamics of Type IIE restriction enzymes NaeI and NarI*, Nucleic acids research, **34** (2006), pp. 167–74.

[18] A.M. van Oijen, P.C. Blainey, D.J. Crampton, C.C. Richardson, T. Ellenberger, and X.S. Xie, *Single-molecule kinetics of lambda exonuclease reveal base dependence and dynamic disorder*, Science, **301** (2003), pp. 1235–8.

[19] F. Vanzi, C. Broggio, L. Sacconi, and F.S. Pavone, *Lac repressor hinge flexibility and DNA looping: single molecule kinetics by tethered particle motion*, Nucl. Acids Res., **34** (2006), pp. 3409–20.

[20] F. Vanzi, S. Vladimirov, C.R. Knudsen, Y.E. Goldman, and B.S. Cooperman, *Protein synthesis by single ribosomes*, RNA, **9** (2003), pp. 1174–9.

[21] M.D. Wang, H. Yin, R. Landick, J. Gelles, and S.M. Block, *Stretching DNA with optical tweezers*, Biophys. J., **72** (1997), pp. 1335–1346.

[22] O.K. Wong, M. Guthold, D.A. Erie, and J. Gelles, *Interconvertable lactose repressor-DNA looped complexes revealed by single-molecule experiments*, 2008. PLoS Biology, in press.

[23] H. Yin, R. Landick, and J. Gelles, *Tethered particle motion method for studying transcript elongation by a single RNA polymerase molecule*, Biophys. J., **67** (1994), pp. 2468–2478.

[24] C. Zurla, A. Franzini, G. Galli, D. Dunlap, D.E.A. Lewis, S. Adhya, and L. Finzi, *Novel tethered particle motion analysis of CI protein-mediated DNA looping in the regulation of bacteriophage lambda*, J. Phys.: Condens. Matter, **18** (2006), pp. S225–S234.

[25] C. Zurla, T. Samuely, G. Bertoni, F. Valle, G. Dietler, L. Finzi, and D.D. Dunlap, *Integration host factor alters LacI-induced DNA looping*, Biophys. Chem., **128** (2007), pp. 245–52.

DIFFERENCE TOPOLOGY: ANALYSIS OF HIGH-ORDER DNA-PROTEIN ASSEMBLIES

MAKKUNI JAYARAM* AND RASIKA HARSHEY*†

Abstract. DNA transactions in biological systems are often carried out by multi-subunit protein assemblies that confer a defined topology on their DNA target sites. A subset of biochemically characterized site-specific recombination reactions and at least one DNA transposition reaction have been subject to extensive topological analysis. It is conceivable that interactions between the replication apparatus and origins of replication or those between transcription machineries and promoters, enhancers and repressor binding sequences also impose precise topological constraints on the path of DNA. Such 'topological filters' are thought to stabilize DNA-protein configurations that are conducive to triggering the chemical steps of the respective reactions. 'Difference topology' is a simple method for deciphering the DNA topology within complex DNA-protein machines that are not readily amenable to standard structural analyses. The logic is to trap the crossings formed by distinct DNA segments by tying them into knots or links by site-specific DNA inversion and deletion, respectively, carried out by a recombinase. The number of such crossings can then be counted by analytical methods such as gel electrophoresis or electron microscopy.

Key words. DNA topology, topological filter, site-specific recombination, DNA transposition.

1. Introduction. Nick Cozzarelli was one of the pioneers in employing topological analyses to describe mechanisms of DNA rearrangements (Cozzarelli et al., 1984; Cozzarelli, 1990; Wasserman and Cozzarelli, 1986). Collaborating first with James White in the nineteen eighties and later with De Witt L. Sumners in the nineties, he was largely responsible for introducing new mathematical ways of thinking about reactions in which DNA strands are broken and joined (Cozzarelli et al., 1984; Sumners et al., 1995). One of us (MJ) was drawn into DNA topology and the topology of DNA recombination by reading an article by Cozzarelli and colleagues in the 1984 volume of the Cold Spring Harbor Symposia series (Cozzarelli et al., 1984). Two simple, if somewhat cheerless, equations describe the conservation of linkage during recombination between two head-to-head or head-to-tail sites located on the same circular substrate molecule that is unknotted to begin with. For the deletion reaction between head-to-tail sites that generates two daughter circles (Figure 1A),

$$^{\mathrm{tra}}\mathrm{Lk}^{\mathrm{s}} + {}^{\mathrm{tra}}\mathrm{Me} = \mathrm{Lk}^{\mathrm{p}} \tag{1}$$

and

$$^{\mathrm{ter}}\mathrm{Wr}^{\mathrm{s}} + {}^{\mathrm{ter}}\mathrm{Me} = \mathrm{Ca}^{\mathrm{p}}. \tag{2}$$

*Section of Molecular Genetics and Microbiology & Institute of Cellular and Molecular Biology, University of Texas at Austin, Austin, TX 78712 (jayaram@icmb.utexas.edu).

†rasika@uts.cc.utexas.edu.

C.J. Benham et al. (eds.), *Mathematics of DNA Structure, Function and Interactions*, The IMA Volumes in Mathematics and its Applications 150, DOI 10.1007/978-1-4419-0670-0_7, © Springer Science+Business Media, LLC 2009

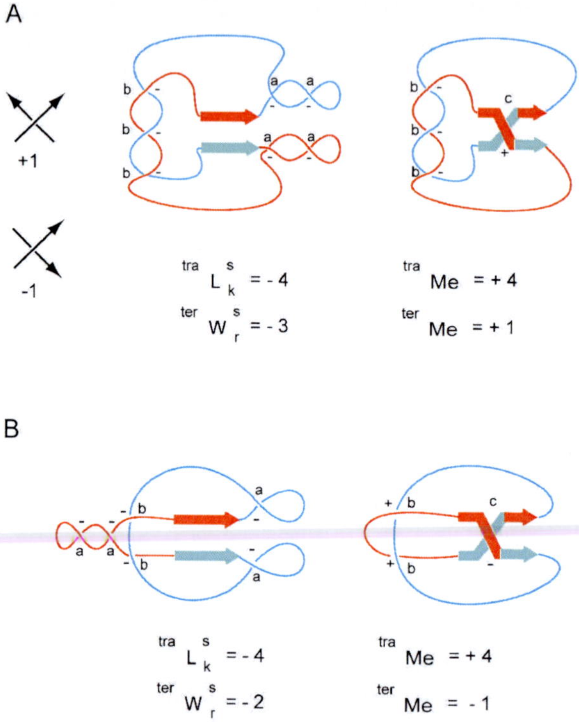

FIG. 1. *Conservation of linkage during recombination reactions: The recombination substrates are negatively supercoiled unknotted circular molecules, divided into two domains (red and blue) by a pair of recombination target sites (represented by the broad arrows). In the deletion substrate (A), the sites are in head-to-tail (direct) orientation; in the inversion substrate (B), they are in head-to-head (inverted) orientation. Intradomainal supercoil crossings (red × red or blue × blue) within the substrate and product(s) are denoted by* ${}^{tra}Lk^{s}$ *and* ${}^{tra}Lk^{p}$ *respectively. Interdomainal crossings (red × blue) within the substrate are referred to as* ${}^{ter}Wr^{s}$*. They contribute to catenane crossings* (Ca^{p}) *in the deletion products or knot crossings* (Kn^{p}) *in the inversion product.* ${}^{ter}Wr^{s}$ *crossings that escape being trapped in knots will end up as interdomainal crossings in the product,* ${}^{ter}Wr^{p}$*. Changes in the intradomainal and interdomainal crossings resulting from the mechanism of action of the recombianse enzyme are denoted by* ${}^{tra}Me$ *and* ${}^{ter}Me$*, respectively. In assigning signs to the DNA crossings, the DNA axis in the substrate circle is given an arbitrary, fixed direction. The conventions for the + and − signs are shown. The continuous and broken arrows correspond to the 'overlying' and 'underlying' axis segments, respectively. For the deletion and inversion substrates shown in A and B, respectively,* ${}^{tra}Lk^{s}(\Sigma a\text{'s}) = -4$, ${}^{tra}Me = +4$ *and* ${}^{tra}Lk^{p} = 0({}^{tra}Lk^{s} + {}^{tra}Me)$*. Within the deletion synapse (A),* ${}^{ter}Wr^{s}(\Sigma b\text{'s}) = -3$*, and* ${}^{ter}Me(c) = +1$ *for strand exchange in the right handed sense.* $Ca^{p} = -2({}^{ter}Wr^{s} + {}^{ter}Me)$*, denoting the formation of a singly linked catenane. For the inversion synapse (B),* ${}^{ter}Wr^{s} = -2$*.* ${}^{ter}Me$ *for the right handed strand exchange is −1 because of the inversion of the blue domain with respect to the red.* $Kn^{p} = 0$ *(indicating an unknotted inversion product) and* ${}^{ter}Wr^{p} = +1$ $(-{}^{ter}Wr^{s} + {}^{ter}Me = Kn^{p} + {}^{ter}Wr^{p})$.

The substrate is divided into two separate domains whose borders are defined by the two recombination sites (or more precisely, by the points of DNA breakage and exchange within the two sites). The first equation represents the redistribution of intradomainal supercoils ($^{tra}Lk^{s}$) of the substrate and the supercoils added (or removed) by the enzyme mechanism (^{tra}Me) into supercoils of the product circles (Lk^{P}). The second one describes the conversion of interdomainal supercoils ($^{ter}Wr^{s}$) of the substrate together with similar crossings introduced by the enzyme mechanism (^{ter}Me) into catenane crossings (Ca^{P}) or links between the product circles. Very similar equations can be written for the inversion reaction between head-to-head recombination sites (Figure 1B).

$$^{tra}Lk^{s} + {}^{tra}Me = Lk^{P} \tag{3}$$

and

$$-{}^{ter}Wr^{s} + {}^{ter}Me = Kn^{P} + {}^{ter}Wr^{P}. \tag{4}$$

The minus sign in (4) denotes the inversion of the DNA axis during the reaction, and the extra term $^{ter}Wr^{P}$ represents the interdomainal supercoils that do not become part of the knot crossings. As explained in Cozzarelli et al. (1984), there are two reasons for this. First, it takes a minimum of three crossings to form the simplest knot. Second, only an odd number of crossings can be trapped into knots when Me and $^{ter}Wr^{s}$ are of the same sign, and only an even number of crossings can be trapped when they are of opposite signs.

The lasting impression made by the straightforward equations of recombination, no doubt, subliminally kindled the logic of the difference topology method (see below). Indeed, Cozzarelli was catholic in his embrace of new analytical tools to gain deeper understanding of complex biological phenomena. At a Cold Spring Harbor Symposium, Frank Stahl (Stahl, 1979) lamented that *'successful in vitro analyses marked the beginning of the end of recombination as a geneticist's playground.' "I can already hear the biochemists circling in the night."* Cozzarelli's response (Cozzarelli, 1986), delivered seven years later at a UCLA Symposium, was: *"I smiled because I am a biochemist. Now I see crystallographers and other physical chemists stacking up in the sky. I say 'Welcome'."*

2. Topology of site-specific recombination: Tn3 or $\gamma\delta$ resolvase reaction and Hin or Gin invertase reaction. Before discussing difference topology *per se*, we shall briefly consider the topology of the closely related resolvase enzymes of Tn3 or $\gamma\delta$ transposons, serine family site-specific recombinases that reduce cointegrate intermediates formed during transposition to simple integrants (Grindley, 2002; Grindley et al., 2006; Li et al., 2005). One of the critical assumptions underlying difference topology analysis is that the resolvase reaction occurs within a unique synapse in which three negative supercoils are sequestered (−3 synapse).

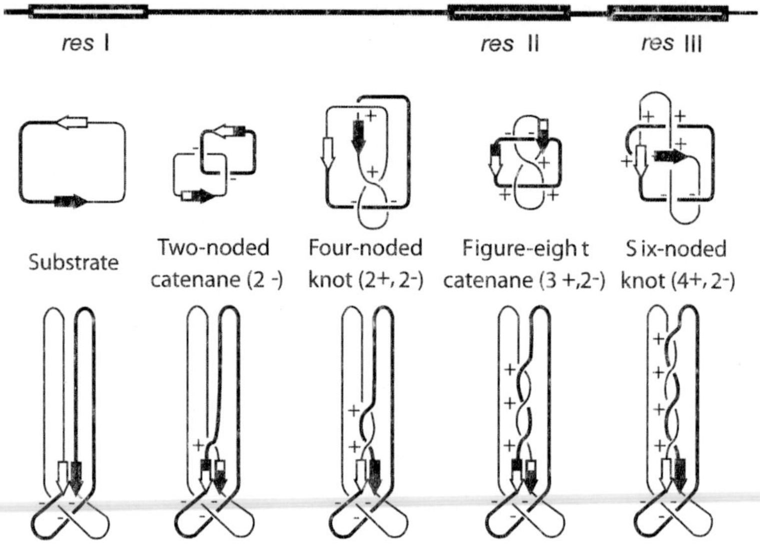

FIG. 2. *The resolvase synapse and the products of processive recombination: The organization of the res site is schematically diagrammed at the top. DNA breakage and exchange occur between the res I sub-sites; res II and res III are accessory sub-sites. The topology of the resolvase recombination conforms to a 3-noded synapse (−3), with a right-handed DNA swap during each exchange event. The res I sites are shown by the unfilled and filled arrows in the substrate, so that the 'hybrid' arrows correspond to an odd number of recombination events. Crossings of DNA axis, here as well as all figures to follow, are assigned + or − signs according to the convention described under Figure 1. During the experimental procedures, supercoils are removed by DNA nicking, leaving behind only the knot and catenane nodes in the recombination products. The figure is adapted from Grindley (2002).*

We shall also touch upon the mechanistically related Hin and Gin DNA inversion systems in which the chemistry of site-specific recombination is triggered within a −2 synapse (harboring two negative supercoils).

The resolvase reaction demands that the recombination sites be arranged in head-to-tail orientation within a negatively supercoiled substrate (Grindley et al., 2006). Each resolvase target site is composed of three sub-sites, *res* I, *res* II and *res* III, each of which can be occupied by a resolvase dimer (Figure 2). The resolvase dimers bound at the accessory sub-sites II and III of the recombination partners organize the −3 synapse, and allosterically activate the resolvase dimers bound at sub-sites I to carry out the chemistry of strand breakage and exchange. The ability of resolvase to perform iterative rounds of recombination without dissociation of the recombination synapse facilitates the topological analysis of the reaction by providing a series of sequential products that can be characterized.

The products of rounds one through four of the resolvase reaction are a two-noded catenane called the Hopf link (2^-), a four-noded knot (2^+, 2^-), a figure-eight catenane (3^+, 2^-) and a six-noded knot (4^+, 2^-), respectively (Figure 2). These product configurations signify a unique synapse in which three negative supercoils are trapped, the *res* I sub-sites are arranged in parallel geometry, and each DNA exchange step occurs in the right-handed sense with the addition of one positive crossing (Grindley, 2002). Results from a series of biochemical experiments together with information from a recently solved resolvase-*res*I co-crystal structure (Kamtekar et al., 2006; Li et al., 2005) agree with a parallel arrangement of the *res* I sites and exchange of the cut strands by a rotary mechanism, as depicted in Figure 2. We therefore choose the -3 resolvase synapse as the standard for deriving topological features of other DNA transactions by difference topology.

Hin and Gin recombinases bind as dimers to their target sites *hix* and *gix*, respectively, arrange the functional synapse with the assistance of the *E. coli* Fis protein bound as a dimer to its cognate enhancer site, and execute DNA inversion. The sequential products for the Hin/Gin reaction for the first four rounds of reaction are an unknotted inversion circle, a three-noded knot (3^-), a four-noded knot (2^+, 2^-) and a five-noded knot (5^-), respectively (Figure 3). A -2 synapse (containing two trapped negative supercoils) with DNA exchange in the right-handed mode (as in the resolvase case) most readily explains the DNA inversion topology (Johnson, 2002; Kanaar et al., 1990) (Figure 3). Since each round of recombination causes relative inversion of the DNA segment between the *hix* or *gix* sites, the sign of the synaptic nodes alternates between $-$ and $+$, as illustrated in Figure 3. So also, the right-handed DNA crossing resulting from strand exchange has a negative sign in the product (contrast this with the corresponding $+$ crossing in the resolvase reaction). Strong biochemical evidence supports the DNA rotation mechanism (Dhar et al., 2004).

To generalize, the synapses arranged by the serine recombinases have a characteristic topological signature. These enzymes impose a parallel geometry on the recombination partners and carry out recombination with DNA rotation in the right-handed sense. Note from Figures 2 and 3 that it takes two and three interdomainal crossings to arrange the head-to-head (*hix* or *gix*) sites and head-to-tail (*res* I) sites, respectively, in parallel fashion. In other words, in inversion substrates, zero or an even number of interdomainal crossings will give rise to parallel synapse; an odd number of such crossings will yield anti-parallel synapse. The reverse is true for deletion substrates. When dealing with the topology of other DNA-protein assemblies (to be discussed later), it is helpful to keep this simple rule in mind.

Strictly, the geometry of the core recombination sites within the synapse can be defined as parallel or antiparallel only if they lie in the same plane. In formulating the logic of difference topology, the co-planarity of recombination sites was a simplifying assumption. The crystal structures

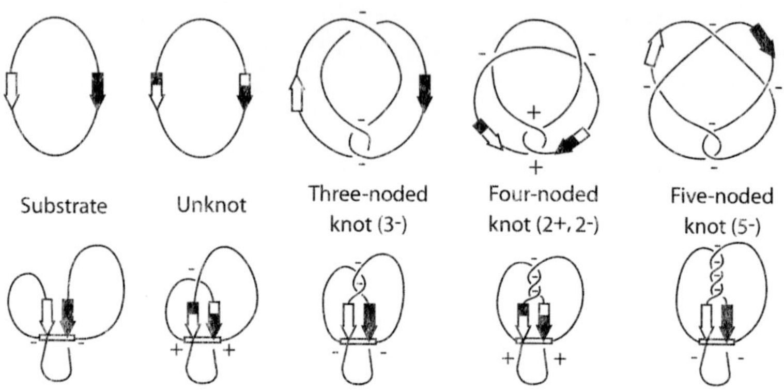

FIG. 3. *The Gin inversion synapse and the products of processive recombination: The Gin recombinase synapse consists of two DNA crossings, arranged with the assistance of the Fis protein and its binding site. Strand exchange by Gin follows a right-handed rotation as in the resolvase reaction. Because each recombination event inverts the DNA segment between the gix sites (changing the direction of the DNA axis; see also Figure 1), the synaptic nodes change in sign with each recombination event. For the same reason, the DNA crossings resulting from strand exchange have − signs as opposed to the + signs in the resolvase reaction (Fig. 2). The figure is adapted from (Johnson, 2002) and (Kanaar et al., 1990).*

of resolvase and of the tyrosine family recombinases Cre and Flp (see below) in association with DNA are consistent with this assumption (Chen et al., 2000; Guo et al., 1997; Li et al., 2005).

3. Difference topology: Trapping DNA crossings within a protein-DNA assembly by Flp or Cre mediated site-specific recombination. Cre and Flp, 'simple' members of the tyrosine family recombinases, can carry out DNA inversion and deletion equally well, depending on the relative orientation of their target sites: *loxP* and *FRT*, respectively (Jayaram, 2002; van Duyne, 2002). They do not require accessory protein factors, are indifferent to the topology of the DNA substrates, and mediate intra- or intermolecular recombination. In principle, they are ideally suited for sealing off, through the act of recombination, the DNA crossings confined within an external synapse. The number of such crossings in the recombination products, knots for inversion and catenanes for deletion, can then be counted by standard analytical procedures such as gel electrophoresis or electron microscopy. However, to derive the topology of the 'unknown synapse', one has to know what the contribution of the recombination reaction itself is to the topology of the final product (the term $^{ter}\mathbf{Me}$ in the Cozzarelli et al. equations for recombination). $^{ter}\mathbf{Me}$ can be derived by answering the following questions. Will Cre (or Flp) introduce a DNA crossing during strand exchange or not? If they do, what is the sign of the crossing?

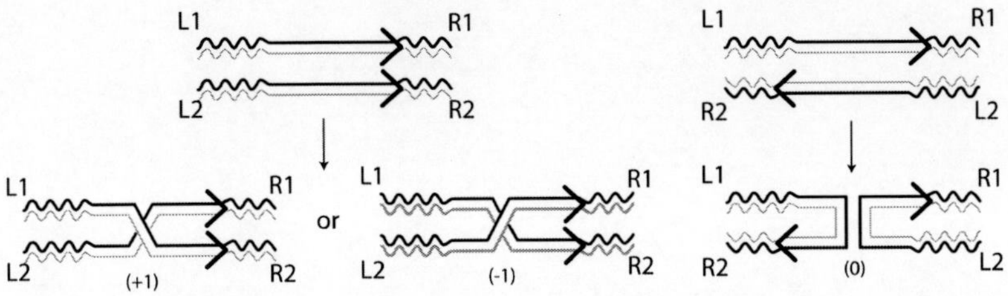

FIG. 4. *Geometry of site alignment during recombination by Flp or Cre: The two partner sites are named L1-R1 and L2-R2 to orient them left to right. In the reaction shown at the left, the two sites are arranged in a parallel fashion (left to right in both cases) in the plane of the paper. If strand exchange occurs with a right-handed rotation of the DNA, the recombinant products (L1-R2 and L2-R1) cross each other to introduce a +1 node. If the sense of rotation is left-handed during exchange, the crossing between L1-R2 and L2-R1 will be −1. In the reaction shown at the right, the partner sites are arranged in the antiparallel orientation (left to right for L1-R1 and right to left for R2-L2). The act of recombination does not introduce a crossing between L1-R2 and L2-R1.*

4. The geometry of site alignment during Flp and Cre recombination. In principle, the *FRT* (or *loxP*) sites may be arranged in a parallel or antiparallel orientation during recombination, provided the reaction is carried out within a planar DNA-protein complex. As noted earlier, the assumption of planarity has been justified by the crystal structures of the Cre and Flp synapses (Chen et al., 2000; Guo et al., 1997). The DNA crossing formed during recombination from the parallel geometry may be denoted by +1 or −1, depending on whether the DNA rotation is right-handed or left-handed (Figure 4). By contrast, recombination from the antiparallel geometry of the recombining sites will not introduce a DNA node (zero crossing). As briefly outlined below, this predicted difference, combined with the known topology of the −3 resolvase synapse, has been exploited to reveal the mode of alignment of the *FRT* and *loxP* sites in their respective synapses (Grainge et al., 2000; Grainge et al., 2002; Kilbride et al., 1999). We describe below the experiments conducted with Flp, but the results hold true for Cre as well.

In the assays performed by Grainge et al., a hybrid *res-FRT* site was constructed by replacing the *res* I sub-site with *FRT* but retaining *res* II and *res* III in their native positions (Figure 5). In a supercoiled plasmid containing two such sites with the normal orientations of *res* II-*res* III, the characteristic synapse with the three negative crossings could be assembled by the addition of resolvase. Subsequently, recombination was performed at the *FRT* sites using Flp, and the product topology was examined. When the *FRT* sites were in head-to-head orientation, the inversion reaction from the hybrid synapse yielded a 3-noded knot with + crossings. When the *FRT* sites were in head-to-tail orientation, the deletion reaction gave a 4-noded catenane.

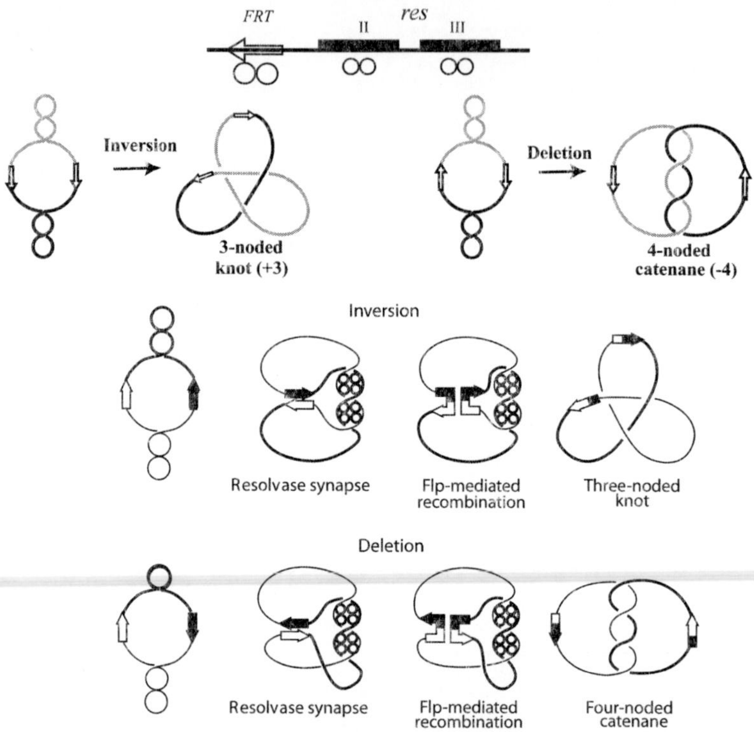

FIG. 5. *Flp-mediated recombination after the assembly of the resolvase synapse: The general organization of a hybrid FRT/res site is diagrammed schematically at the top. The large twin circles below FRT represent two Flp monomers; the smaller ones below res represent resolvase dimers. The res II and res III accessory sites together with the bound resolvase subunits from recombination partners will establish the well characterized* -3 *synapse (see Fig. 2). The res I site is replaced by an FRT site. In the two hybrid FRT/res sites present on a plasmid, the res II/III sites are arranged in their native orientation. The FRT sites are present in a head-to-head orientation in the inversion substrate (left) and a head-to-tail orientation in the deletion substrate (right). Recombination after preincubation with resolvase (to establish the 3-noded synapse) enriches the 3-noded knot for inversion and the 4-noded catenane for deletion). These product topologies are consistent with antiparallel alignment of the FRT sites (see text).*

The outcomes from the *res-FRT* hybrid site recombination reactions are consistent with an antiparallel geometry for the *FRT* sites within the Flp synapse, with no DNA crossing being added during recombination (Figure 5). The three external DNA crossings will synapse the head-to-head *FRT* sites in the antiparallel mode. Recombination will preserve these three nodes in the inversion knot. For the head-to-tail *FRT* sites, a fourth node from the negatively supercoiled DNA must be trapped to orient them in the antiparallel fashion. Hence recombination results in the 4-noded catenane. The antiparallel geometry of the *FRT* and *loxP* sites deduced by the topology method agrees with the arrangements of these sites in the Flp and Cre crystal structures, respectively.

The arguments summarized in Table 1 rule out recombination in the parallel mode with either a +1 crossing or a −1 crossing. For the +1 mode of strand exchange from the parallel sites (top panel in Table 1), the predicted products are a three-noded knot for inversion and a two-noded catenane for deletion. For the −1 mode of strand exchange (bottom panel), the corresponding products are a five-noded knot and a four-noded catenane. The experimentally observed pair of products, a three-noded knot and a four-noded catenane (see Figure 5), disagree with the topological predictions for parallel site alignment, and strongly support antiparallel synapsis of the *FRT* sites. Note that the 'additional' extraneous negative crossing for the deletion reaction (−4 compared to −3 for the inversion reaction) represents a negative supercoil from the plasmid substrate that must be trapped to bring the *FRT* sites in parallel geometry. Note also that the signs of all the crossings change during inversion (rows 1 and 3), and hence the knot nodes are '+'.

In the vast of majority of difference topology assays that we have performed, the crossing numbers in the knot and catenane products resulting from recombination were inferred by gel electrophoresis. However, in those cases where these products were further examined by electron microscopy, they were all torus knots and catenanes. The knot crossings were all plus, and the catenanes were all right handed. We have assumed that these toplogical features of the recombination products are constant in all the reactions described.

5. Deciphering the DNA path within an unknown synapse. Having established that Flp and Cre do not introduce a DNA crossing during strand exchange, deriving the number of DNA crossings trapped in an 'unknown' synapse formed by, say, a recombination, transposition, replication or transcription complex becomes straightforward. Any given assay measures the number of plectonemic wraps formed between two interacting DNA sites, separated into two domains by suitably positioned recombination sites. As described with resolvase, the unknown synapse is first assembled in two matched plasmid substrates that differ only in the relative orientation of the recombination sites (head-to-head in one case and head-to-tail in the other), after which the deletion and inversion reactions are carried out and the catenane and knot crossings counted. The relevant Cozzarelli et al. equations may be rewritten as

$$[{}^{\text{ter}}\text{Wr}^{\text{us}} + (0 \text{ or } -1)] + {}^{\text{ter}}\,\text{Me}^{\text{Flp or Cre}} = \text{Ca}^{\text{p}}, \quad \text{for the deletion reaction} \tag{5}$$

and

TABLE 1

Predictions on recombination topology for parallel geometry of FRT sites: Predicted topologies of inversion knots and deletion catenanes during resolvase-assisted Flp recombination if the FRT sites have a parallel geometry. In the top panel, strand rotation during Flp-mediated cross-over is assumed to be right-handed (+1); in the bottom panel, the rotation is assumed to be left-handed (−1). For the inversion and deletion reactions, the external resolvase synapse has a fixed topology (−3 crossings). An additional negative supercoil (−1) is trapped in the inversion substrate (rows 1 and 3) for parallel alignment of the FRT sites. The signs of the DNA crossings in the substrate and product are reversed when recombination inverts DNA (+3 knot in row 1 and +5 knot in row 3). Flp recombination with FRT sites arranged in parallel fashion predicts either the +3 knot/−2 catenane (top panel) or the +5 knot/−4 catenane (bottom panel) pair of inversion and deletion products. The experimental results, yielding the +3 knot/−4 catenane combination, contradict these predictions and rule out parallel site alignment.

	Resolvase synapse	Additional crossing	Crossing during recombination	Product topology
Inversion	-3	-1	+1	Kn_3
Deletion	-3	0	+1	Ca_2
Inversion	-3	-1	-1	Kn_5
Deletion	-3	0	-1	Ca_4

$$-[{}^{ter}Wr^{us} + (0 \text{ or } -1)] + {}^{ter}Me^{Flp \text{ or } Cre} = Kn^{p} + {}^{ter}Wr^{p}, \tag{6}$$

for the inversion reaction.

Here ${}^{ter}Wr^{us}$ represents the interdomainal negative supercoils within the unknown synapse. As demonstrated in the previous section ${}^{ter}Me^{Flp \text{ or } Cre}$ is equal to 0. ${}^{ter}Wr^{us}$ may be odd or even, depending on the particular system being studied. If it is odd, there is no need for an extra supercoil for aligning head-to-head recombination sites in antiparallel geometry (${}^{ter}Wr^{us} + 0$); if it is even, an extra negative supercoil from the substrate is needed to establish this geometry (${}^{ter}Wr^{us} - 1$). The converse is true for recombination sites in the head-to-tail orientation. The additional supercoil is needed when ${}^{ter}Wr^{us}$ is odd.

The prediction for the experimental outcome is that the number of crossings in the inversion knot and those in the deletion catenane will dif-

TABLE 2

Recombination products predicted by difference topology for fixed topologies of the outside synapse. For a series of extraneous synapses containing a fixed number of negative supercoils, the crossing numbers in the Flp or Cre mediated inversion and deletion products expected from difference topology are listed. When no supercoils are trapped, the inversion circle will be unknotted, and the deletion circles will be unlinked. When only one supercoil is trapped, the inversion product will be unknotted but the deletion circles will be catenated. For any other value for the trapped supercoils, the inversion knot and the deletion catenane will differ in crossing number by one.

Extraneous synapse	Predicted Flp or Cre recombination product	
Number of negative supercoils	**Inversion**	**Deletion**
0	Unknot	Unlinked circles
1	Unknot	Ca_2
2	Kn_3	Ca_2
3	Kn_3	Ca_4
4	Kn_5	Ca_4
5 | | | n (odd)	Kn_5 | | | Kn_n	Ca_6 | | | $Ca_{(n+1)}$
n+1	$Kn_{(n+2)}$	$Ca_{(n+1)}$

fer by one, except for the cases in which the synapse contains no interdomainal crossings or a single crossing (see Table 2). The smaller of the two numbers gives the DNA crossings within the unknown synapse. Again, the additional crossing in one of the products reflects the need to keep the *loxP* or *FRT* sites antiparallel for recombination. When no outside crossings are trapped, recombination will result in topologically simple products: the unknot during inversion and unlinked circles during deletion. When there is only one outside crossing, the inversion product will not be knotted; the deletion product will be a 2-noded catenane.

Two conditions must be met for this analysis to work. The extraneous synapse must be stable enough to preserve the DNA crossings within it in the context of the Cre/Flp recombination synapse. Furthermore, the recombination sites must be placed sufficiently close to a DNA region of interest to avoid random entrapment of interdomainal nodes. For a two-site interaction resulting in non-integral DNA crossings, the difference topology

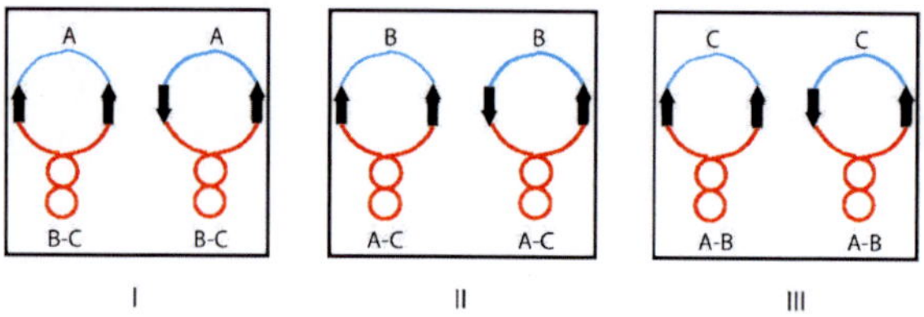

FIG. 6. *Deciphering the DNA crossings established by the bridging interactions of three DNA sites. In (I), the site A (blue domain) is separated from sites B-C (red domain) by recombination sites arranged in head-to-head and head-to-tail orientations. In (II) and (III), the blue domain is constituted by sites B and C, respectively; the corresponding red domains are A-C and A-B. As described in the text, the difference topology assays for (I), (II) and (III) will give the crossings made by (i) A with B and C, (ii) B with C and A and (iii) C with A and B.*

assay would reveal this number as the next higher integer (assuming that the synapse is stable). If the partial turn can be unwound during the assembly of the Flp or Cre recombination synapse, the result would be the next lower integer. Thus, an inherent limitation of difference topology is that the output is quantized, the DNA intertwining of interest being rounded off to the nearest higher or lower integer.

6. Application of difference topology to a three DNA site interaction system. Imagine a DNA protein assembly in which three distant DNA sites A, B and C are brought together with a unique topology by DNA-protein and protein-protein interactions. To arrive at the DNA topology within this synapse, a minimum of three assays, each using the matched inversion and deletion substrates must be performed: one in which the A domain is separated from the B-C domains by recombination sites close to and on either side of A, and the other two in which the B and C domains are similarly separated from the C-A and A-B domains, respectively (Figure 6). The first yields the number of crossings 'N1' that A makes with B and C, the other two the number of crossings N2 and N3 that B makes with A and C and C makes with A and B, respectively.

$$(A \times B) + (A \times C) = N1 \tag{7}$$

$$(B \times C) + (B \times A) = N2 \tag{8}$$

$$(C \times A) + (C \times B) = N3. \tag{9}$$

From (7) and (8),

$$(A \times C) - (B \times C) = N1 - N2. \tag{10}$$

From (9) and (10),

$$(A \times C) = \frac{1}{2}(N1 - N2 + N3). \tag{11}$$

Substituting this value for $(A \times C)$ in (7) and (9),

$$(A \times B) = \frac{1}{2}(N1 - N3 + N2) \tag{12}$$

$$(B \times C) = \frac{1}{2}(N2 - N1 + N3). \tag{13}$$

The DNA transposition reaction carried out by the bacteriophage Mu has been subjected to a detailed analysis by difference topology (Grainge et al., 2002; Harshey and Jayaram, 2006; Pathania et al., 2002). The reaction requires negative supercoiling of the DNA substrate plus the interaction of three separate DNA sites, the left and right ends of Mu (attL and attR, referred to here as L and R, respectively) and the enhancer element (E), mediated by the transposase protein MuA. Do these interactions sequester a fixed number of supercoils within the transpososome? And if so, how are the crossings distributed among the three sites?

The key results for the Mu system and the three-site, five-noded synapse that they signify are assembled in Figure 7. The experimental observations were:

$$(E \times L) + (E \times R) = 3 \tag{14}$$

$$(L \times E) + (L \times R) = 3 \tag{15}$$

$$(R \times E) + (R \times L) = 4. \tag{16}$$

Solving these equations yields, $(E \times L) = 1$; $(E \times R) = 2$; and $(L \times R) = 2$. The number of crossings between L and R were further verified by an alternative strategy. Here, the enhancer was deleted from the substrate and provided as a linear DNA fragment in *trans*, thus topologically dissociating it from the synapse. In agreement with the two DNA crossings between L and R, the inversion product in this case was a three-noded knot, and the deletion product was a two-noded catenane. Similarly, the number of crossings between E and R could also be confirmed as two by deleting L from the substrate. The products in this case were a two-noded catenane and a three-noded knot. A similar analysis of the crossings between E and L was not possible, because a stable E-L MuA complex cannot be assembled in the absence of R. A mathematical analysis of the difference topology data for the Mu transpososome using tangle calculus has established the transposition synapse diagrammed in Figure 7 as the biologically most plausible solution (I. Darcy, J. Luecke, M. Vasquez, "Tangle analysis of difference topology experiments: Applications to a Mu protein-DNA complex", preprint; arXiv:0710.4150v1 [math.GT], http://xxx.lanl.gov/abs/0710.4150v1).

In principle, the difference topology analysis is applicable to systems that may require bridging interactions among more than three DNA sites. For a four site interaction involving sites A-D, one can derive four crossing numbers N1-N4 as

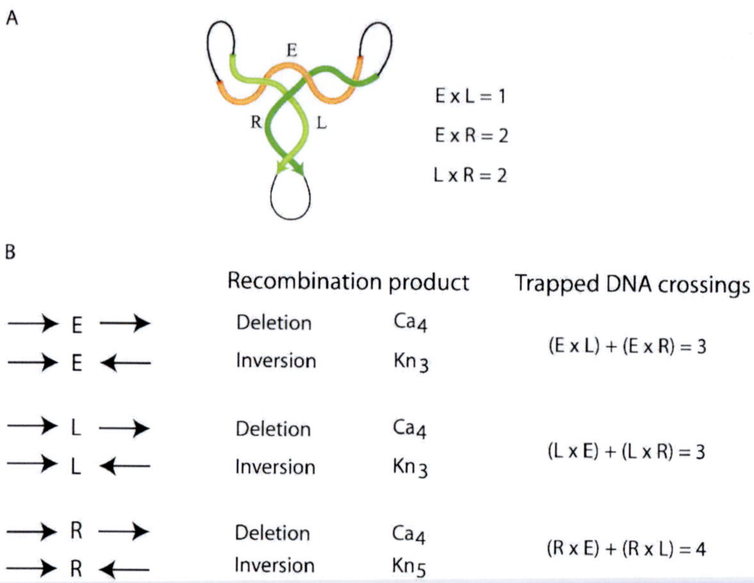

FIG. 7. *The topology of the Mu transposition synapse: DNA crossings trapped by MuA interactions with attL, attR and the enhancer:*
A. The five DNA crossings trapped by the left (L) and right (R) ends of Mu and the transposition enhancer (E) when the transposase protein bridges these three sites are schematically depicted ($E \times L = 1; E \times R = 2; L \times R = 2$).
B. The crossing numbers between each of the three sites and its two partners were deduced by the difference topology method. For each experiment, the site that was sequestered by flanking recombination target sites is shown at the left. The arrows represent the recombination sites and their relative orientation.

$$\begin{aligned}(A \times B) + (A \times C) + (A \times D) &= N1 \\ &\text{[site A in one domain and BCD in the other]}\end{aligned} \tag{17}$$

$$\begin{aligned}(B \times C) + (B \times D) + (B \times A) &= N2 \\ &\text{[site B in one domain and CDA in the other]}\end{aligned} \tag{18}$$

$$\begin{aligned}(C \times D) + (C \times A) + (C \times B) &= N3 \\ &\text{[site C in one domain and DAB in the other]}\end{aligned} \tag{19}$$

$$\begin{aligned}(D \times A) + (D \times B) + (D \times C) &= N4 \\ &\text{[site D in one domain and ABC in the other].}\end{aligned} \tag{20}$$

These results alone are insufficient to derive all six crossing numbers for the pairwise combinations of these sites. However, if one of the sites can be deleted or supplied in *trans* without affecting the topological interactions among the others (as is the case with the Mu transposition system), the corresponding three pairwise crossing numbers can be estimated. Substitution of these values in Equations 17–20 will yield the remaining three crossing numbers.

7. Potential wider applications of difference topology. The topological features of the interactions of DNA sites in replication, transcription and repair complexes should be amenable to the types of analyses described for the Mu transposition system. As explained above, the number of crossings made by a given site with the rest of the sites can be determined after isolating it into one DNA domain with the recombination target sites placed on either side of it. The procedure is then iterated for each of the rest of the sites. For a subset of these sites, it is often possible to topologically unlink that particular site by deleting it or by supplying it in *trans*. The subtopologies thus obtained can then be integrated into the final composite topology.

Difference topology should be generally useful in following changes in DNA topology as a DNA-protein complex matures through intermediate stages to its functional form. Of course, an intermediate has to be stable enough to prevent the DNA configuration within it from being altered by the assembly of Flp or Cre recombination synapse. It must also be able to withstand isolation for post-recombination analysis of DNA topology. The Mu transpososome progresses through three sequential states of increasing stability called the LER, Type0 and TypeI complexes (Chaconas, 2002) (Figure 8). The conversion of Type0 to the TypeI complex is coincident with the single stranded cleavages at the L and R ends of Mu. Analysis of each of these three complexes reveals that the five-noded synapse topology is established as early as the LER complex and preserved through the TypeI complex; however, a unique E-R complex with 2 DNA crossings precedes LER (Pathania et al., 2002; Pathania et al., 2003). The TypeI complex carries out joining of the cleaved strands to the target DNA to yield the TypeII complex.

Another potential use of difference topology is in dissecting the contributions of individual DNA-protein interactions towards the topology of the final assembly. For example, the L-E-R interactions during Mu transposition are mediated by the MuA protein which has two DNA binding specificities. A MuA monomer can associate with one of three sub-sites at each Mu end, L1, L2 and L3 at the left and R1, R2 and R3 at the right (Figure 9). It can potentially interact with one of three sub-sites within the enhancer designated O1, O2 and O3, as these are also 'operator' sites to which a Mu coded repressor protein binds to negatively regulate transcription of the phage genome. Although the assembly of the transpososome is initiated with six MuA monomers (consistent with the three binding sub-sites each at L and R), only four monomers form part of the stable chemically competent transpososome, and the other two can be stripped off from the complex without affecting the reaction. By individually ablating the sub-sites at L, R or E, it is possible, in principle, to distinguish

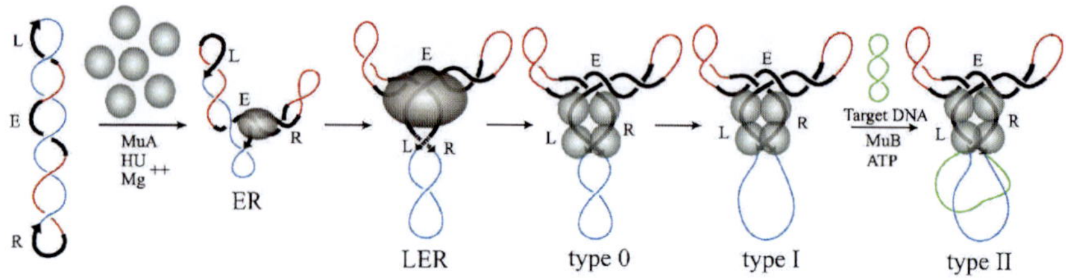

FIG. 8. *Order and dynamics of Mu transpososome assembly: Assembly of the Mu transpososome is initiated by MuA mediated interaction of E and R segments, followed by the capture of L. The process is assisted by the E. coli HU protein and divalent cations. The resulting LER complex, which traps five negative supercoils (see Fig. 7), progresses through Type0 and TypeI complexes to the TypeII complex. The 5-noded DNA topology is maintained through strand cleavage (TypeI), and likely through strand transfer (TypeII). The strand transfer step is promoted by the MuB protein and ATP. The assembly of the transpososome starts with six MuA monomers; however, a tetrameric form of MuA, stabilized in the Type0 complex, is responsible for strand cutting. The other two MuA monomers appear to remain in the complexes in a loosely associated form. The complexes become progressively more stable, TypeII being the most stable one.*

sub-sites that are essential for the native topology of the transpososome from those that are dispensable. To rephrase the issue, does the removal of a sub-site lead to an altered transpososome lacking one or more DNA crossings? To illustrate the point, the results of the difference topology assay for the Type0 complex organized with a substrate lacking the R3 sub-site are displayed in Figure 9 (Yin et al., 2007). The number of crossings E makes with L and R decreases by one, from 3 to 2, when R3 is absent; and so does the number of crossings R makes with E and L (from 4 to 3). By contrast the number of crossings L makes with E and R remains unchanged (3). Thus lack of R3 does not arrest assembly, but gives rise to a four-noded transpososome that lacks one of the two E-R crossings present in the normal five-noded transpososome.

Finally, difference topology can also probe whether an accessory DNA site leaves the synapse once the assembly process is completed or stays associated with it during the chemical steps of the reaction. An earlier analysis of the Mu transpososome by electron microscopy had suggested that the enhancer is released once the Type0 complex is formed (Watson and Chaconas, 1996). However, as the topological assays revealed, the enhancer remains intertwined with the left and right Mu ends in the LER, Type0 and TypeI complexes. In fact, the enhancer is stably trapped within the synapse even after the cleaved Mu ends have been joined to the target DNA in the TypeII complex (Pathania et al., 2003). When the enhancer is flanked by the recombination sites in head-to-tail orientation, the predicted topological outcomes of the deletion reaction for the 'enhancer-associated' and 'enhancer-dissociated' synapses are diagrammed in Figure 10. In the former case, the two product circles, one of which contains the enhancer,

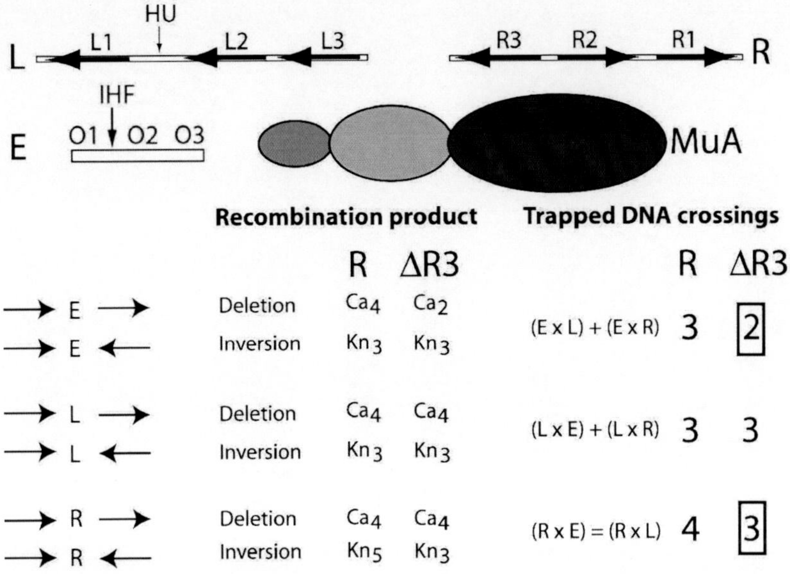

FIG. 9. *Dissecting individual interactions that shape the final topology of a DNA-protein assembly:*
A. The three sub-sites within the left and right ends of Mu (L1-L3 and R1-R3) and their relative orientations are schematically shown. Similarly, the three sub-sites (O1-O3) within the enhancer are also indicated. The binding sites for the E. coli proteins HU (at L) and IHF (at E) are denoted by vertical arrows. In the schematic representation of the MuA protein, the left and middle ovals represent the enhancer- and end-binding domains, respectively.
B. Results from the difference topology assays using plasmid substrates containing the native R end or the $\Delta R3$ end are assembled. Each of the individual sites E, L and R (or $\Delta R3$) isolated into a domain by closely flanking recombination sites (horizontal arrows in head-to-head or head-to-tail orientation) is depicted at the left. Note that the lack of R3 causes a deficit of one DNA crossing between E and L-R or between R and E-L, but leaves unaffected the DNA crossings between L and E-R.

will be topologically linked; freeing the enhancer circle will require linearizing the partner circle by restriction enzyme digestion. In the latter, deletion products will be unlinked circles.

This logic may be applied to interrogate the transience or persistence of accessory sites within other recombination synapses, for example those assembled by the Hin and Gin invertases.

8. Concluding remarks. Through his brilliance, foresight and power of persuasion, Nick Cozzarelli left his mark on several exciting areas of biology. His keen analytical mind influenced many scientists, even those who did not directly interact with him. He was tough, tenacious and gracious. One of us recalls how Cozzarelli once publically declared that "these

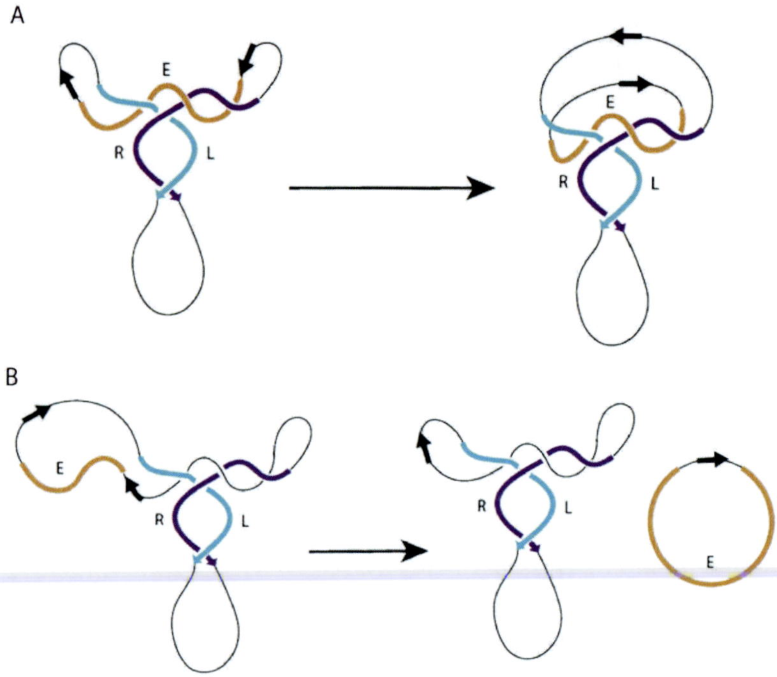

FIG. 10. *Probing the longevity of the association of an accessory site with the fully assembled form of a protein-DNA complex: In the example of the Mu transpososome shown here, one can envisage two scenarios: (A) the enhancer remains stably associated with the complex during the chemical steps of transposition or (B) the enhancer diffuses away from the complex once the assembly is completed but prior to strand breakage. Site-specific recombination between two head-to-tail target sites placed close to and on either side of the enhancer permits the distinction between the two alternatives. For A, the deletion circle containing the enhancer is predicted to be topologically linked to the larger deletion circle. For B, the deletion products are expected to be free (unlinked) circles. An experimental test based on this logic supports A.*

elegant experiments with Flp were possible only because the system is so simple." He was generous in applauding the Mu topological analysis- "for having nailed the system." Cozzarelli had a flair for identifying and solving complex problems, and then explaining the solutions with a simplicity and clarity that would 'gratify most and astonish the rest.'

Acknowledgements. Research in our laboratories covered by this article is supported by the National Institutes of Health (RMH, GM33247; MJ, GM35654 and GM62167) and the Robert Welch Foundation (RMH, F-1351; MJ, F-1274). We thank Jennifer Mann for critically reading the manuscript and the referee for suggesting an additional figure for improving clarity. We acknowledge the assistance of Chien-Hui Ma in preparing Figure 1.

REFERENCES

[1] Chaconas G. and Harshey R.M., Transposition of phage Mu DNA. In Mobile DNA II, Eds. Craig. N.L., Craigie R., Gellert M., and Lambowitz A.M. (Washington DC: ASM Press). pp. 384–402, 2002.

[2] Chen Y., Narendra U., Iype L.E., Cox M.M., and Rice P.A., Crystal structure of a Flp recombinase-Holliday junction complex: assembly of an active oligomer by helix swapping. Mol. Cell. **6**: 885–897, 2000.

[3] Cozzarelli N.R., UCLA Symp Mol. Cell. Biol. **47**: 1, 1986.

[4] Cozzarelli N.R., Krasnow M.A., Gerrard S.P., and White J.H., A topological treatment of recombination and topoisomerases. Cold Spring Harb. Symp. Quant. Biol. **49**: 383–400, 1984.

[5] Cozzarelli N.R., Krasnow M.A., Gerrard S.P., and White J., Primer on the topology and geometry of DNA supercoiling: In DNA topology and its biological effects. Cold Spring Harbor Laboratory Press, Cold Spring Harbor, NY, 1990.

[6] Dhar G., Sanders E.R., and Johnson R.C., Architecture of the Hin synaptic complex during recombination: the recombinase subunits translocate with the DNA strands. Cell. **119**: 33–45, 2004.

[7] Grainge I., Buck D., and Jayaram M., Geometry of site alignment during int family recombination: antiparallel synapsis by the Flp recombinase. J. Mol. Biol. **298**: 749–764, 2000.

[8] Grainge I., Pathania S., Vologodskii A., Harshey R.M., and Jayaram M., Symmetric DNA sites are functionally asymmetric within Flp and Cre site-specific DNA recombination synapses. J. Mol. Biol. **320**: 515–527, 2000.

[9] Grindley N.D., The movement of Tn3 like elements: Transposition and cointegrate resolution. In Mobile DNA II, Eds. Craig N.L., Craigie R., Gellert M., and Lambowitz A.M. (Washington DC: ASM Press). pp. 272–302, 2002.

[10] Grindley N.D., Whiteson K.L., and Rice P.A., Mechanisms of site-specific recombination. Annu. Rev. Biochem. **75**: 567–605, 2006.

[11] Guo F., Gopaul D.N., and van Duyne G.D., Structure of Cre recombinase complexed with DNA in a site-specific recombination synapse. Nature **389**: 40–46, 1997.

[12] Harshey R.M. and Jayaram M., The Mu transpososome through a topological lens. Crit. Rev. Biochem. Mol. Biol. **41**: 387–405, 2006.

[13] Jayaram M., Tribble G., and Grainge I., Site-specific recombination by the Flp protein of *Saccharomyces cerevisiae*. In Mobile DNA II, Eds. Craig N.L., Craigie R., Gellert M., and Lambowitz A.M. (Washington DC: ASM Press). pp. 192–218, 2002.

[14] Johnson R., Bacterial site-specific DNA inversion systems. In Mobile DNA II, Eds. Craig N.L., Craigie R., Gellert M., and Lambowitz A.M. (Washington DC: ASM Press). pp. 230–271, 2002.

[15] Kamtekar S., Ho R.S., Cocco M.J., Li W., Wenwieser S.V., Boocock M.R., Grindley N.D., and Steitz T.A., Implications of structures of synaptic tetramers of $\gamma\delta$ resolvase for the mechanism of recombination. Proc. Natl. Acad. Sci. USA **103**: 10642–10647, 2006.

[16] Kanaar R., Klippel A., Shekhtman E., Dungan J.M., Kahmann R., and Cozzarelli N.R., Processive recombination by the phage Mu Gin system: Implications for the mechanisms of DNA strand exchange, DNA site alignment, and enhancer action. Cell **62**: 353–366, 1990.

[17] Kilbride E., Boocock M.R., and Stark W.M., Topological selectivity of a hybrid site-specific recombination system with elements from Tn3 res/resolvase and bacteriophage P1 loxP/Cre. J. Mol. Biol. **289**: 1219–1230, 1999.

[18] Li W., Kamtekar S., Xiong Y., Sarkis G.J., Grindley N.D., and Steitz T.A., Structure of a synaptic $\gamma\delta$ resolvase tetramer covalently linked to two cleaved DNAs. Science **309**: 1210–1215, 2005.

[19] PATHANIA S., JAYARAM M., AND HARSHEY R.M., Path of DNA within the Mu transpososome. Transposase interactions bridging two Mu ends and the enhancer trap five DNA supercoils. Cell **109**: 425–436, 2002.

[20] PATHANIA S., JAYARAM M., AND HARSHEY R.M., A unique right end-enhancer complex precedes synapsis of Mu ends: the enhancer is sequestered within the transpososome throughout transposition. EMBO J. **22**: 3725–3736, 2003.

[21] STAHL F.W., Symposium on DNA replication and recombination. Summary. Cold Spring Harb. Symp. Quant. Biol. **43**(2): 1353–1356, 1979.

[22] SUMNERS D.W., ERNST C., SPENGLER S.J., AND COZZARELLI N.R., Analysis of the mechanism of DNA recombination using tangles. Q. Rev. Biophys. **28**: 253–313, 1995.

[23] VAN DUYNE G.D., A structural view of tyrosine recombinase site-specific recombination, In Mobile DNA II, Eds. Craig N.L., Craigie R., Gellert M., and Lambowitz A.M. (Washington DC: ASM Press). pp. 93–117, 2002.

[24] WASSERMAN S.A. AND COZZARELLI N.R., Biochemical topology: Applications to DNA recombination and replication. Science **232**: 951–960, 1986.

[25] WATSON M.A. AND CHACONAS G., Three-site synapsis during Mu DNA transposition: a critical intermediate preceding engagement of the active site. Cell **85**: 435–445, 1996.

[26] YIN Z., SUZUKI A., LOU Z., JAYARAM M., AND HARSHEY R.M., Interactions of phage Mu enhancer and termini that specify the assembly of a topologically unique interwrapped transpososome. J. Mol. Biol. **372**: 382–396, 2007.

USEFUL INTRUSIONS OF DNA TOPOLOGY INTO EXPERIMENTS ON PROTEIN-DNA GEOMETRY

JASON D. KAHN*, JAMES R. JENSSEN†, AND VASAVI VITTAL‡

Abstract. Small DNA minicircles are useful for characterizing protein-induced DNA bending and twisting, because obfuscating effects of DNA flexibility are less important than in larger DNA. Our work on DNA geometry and flexibility in protein-DNA complexes has employed T4 ligase-mediated DNA cyclization to make minicircles. Experiments can be carried out as forward ligations, or equivalently protein binding to minicircles can be characterized. In every case we have studied, topological characterization of minicircle synthesis or properties has led to unexpected geometric or mechanistic conclusions. Examples concerning the catabolite activator protein, *E. coli* RNA polymerase, the Lac repressor, and the TATA-box binding protein are discussed. Topological results have the experimental advantages that they are qualitatively unmistakable and internally controlled: new topoisomers are readily identified even in small amounts, and they are formed in the same reaction as relaxed products. Simulations of topoisomer distributions are quite sensitive to geometrical and flexibility parameters, which helps set stringent constraints on possible structural/dynamic models. However, the disadvantage of a topological measurement is it is consistent with any combination of writhe and twist that sums to the observed ΔLk, so it is difficult to be confident that a structural/dynamic model is a unique solution.

Key words. Minicircle, DNA bending, DNA cyclization, RNA polymerase, TBP, Lac repressor, CAP.

AMS(MOS) subject classifications. 92C40, 74K10.

Abbreviations. CAP, catabolite activator protein; RNAP, RNA polymerase; TBP, TATA box binding protein; LacI, Lactose operon repressor protein; bp, base pair; EMSA, electrophoretic mobility shift assay; FRET, fluorescence resonance energy transfer; RPo, RNAP:DNA open complex; TEC, RNAP:DNA:RNA ternary elongation complex.

1. Introduction.

DNA minicircles (closed circular double-stranded DNA of < 1000 bp, typically 150–500 bp) are a useful experimental system for studying intrinsic or ligand-induced DNA bending and twisting. Minicircles of < 200 bp are strongly bent, so they also offer models for the study of constrained DNA, as in chromatin, DNA loops, or supercoiled environments like the bacterial chromosome. Experiments on minicircles can be carried out in several ways. The classic Shore and Baldwin experiment monitors the rate of minicircle formation upon T4 DNA ligase-mediated ligation of linear restriction fragments [1]. Typically the cyclization rate is referenced against the rate of bimolecular ligation to give the J factor, a measure of the effective concentration of one properly-aligned and torsionally-phased end in the neighborhood of the other [2]. J factors

*Department of Chemistry and Biochemistry, University of Maryland, College Park, MD 20742-2021 (jdkahn@umd.edu). The work was funded by an NSF Career Award and an NIH grant.

†Collagen Matrix, Inc., 509 Commerce Street, Franklin Lakes, NJ 07417.

‡Atlantic Equities LLP, 15 St. Helen's Place, London EC3A 6DE, United Kingdom.

C.J. Benham et al. (eds.), *Mathematics of DNA Structure, Function and Interactions*, The IMA Volumes in Mathematics and its Applications 150,
DOI 10.1007/978-1-4419-0670-0_8,

can range from sub-nanomolar to millimolar, an unusually large dynamic range for a simple solution kinetics assay. In a complementary experiment, purified minicircles can be used as binding partners for proteins. DNA deformation can be assessed directly by electrophoretic mobility or nuclease treatment. There is a rich body of theory on prediction of J factors and minicircle shape [3–6]. In all of these applications, it is important to measure or control the topology of the DNA minicircles.

Our laboratory uses minicircle DNA as one tool to study DNA bending, looping, twisting, and flexibility in protein-DNA complexes. In these experiments, it is useful to use families of DNA with intrinsic bends, because constructs with different helical phasing between a known bend and an unknown deformation can be used to identify the direction and magnitude of the unknown bend or hinge [7-9]. Also, the use of pre-bent substrates minimizes the size of the DNA required for efficient minicircle formation, which means that the molecules are not so flexible that local bending and twisting effects are washed out.

A population of small minicircles at equilibrium tends to have a single linking number Lk, or at most two linking numbers if Lk_0 = (number of bp)/(average helical repeat) is close to half-integral. Therefore, the appearance of a new topoisomer upon protein binding and cyclization can offer a sensitive and qualitatively obvious signal that the protein has introduced a change in twist or writhe. In keeping with the theme of the meeting and this volume, we will focus here on the insights gained from minicircle topology, as applied to several classic systems: the catabolite activator protein (CAP), the TATA box binding protein (TBP), *E. coli* RNA polymerase (RNAP), and the lactose operon repressor (Lac repressor, LacI). In the first two systems, the appearance of new topoisomers was unexpected (hence the title of the paper), and in all four the topology provided essential constraints on possible models for the structure and flexibility of the protein-DNA complexes.

2. Materials and methods. The CAP protein was a gift from the Steitz laboratory, Yale University. TATA box binding protein (TBP) was a gift from Stephen Burley, then at Rockefeller University. For later experiments, yeast TBP was expressed from plasmids provided by Michael Brenowitz, Albert Einstein. *E. coli* RNA polymerase was a generous gift from the Darst laboratory, Rockefeller. Lac repressor protein as well as expression constructs for it were obtained from Michael Brenowitz, and the protein was purified as described [10].

All DNA cyclization and FRET substrates were PCR products derived from plasmid clones whose sequences were constructed from synthetic oligonucleotides by standard methods. Sequences are available in the papers describing each system or on request. The three $\sim$ 250 bp *lac* UV5 promoter–bent DNA phasing constructs were prepared as described [11].

For RNA polymerase studies, electrophoretic mobility shift (EMSA), abortive initiation, and transcription elongation assays were performed by standard methods [11, 12]. Promoter escape was measured as the production of a 20 nt RNA in a reaction mixture containing only ApA, ATP, UTP, and GTP, from a template coding for C at position 21 [13]. Topoisomers were separated on one-dimensional polyacrylamide gels containing the intercalator chloroquine; since there were at most three topoisomers present there was no need for 2-D gel separations.

3. Results and discussion: CAP: Bend phasing can cause ΔLk. Figure 1 sketches the application of DNA cyclization kinetics to DNA bending by the CAP protein, an *E. coli* transcription factor that is the prototypical DNA bending protein [14]. The use of a sequence-directed phased A-tract bend as an internal standard allowed us to determine the direction of the protein-induced bend entirely based on solution measurements of ring closure rates, thereby avoiding contentious issues surrounding the interpretation of electrophoretic mobilities for bent DNA [8]. (In this case, the bend direction for the CAP protein was previously known; the goal was to establish the cyclization assay and to use it to make a quantitative solution estimate of the CAP-induced bend angle.) The cyclization assay is more complicated than an electrophoretic phasing assay because the lengths of two different linkers must be varied: one (B in Figure 1) controls the helical phasing of the bends and the other (A in Figure 1) the overall length. The cyclization efficiency is a sinusoidal function of length (actually a sum of Gaussians), and to measure a bend angle the effect of the torsional variation must be disentangled from the effect of bend phasing.

When the two bends are in-phase and the overall length of the DNA is a near-integral multiple of the helical repeat, cyclization efficiency is dramatically enhanced. The CAP protein does not introduce significant twist change, so initially we did not anticipate protein-induced changes in topology. At the time it was surprising to observe topological changes when out-of-phase constructs are cyclized. The observed inefficient cyclization must require some combination of inversion of the A tract bend, loss of protein binding, or twist changes in the phasing linker that bring the two bends into phase (linker B in Figure 1). These twist changes could be either overtwisting or undertwisting depending on the length of the linker. If the length of the second linker (A in Figure 1) gives an integral number of turns between the bends in the cyclized product, then cyclization will not require twist changes in the second linker, and the product will have a change in linking number ΔLk due to the twist change (ΔTw) in the B linker. The observation of a new topoisomer (in this case $\Delta Lk = -1$) strongly suggests that the mechanism of cyclization goes through linker undertwisting as opposed to bend inversion, and since the new topoisomer is dependent on the presence of CAP it is clear that it is not due to cyclization of free DNA [4]. Finally, if the length of linker A is such that it must also overtwist

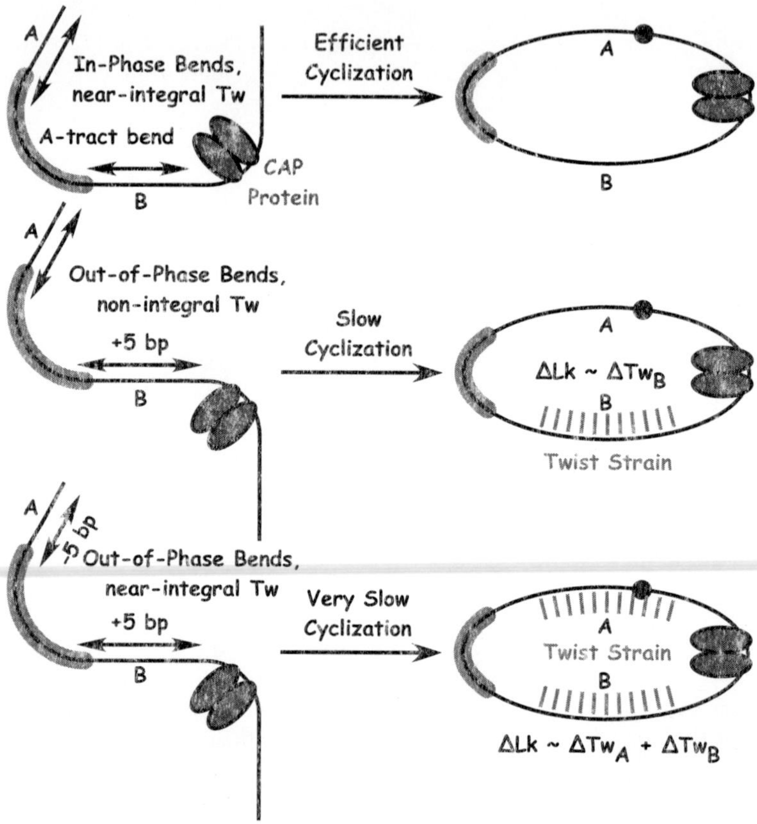

FIG. 1. *Mechanisms of DNA cyclization for different bend phasing constructs. A and B represent linkers of variable length. Linker B controls the helical phasing of the sequence-directed A-tract bend and the protein-induced CAP bend. Linker A independently controls the total length of the DNA. Cyclization of out-of-phase constructs is proposed to occur via twist changes in the linkers, which can be detected experimentally if they lead to new topoisomers. Since these constructs are only about* ~ 150 *bp in size, the minicircle products have very little writhe, so the linking number change* ΔLk *is approximately the sum of the twist changes required in the linkers.*

or undertwist to allow cyclization, then the topological change will be a sum of the two twist changes, and this can either increase or conceal the effect. In practice, the probability of large twist changes occurring in both linkers at once is so low that the J factor may be too small to measure.

In the case of CAP, the appearance of negative topoisomers was predicted by Monte Carlo simulations of the cyclization process, and the theoretical results prompted us to reexamine our experimental results to identify the topoisomers. Their presence correspondingly increased our confidence in the simulations, which had also determined a much lower value for the DNA torsional modulus C than was then accepted

($1.7 \pm 0.2 \times 10^{-19}$ erg cm vs. the previous 2.0–3.4 $\times 10^{-19}$ erg cm [15]). Cyclization has contributed to a long-running debate about the torsional flexibility of DNA. Subsequent cyclization work analyzed by an independent approach has also been consistent with the lower end of the range for C [16]. Recent controversial results on the cyclization of very short DNA molecules ($\sim$ 94 bp) suggests that anharmonicity in the twist potential may depress C further [17]. Single-molecule studies have demonstrated subtle twist anharmonicity, though they arrive at a much larger value of C, $4.1 \pm 0.3 \times 10^{-19}$ erg cm [18]. Our results on these $\sim$ 150 bp molecules did not require consideration of anharmonic effects. Finally, estimates for C obtained from in vivo repression measurements are typically much lower, probably due to stochastic twist fluctuations due to nonspecific DNA binding proteins that alter twist [19].

4. TBP: Strain-responsive protein-induced writhe. The TATA box binding protein (TBP) binds to TATAAAAG and related sequences upstream of the transcription initiation sites for eukaryotic RNA polymerase II. It is an essential part of the general transcription machinery for all three eukaryotic RNA polymerases. The X-ray co-crystal structure of TBP bound to DNA shows dramatic DNA unwinding and a widened minor groove in the center of the TATA box, suggesting a ΔTw of about $-100°$. There are two 50° roll kinks introduced by pairs of intercalating phenylalanine residues that confer out-of-plane bending, leading to a right-handed helical writhe of about 1/3. Therefore, writhe and twist changes cancel, and there is little or no ΔLk when TBP binds plasmid DNA [20].

We carried out cyclization experiments on DNA constructs derived from those above, with TATA box DNA replacing the CAP binding sites [9]. The goal was to characterize the protein-DNA geometry, specifically to assess whether the solution the crystal structures are the same. This would also provide a platform for studying any structural changes induced upon the assembly of larger transcription complexes.

The cyclization experiments provided three main conclusions. First, the TATA box itself provides unusual anisotropic flexibility, in a direction opposite to the TBP-induced bend. This was confirmed by subsequent cyclization experiments [21] and is in accord with selection-amplification results on the high prevalence of TA dinucleotides in sequences that form nucleosomes with high affinity [22]. Second, the geometry of the TBP-DNA complex, inferred from the linker lengths needed for optimal cyclization and the magnitudes of the J factors, is in agreement with the high-resolution structures. Third, most surprisingly and most relevant here, in the minicircle context TBP clearly induces $\Delta Lk = -0.3$, apparent as a $\sim$ 3 bp increase in the optimal length for cyclization. Since the increase appears even for in-phase constructs, the cause is not bend alignment as for CAP, but instead must be due to a local change induced by TBP. When the number of helical turns in the molecule is between $n + 0.5$ and $n + 0.8$

(where n in an integer), then cyclization also gives a qualitatively obvious $\Delta Lk = -1$ topoisomer, as the cyclization process selects the integer nearest to the twist of the linear starting material. As expected, the efficiency of cyclization is also low because it requires additional untwisting of the linkers. The bend alignment mechanism for out-of-phase constructs described for CAP also operates, and in fact in this case provided a positively supercoiled product for one phasing isomer. It requires a large number of constructs (18 in this case) to disentangle these effects.

Our work on CAP had established a thermodynamic linkage between the decreased binding free energy of a protein to pre-strained minicircle DNA and the decreased cyclization free energy conferred by protein-induced bending [8]. This means that we can interpret cyclization and binding experiments in the same framework, and infer the results of experiments that for one reason or another are impossible. In this case, the induction of negative supercoils by TBP strongly suggests that TBP should bind more tightly to negatively supercoiled minicircles than to relaxed circles or linear DNA. We have verified this experimentally (J. Byun and JDK, unpublished), although there may be kinetics issues that make the enhanced binding less dramatic than expected.

The origin of the ΔLk introduced by TBP was a puzzle, as the protein had previously been shown not to introduce supercoiling in plasmid relaxation assays [23]. The coincidence between the ΔLk observed in our experiments and the ΔTw in the crystal structure suggested to us that the positive writhe introduced by the protein was being removed [24]. Examination of the structure suggests that the simplest mechanism would be the removal of the two kinks bracketing the TATA box. Setting the two roll kink angles to zero would give a structure with nearly the same overall bend angle as the crystal structure, but the proposed structure is almost perfectly planar and therefore cannot contribute to minicircle writhe. This proposal also explains why the effect is seen only in small minicircles: the flattening is stabilized by the relief of bending strain in the remainder of the molecule. This is equivalent to stating that the free energy of the circle is decreased by flattening, and therefore that the J factor is much larger for the flattened form. Of course, to be observed in the forward ligation experiment the flattened form must be populated to some extent in the absence of strain. Comprehensive FRET analysis of the TBP-DNA complex [25], as well as recent tethered particle microscopy experiments [26], have identified multiple TBP-DNA conformations, and extending these studies may be able to tell us whether the proposed flattened form is one of them.

The proposed strain-induced flattening mechanism predicts that the supercoiling should decrease as the size of the DNA increases, and indeed the observed ΔLk disappears for DNA of greater than 600 bp (S. Magee and JDK, unpublished). Also, the model predicts that TBP mutants lacking the intercalating phenylalanines should bind only in the flattened mode. We have demonstrated that such mutants are defective in binding linear

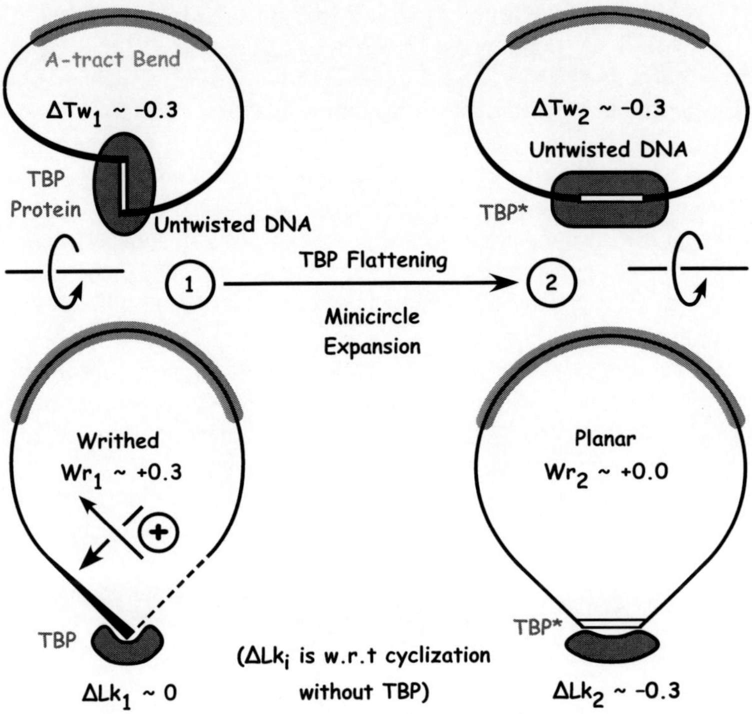

FIG. 2. *Proposed strain-induced TBP-DNA flattening suggested by topological results. Structure 1 on the left, sketched in perspective and from above, represents the TBP-DNA cocrystal structure embedded in a minicircle. Positive Wr and negative* ΔTw *sum to give* $\Delta Lk_1 \sim 0$*. Removal of the two phenylalanine kinks gives the flattened Structure 2 on the right, which has zero writhe and therefore* $\Delta Lk_2 \sim \Delta Tw = -0.3$*. The flattened form is stabilized in a small minicircle because the product has decreased writhe and curvature, hence decreased bending strain.*

DNA but that they can bind stably to supercoiled minicircles (J. Byun and JDK, unpublished). Intermediate degrees of flattening are possible, but we have no evidence regarding them.

The TBP flattening mechanism is of interest for two reasons. First, it would be the simplest example to date of a protein-DNA complex whose geometry responds to mechanical strain. Second, it may have relevance to the chromatin environment. The mechanism predicts that remodeling chromatin, which introduces local superhelical strain, could drive TBP binding. This would act to accelerate transcription of genes upon activation [24]. Local dynamic supercoiling can also result from the torque of transcription [27]. This supercoiling-dependent TBP "hot spot" proposal should be testable using chromatin immunoprecipitation to monitor TBP occupancy during remodeling or pulses of transcription.

5. RNAP: Invariant topology suggests minimal DNA bending. *E. coli* RNA polymerase (RNAP) introduces a ΔLk of –1 to –1.5 in either the binary RNAP:DNA open complex (RPo), in which promoter DNA is melted but transcription has not initiated, or the RNAP:DNA:RNA ternary elongation complex (TEC) [28–30]. We set out to measure the extent of DNA bending by RNAP using approaches similar to those above. We anticipated topological changes, and it was a surprise that essentially invariant topologies emerged as the most significant result [11]. The phasing constructs used are sketched in Fig. 3. The results of the cyclization experiments of Fig. 4 and experiments not shown are summarized at the bottom of Fig. 3.

Fig. 4 shows that the cyclization efficiency of phasing constructs bearing RPo varies for the different phasing linkers, but only several-fold. This suggests a relatively small net bend ($< 50°$), in accord with our previous results on RNAP binding to supercoiled minicircles that lack bent segments [12]. The bend direction is consistent with the model shown in Fig. 3, with the DNA curving around the surface of the protein to contact the two α subunits on the left side of the structure. The cyclization efficiency of TEC constructs is nearly independent of bend phasing. This result suggests that the bend angle induced by RNAP is $n \times 180°$, because the DNA end would be roughly in the same place for any such bend. However, a 180° bend would predict that different phasing constructs would yield different topoisomers, as in Fig. 3, because cyclization would give figure eight molecules with different node signs. Since only the $\Delta Lk = -1$ topoisomer characteristic of the polymerase is observed, we conclude that the bend angle is $n \times 360°$. A 360° bend is not excluded by our data, but a complete DNA wrap seems quite unlikely given the crystal structure and the fact that only about 20 bp are protected from nucleases in the TEC. Therefore, we suggest that the TEC induces essentially no net bend. The crystal structure [31, 32] shows roughly a 90° bend, but the upstream DNA is not present; we suggest the DNA must bend sharply again to contact the surface of the protein just upstream of the transcription bubble. This proposal could be tested using FRET, as has been done for T7 RNA polymerase [33].

We found it ironic that in the one system where we expected informative topological changes, the product topologies were in fact invariant. However, this apparently unsatisfying result, in combination with other observations (data not shown), allows us to draw conclusions about the energetics of transcription initiation and promoter escape. The additional experiments employ pre-formed $\Delta Lk = -1$ minicircles as transcription templates. As judged by either EMSA or by chase experiments using radiolabeled rNTPs, RPo formed on linear DNA converts to the TEC much more efficiently than RPo formed on cyclized constructs 7–16 and 11–14. Circular construct 9–12 does not readily form open complex, but once formed it is readily converted to TEC. The free energy reaction coordinate diagram

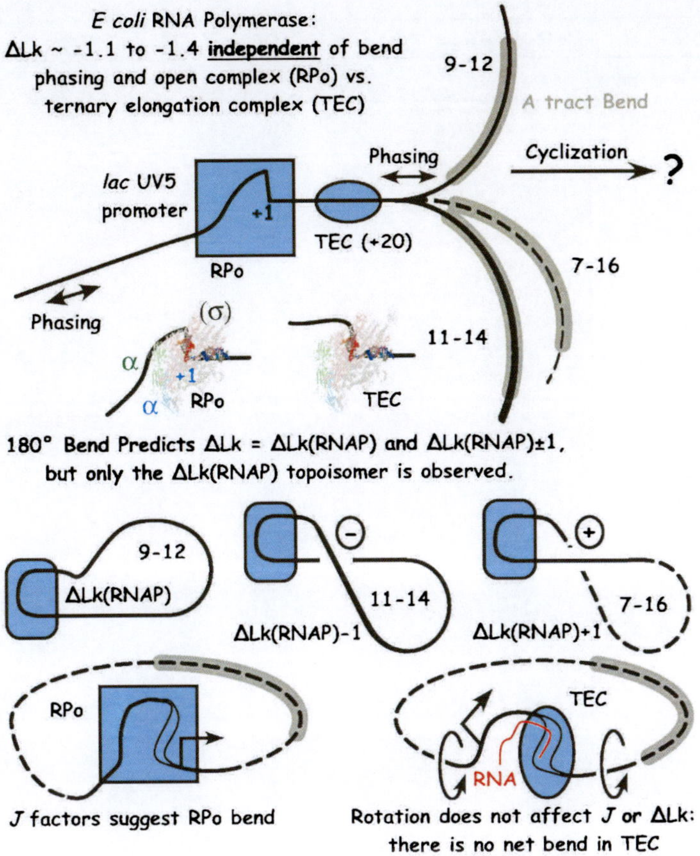

FIG. 3. *Summary of cyclization results on RNAP:DNA open complexes (RPo) and ternary elongation complexes (TEC). The top sketch shows the three phasing constructs studied, with the blue boxes indicating the DNA bound either in RPo or the TEC, approximately to scale. The structures below are from PDB file 2o5I, of the T. thermophilus elongation complex, with the probable location of the σ factor in RPo indicated [31, 32]. The upstream DNA is not present in the structure but it is known to interact with the α subunit in RPo. The diagrams at the bottom summarize the results of Fig. 4 and experiments not shown: the observed topology of cyclization products and the J factors for TECs are independent of bend phasing, suggesting there is only a small bend in RPo and no net bending in the TEC. This might allow transcription with minimal movement of the neighboring DNA, as opposed to the disruption that would be caused by a large moving DNA bend.*

of Fig. 5 synthesizes the cyclization, EMSA, and RNA transcription results to give a consistent qualitative ordering of the free energies of complexes formed on linear DNA or on the 7–16, 9–12, and 11–14 minicircles. The 7–16 and 11–14 minicircles are stabilized relative to RPo on linear DNA. Circular substrates have the added advantage that end-binding is impos-

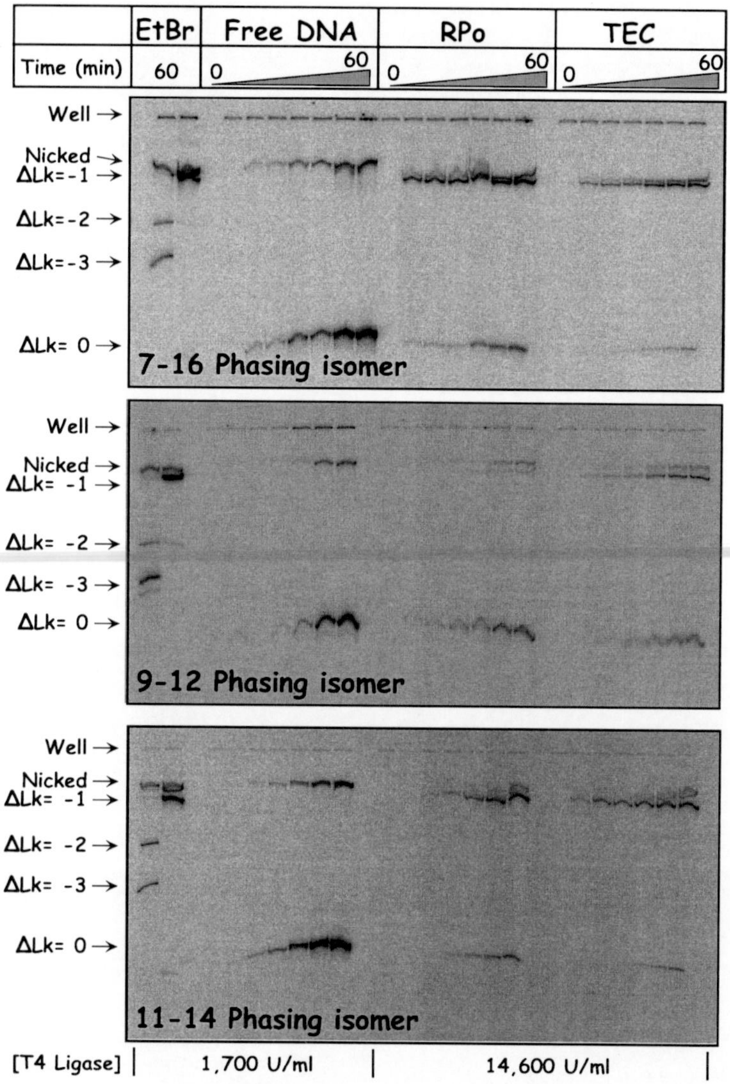

FIG. 4. *Cyclization experiments on RPo and TEC complexes. The Phosphorimages of native chloroquine-containing polyacrylamide gels show that variation among phasing constructs is much more pronounced for RPo than for TEC, with the 9–12 isomer RPo cyclizing the most poorly; in fact, the RPo lanes for 9–12 show that residual free DNA appears to give most of the circular product. All of the TECs cyclize with similar efficiency to each other, and all of the RNAP complexes give the same topology,* $\Delta Lk = -1$ *(experiments not shown with length variants suggest that* ΔLk *in the minicircle context is* -1.1 *to* -1.4*). RNAP inhibits cyclization, perhaps due to binding at DNA ends or to steric interference with T4 DNA ligase; cyclization requires much higher ligase concentration than for free DNA. EtBr indicates cyclization in the presence of ethidium bromide to provide markers for negative supercoils.*

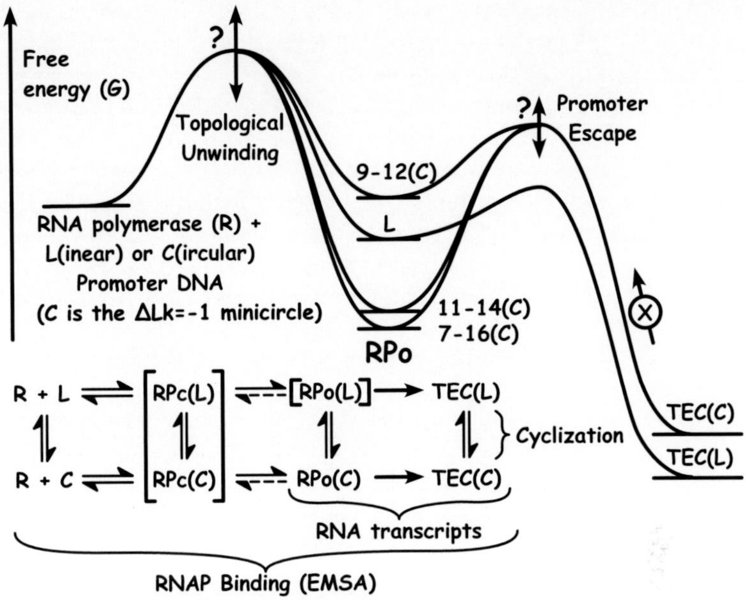

FIG. 5. *Conceptual free energy reaction coordinate diagram for open complex formation and promoter escape on linear and supercoiled minicircle templates. The relative free energies of the RPo complexes formed on different templates are sensitive to bend phasing, as explained in the text. Promoter escape from RPo to form the TEC is effectively irreversible, but since the efficiency of conversion is inversely related to the stability of the RPo, the transition state appears to be similar for all the constructs. This suggests that the measured relative free energies of the TEC(C) states, which are all comparable, apply as well to the transition state for promoter escape. The diagram at the bottom left illustrates the experiments that apply to each interconversion. The brackets indicate states that are not observed directly.*

sible: under our conditions linear DNA did not give well-resolved EMSA bands, but all of the minicircles did. The relative rates of conversion of RPo to TEC correlate inversely with the stability of the open complexes, suggesting that there is a transition state of similar free energy for all of the constructs. The DNA bending in this transition state presumably has substantial product character: if the RNAP-bound DNA in the transition state is not bent, then all of the phasing isomers will be of similar free energy, like the TEC. The transition to elongation probably entails breaking contacts between the DNA and the α subunits. Breaking or weakening contacts to the σ subunit would also be necessary, but would not be correlated as obviously to the change in DNA bending.

In summary, the RNA polymerase results show that even an essentially negative topological result, that no changes are observed under any circumstances, sets useful constraints on possible models for the geometry of a protein-DNA complex. We have shown that the TEC has very little net DNA bending, and that there is a net change in DNA bending during the transition to elongation even though there is no change in topology. Recent single-molecule experiments demonstrate that this occurs via "DNA scrunching," in which contacts at the upstream end of the polymerase are maintained as the front end of the protein draws in downstream DNA and the DNA is transiently more untwisted [30]. Loss of the upstream contacts and hence loss of bending appear to be part of the rate limiting step in the transition to a stable TEC. These mechanistic properties suggest that *E. coli* RNAP has evolved to initiate transcription and travel along the DNA without requiring dramatic long-range conformational change.

6. LacI: Topology suggests geometry, but ambiguity remains. The Lac repressor tetramer is the prototypical transcriptional repressor. The efficiency of repression is increased by DNA looping, whereby binding to one operator site by one dimeric DNA binding domain increases the local concentration of the other DNA binding domain in the neighborhood of a second operator [34]. Designed molecules in which *lac* operators are phased against a sequence-directed bend [35] were used to ask whether there is a tight DNA loop anchored by a V-shaped protein [36] or a U-shaped DNA loop anchored by an extended or open form of the protein, as diagrammed in Fig. 6 [37].

EMSA and DNA footprinting assays demonstrated that the both the "9C14" and the "11C12" molecules illustrated in Fig. 6 form hyperstable LacI-DNA loops, with half-lives of days [35]. This suggested that the protein could adopt two different conformations depending on the DNA shape. The critical supporting experiment was the observation of a $\Delta Lk = +1$ topoisomer (as well as relaxed and a trace of $\Delta Lk = -1$ topoisomers) upon cyclization of extended 9C14 constructs, but only relaxed and a trace of $\Delta Lk = -1$ topoisomers from 11C12. The presence of a crossover node in 9C14 and its absence in 11C12 were confirmed by efficient fluorescence resonance energy transfer (FRET) between donor and acceptor fluorophores inserted near the 9C14 operator sites. 11C12 loops gave much less efficient FRET [38].

There are several possible "loop topologies" for LacI loops, with different orientations of operators with respect to each other or the asymmetric LacI protein [39]. One example is the $\Delta Lk = -1$ topoisomer illustrated for 11C12 in Fig. 6, which is an example of an antiparallel loop, as opposed to a the parallel loop proposed for 9C14. We discounted this geometry because it appeared to predict much more negative writhe than an open form loop, but this may not be correct [40].

The designed hyperstable loops have been useful in several subsequent studies. Single-molecule FRET experiments confirmed the existence of a loop in which FRET efficiency is nearly 100 %, but discrepancies between the single-molecule results and the earlier experiments suggest unexpected roles for the flanking DNA in determining conformation [41]. Monte Carlo

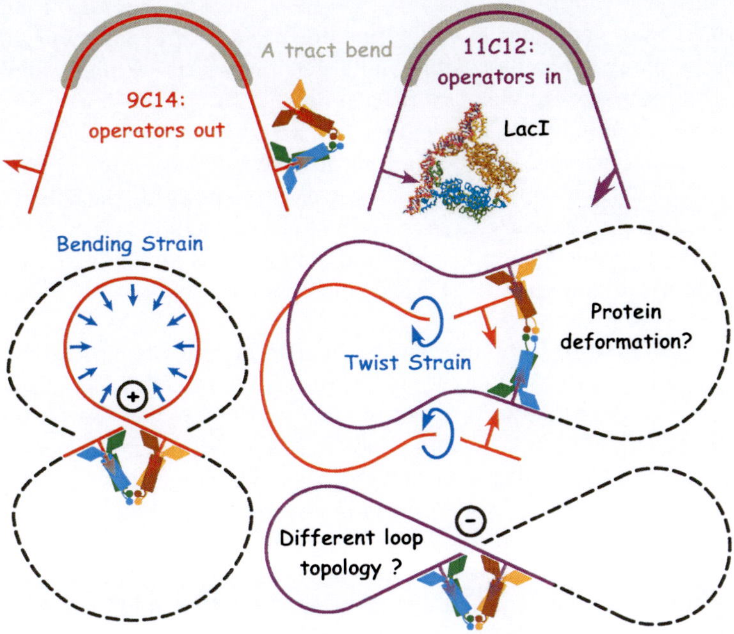

FIG. 6. *Hyperstable LacI-DNA loop constructs. The 9C14 and 11C12 constructs differ in the helical phasings between lac operators and A tract bends. The arrows indicate the symmetry axes of dyad-symmetric idealized lac operators, so they are directed down the center of LacI dimers as shown. The 9C14 molecule can form the* $\Delta Lk = +1$ *parallel loop at the bottom left by bending the DNA to tighten its curvature. Some of our results suggest that twist strain that rotates the operators inward could also allow 9C14 to adopt an open form. Based on topology and FRET results, 11C12 is proposed to adopt the open form at the right, with the tetramer held together by a C-terminal 4-helix bundle attached to the core of the protein by flexible protein loops. A V-shaped repressor in an antiparallel loop would give a circle with* $\Delta Lk = -1$*; a small amount of this topoisomer is observed but we do not know whether it is due to twist or writhe changes. The dashed black lines indicate the DNA closing the circles.*

simulations of the cyclization reactions confirmed that a DNA-directed distribution of open and closed forms could explain our observed topoisomer distributions and J factors [42]. The analysis also suggested a strong intrinsic preference for the open form on the part of the LacI protein with occupied DNA binding sites. The open and closed forms have recently been confirmed by X-ray scattering on LacI bound to two double-stranded DNA oligonucleotides [43].

There have been many modeling studies on Lac repressor that touch on loop geometry and topology. Molecular dynamics simulations emphasized the importance of head-group rotations as opposed to large scale opening of the core of the protein; our FRET results were simulated in this work but it did not consider the topological results [44]. Treatments based on

DNA elasticity [45] or on the statistical mechanics of cyclization [46, 47] confirmed that open and closed forms are needed to explain our results as well as many other biochemical and genetic results. Finally, sophisticated rod mechanics modeling has been applied to our constructs [40, 48]. Surprisingly, many of the topological and FRET results can be simulated without invoking the LacI open form; the 11C12 results are explained in terms of an antiparallel loop and larger twist changes in the intervening DNA than in our models. These issues will be resolved by systematic mapping of the linker length landscape and comparisons among experimental and theoretical loop stability, FRET, and topology results.

7. Conclusions. Quantitative systems biology of transcription networks requires that we understand protein-DNA interaction is sufficient detail to be able to predict protein binding and DNA shape as conditions in the cell change [49, 50]. Complete models must consider dynamic supercoiling and the chromatin environment, and therefore must consider the effects of DNA topology and protein-DNA structure on each other [51]. The structure and dynamics of DNA loops are especially sensitive to DNA stiffness and topology, and looping can in turn be used to control the local DNA shape. The combination of biochemical model systems for strained DNA (like minicircles), in vivo characterization of gene expression patterns as a function of DNA shape [52], and new theoretical approaches may provide a comprehensive view of the role of DNA topology in the control of gene expression.

DNA topology is uniquely useful in studying protein-DNA geometry because the linking number is a discrete variable. The appearance of topological changes is a qualitatively obvious change that cannot be ignored, and yet the relative amounts of topoisomers can be analyzed quantitatively. Ratios of topoisomers are internally referenced and often more reliable than J factors. We have found that correct predictions of topological distributions provides a stringent constraint on possible models for the geometry and flexibility of a protein-DNA complex. Initially the dominance of topology in our thinking came as surprise to us; one might expect that the J factor for cyclization would be a more useful measure of geometric properties. Unfortunately, J is difficult to measure with absolute accuracy, and often more than one model can explain any J factor measurement. This situation might be improved with more use of high-throughput cyclization measurements, but these high-throughput measurements still require preparation of a large number of individual constructs [53].

The unifying theme of our studies is that the proteins we have studied have evolved to bind to DNA in a wide range of conditions, without requiring inadvertent changes to the DNA. CAP-induced DNA bending is necessary to enable protein-protein interaction, but by virtue of CAP's homodimeric state it can act at several CAP site-promoter spacings. The proposed TBP flattening mechanism may allow TBP to respond to variable levels of supercoiling and to buffer the induced bending strain. RNA polymerase introduces minimal DNA bending and can convert between dramatically different shapes without additional topological change. LacI-DNA loops can exist in a variety of shapes and sizes by virtue of protein flexibility. These remarkably adaptable proteins have evolved to work with inherently stiff or otherwise constrained DNA.

Acknowledgements. We are grateful to members of the Kahn laboratory and other coauthors for reagents and encouragement and to John Maddocks and Noel Perkins for discussion. Work on the CAP protein and preliminary work on RNA polymerase was done with the guidance and inspiration of Donald Crothers.

REFERENCES

[1] D. Shore, J. Langowski, and R.L. Baldwin, *DNA flexibility studied by covalent closure of short fragments into circles*, Proc. Natl. Acad. Sci. USA, **78**: 4833–4837, 1981.

[2] P.J. Flory, U.W. Suter, and M. Mutter, *Macrocyclization Equilibria. I. Theory*, J. Am. Chem. Soc., **98**: 5733–5739, 1976.

[3] S.D. Levene and D.M. Crothers, *Ring Closure Probabilities for DNA Fragments by Monte Carlo Simulation*, J. Mol. Biol., **189**: 61–72, 1986.

[4] J.D. Kahn and D.M. Crothers, *Measurement of the DNA bend angle induced by the catabolite activator protein using Monte Carlo simulation of cyclization kinetics*, J. Mol. Biol., **276**: 287–309, 1998.

[5] Y. Zhang and D.M. Crothers, *Statistical mechanics of sequence-dependent circular DNA and its application for DNA cyclization*, Biophys. J., **84**: 136–153, 2003.

[6] R.S. Manning, J.H. Maddocks, and J.D. Kahn, *A continuum rod model of sequence-dependent DNA structure*, J. Chem. Phys., **105**: 5626–5646, 1996.

[7] S.S. Zinkel and D.M. Crothers, *DNA bend direction by phase sensitive detection*, Nature, **328**: 178–181, 1987.

[8] J.D. Kahn and D.M. Crothers, *Protein-induced bending and DNA cyclization*, Proc. Natl. Acad. Sci. USA, **89**: 6343–6347, 1992.

[9] N.A. Davis, S.S. Majee, and J.D. Kahn, *TATA box DNA deformation with and without the TATA box-binding protein*, J. Mol. Biol., **291**: 249–265, 1999.

[10] M. Brenowitz, A. Pickar, and E. Jamison, *Stability of a Lac repressor mediated "looped complex"*, Biochemistry, **30**: 5986–5998, 1991.

[11] J.R. Jenssen, *The geometry and topology of DNA in binary and ternary complexes with E. coli RNA polymerase*, Dissertation, University of Maryland, College Park, MD, 2001

[12] J.D. Kahn and D.M. Crothers, *DNA Bending in Transcription Initiation*, Cold Spring Harbor Symp. Quant. Biol., **58**: 115–122, 1993.

[13] J.R. Levin, B. Krummel, and M.J. Chamberlin, *Isolation and Properties of Transcribing Ternary Complexes of Escherichia coli RNA Polymerase Positioned at a Single Template Base*, J. Mol. Biol., **196**: 85–100, 1987.

[14] S.C. Schultz, G.C. Shields, and T.A. Steitz, *Crystal structure of a CAP-DNA complex: the DNA is bent by 90 degrees*, Science, **253**: 1001–1007, 1991.

[15] D.M. Crothers, J. Drak, J.D. Kahn, and S.D. Levene, *DNA Bending, Flexibility, and Helical Repeat by Cyclization Kinetics*, Methods Enzymol., **212B**: 1–29, 1992.

[16] Y. Zhang, Z. Xi, R.S. Hegde, Z. Shakked, and D.M. Crothers, *Predicting indirect readout effects in protein-DNA interactions*, Proc. Natl. Acad. Sci. USA, **101**: 8337–8341, 2004.

[17] T.E. Cloutier and J. Widom, *DNA twisting flexibility and the formation of sharply looped protein-DNA complexes*, Proc. Natl. Acad. Sci. USA, **102**: 3645–3650, 2005.

[18] Z. Bryant, M.D. Stone, J. Gore, S.B. Smith, N.R. Cozzarelli, and C. Bustamante, *Structural transitions and elasticity from torque measurements on DNA*, Nature, **424**: 338–341, 2003.

[19] N.A. Becker, J.D. Kahn, and L.J. Maher, 3rd, *Effects of nucleoid proteins on DNA repression loop formation in Escherichia coli*, Nucleic Acids Res., **35**: 3988–4000, 2007.

[20] J.L. Kim and S.K. Burley, *1.9 Å resolution refined structure of TBP recognizing the minor groove of TATAAAAG*, Nature Struct. Biol., **1**: 638–653, 1994.

[21] M. Roychoudhury, A. Sitlani, J. Lapham, and D.M. Crothers, *Global structure and mechanical properties of a 10-bp nucleosome positioning motif*, Proc. Natl. Acad. Sci. USA, **97**: 13608–13613, 2000.

[22] A. Thåström, P.T. Lowary, H.R. Widlund, H. Cao, M. Kubista, and J. Widom, *Sequence motifs and free energies of selected natural and non-natural nucleosome positioning DNA sequences*, J. Mol. Biol., **288**: 213–229, 1999.

[23] Y. Lorch and R.D. Kornberg, *Near-Zero Linking Difference upon Transcription Factor IID Binding to Promoter DNA*, Mol. Cell. Biol., **13**: 1872–1875, 1993.

[24] J.D. Kahn, *Topological effects of the TATA box binding protein on minicircle DNA and a possible thermodynamic linkage to chromatin remodeling*, Biochemistry, **39**: 3520–3524, 2000.

[25] K.M. Parkhurst, R.M. Richards, M. Brenowitz, and L.J. Parkhurst, *Intermediate Species Possessing Bent DNA are Present Along the Pathway to Formation of a Final TBP-TATA Complex*, J. Mol. Biol., **289**: 1327–1341, 1999.

[26] S.F. Tolic-Norrelykke, M.B. Rasmussen, F.S. Pavone, K. Berg-Sorensen, and L.B. Oddershede, *Stepwise bending of DNA by a single TATA-box binding protein*, Biophys. J., **90**: 3694–3703, 2006.

[27] F. Kouzine, S. Sanford, Z. Elisha-Feil, and D. Levens, *The functional response of upstream DNA to dynamic supercoiling in vivo*, Nat. Struct. Mol. Biol., **15**: 146–154, 2008.

[28] M. Amouyal and H. Buc, *Topological Unwinding of Strong and Weak Promoters by RNA Polymerase: A Comparison Between the lac Wild-type and the UV5 Sites of Escherichia coli*, J. Mol. Biol., **195**: 795–808, 1987.

[29] H.B. Gamper and J.E. Hearst, *A Topological Model for Transcription Based on Unwinding Angle Analysis of E. coli RNA Polymerase Binary, Initiation and Ternary Complexes*, Cell, **29**: 81–90, 1982.

[30] A. Revyakin, C. Liu, R.H. Ebright, and T.R. Strick, *Abortive initiation and productive initiation by RNA polymerase involve DNA scrunching*, Science, **314**: 1139–1143, 2006.

[31] K.S. Murakami, S. Masuda, E.A. Campbell, O. Muzzin, and S.A. Darst, *Structural basis of transcription initiation: an RNA polymerase holoenzyme-DNA complex*, Science, **296**: 1285–1290, 2002.

[32] D.G. Vassylyev, M.N. Vassylyeva, A. Perederina, T.H. Tahirov, and I. Artsimovitch, *Structural basis for transcription elongation by bacterial RNA polymerase*, Nature, **448**: 157–162, 2007.

[33] G.Q. Tang and S.S. Patel, *T7 RNA polymerase-induced bending of promoter DNA is coupled to DNA opening*, Biochemistry, **45**: 4936–4946, 2006.

[34] J. Müller, S. Oehler, and B. Müller-Hill, *Repression of lac promoter as a function of distance, phase and quality of an auxiliary lac operator*, J. Mol. Biol., **257**: 21–29, 1996.

[35] R.A. Mehta and J.D. Kahn, *Designed hyperstable Lac repressor-DNA loop topologies suggest alternative loop geometries*, J. Mol. Biol., **294**: 67–77, 1999.

[36] M. Lewis, G. Chang, N.C. Horton, M.A. Kercher, H.C. Pace, M.A. Schumacher, R.G. Brennan, and P. Lu, *Crystal structure of the lactose operon

repressor and its complexes with DNA and inducer, Science, **271**: 1247–1254, 1996.

[37] A.M. Friedman, T.O. Fischmann, and T.A. Steitz, *Crystal structure of lac repressor core tetramer and its implications for DNA looping*, Science, **268**: 1721–1727, 1995.

[38] L.M. Edelman, R. Cheong, and J.D. Kahn, *Fluorescence Resonance Energy Transfer over ∼130 Basepairs in Hyperstable Lac Repressor-DNA Loops*, Biophys. J., **84**: 1131–1145, 2003.

[39] S. Semsey, K. Virnik, and S. Adhya, *A gamut of loops: meandering DNA*, Trends Biochem. Sci., **30**: 334–341, 2005.

[40] S. Goyal, T. Lillian, S. Blumberg, J.C. Meiners, E. Meyhofer, and N.C. Perkins, *Intrinsic curvature of DNA influences LacR-mediated looping*, Biophys. J., **93**: 4342–4359, 2007.

[41] M.A. Morgan, K. Okamoto, J.D. Kahn, and D.S. English, *Single-molecule spectroscopic determination of lac repressor-DNA loop conformation*, Biophys. J., **89**: 2588–2596, 2005.

[42] J.D. Kahn, R. Cheong, R.A. Mehta, L.M. Edelman, and M.A. Morgan, *Flexibility and Control of Protein-DNA Loops*, Biophysical Reviews and Letters (BRL), **1**: 327–341, 2006.

[43] M. Taraban, H. Zhan, A.E. Whitten, D.B. Langley, K.S. Matthews, L. Swint-Kruse, and J. Trewhella, *Ligand-induced conformational changes and conformational dynamics in the solution structure of the lactose repressor protein*, J. Mol. Biol., **376**: 466–481, 2008.

[44] E. Villa, A. Balaeff, and K. Schulten, *Structural dynamics of the lac repressor-DNA complex revealed by a multiscale simulation*, Proc. Natl. Acad. Sci. USA, **102**: 6783–6788, 2005.

[45] D. Swigon, B.D. Coleman, and W.K. Olson, *Modeling the Lac repressor-operator assembly: The influence of DNA looping on Lac repressor conformation*, Proc. Natl. Acad. Sci. USA, **103**: 9879–9884, 2006.

[46] Y. Zhang, A.E. McEwen, D.M. Crothers, and S.D. Levene, *Statistical-mechanical theory of DNA looping*, Biophys. J., **90**: 1903–1912, 2006.

[47] Y. Zhang, A.E. McEwen, D.M. Crothers, and S.D. Levene, *Analysis of In-Vivo LacR-Mediated Gene Repression Based on the Mechanics of DNA Looping*, PLoS ONE, **1**: e136, 2006.

[48] S. Goyal, N.C. Perkins, and C.L. Lee, *Nonlinear dynamics and loop formation in Kirchhoff rods with implications to the mechanics of DNA and cables*, J. Comp. Phys., **209**: 371–389, 2005.

[49] L. Bintu, N.E. Buchler, H.G. Garcia, U. Gerland, T. Hwa, J. Kondev, T. Kuhlman, and R. Phillips, *Transcriptional regulation by the numbers: applications*, Curr. Opin. Genet. Dev., **15**: 125–135, 2005.

[50] L. Saiz and J.M. Vilar, *Multilevel deconstruction of the in vivo behavior of looped DNA-protein complexes*, PLoS ONE, **2**: e355, 2007.

[51] A. Travers, *DNA topology: Dynamic DNA looping*, Current Biology, **16**: R838–R840, 2006.

[52] N.A. Becker, J.D. Kahn, and L.J. Maher, 3rd, *Bacterial repression loops require enhanced DNA flexibility*, J. Mol. Biol., **349**: 716–730, 2005.

[53] Y. Zhang and D.M. Crothers, *High-throughput approach for detection of DNA bending and flexibility based on cyclization*, Proc. Natl. Acad. Sci. USA, **100**: 3161–3166, 2003.

TOPOLOGICAL ANALYSIS OF DNA-PROTEIN COMPLEXES

SOOJEONG KIM* AND ISABEL K. DARCY*

Abstract. A tangle consists of strings properly embedded in a 3-dimensional ball. Tangles have been used to model protein-bound DNA. The protein is represented by the 3D ball and the protein-bound DNA is represented by the strings embedded in the 3D ball. We review tangle analysis of protein-DNA complexes involving three or four segments of DNA.

Key words. Site-specific recombination, difference topology, tangle.

AMS(MOS) subject classifications. Primary 57M25, 92C40, 92E10.

1. Introduction. An n-string tangle is a three dimensional ball with n-strings properly embedded in it. Tangles were studied by Conway in the 1960's [3]. In the 1980's, Ernst and Sumners introduced a mathematical tangle model for protein-bound DNA complexes [8]. In this model, the protein is modeled by a three dimensional ball and the protein-bound DNA is modeled by strings. They used a 2-string tangle model to analyze experimental results for Tn3 resolvase and phage λ integrase.

This work was motivated by Nick Cozzarelli [7, 16, 19, 21]. Some proteins can break and rejoin DNA segments and will knot circular DNA molecules. The knot types of the products can be used to determine information regarding how these proteins act. Nick Cozzarelli also used such proteins to study other protein-DNA complexes [13]. Type II topoisomerases will knot circular DNA by cutting the DNA, allowing a segment of DNA to pass through the break before resealing the DNA. In order to study the protein 13S condensin, DNA was first incubated with 13S condensin allowing the condensin to bind the DNA. Topoisomerase was then added. A spectrum of knots resulted which was different than that when topoisomerase acts on DNA in the absence of condensin. The difference in the knot spectrum in the presence versus absence of condensin was used to determine the manner in which 13S condensin is bound to DNA.

Pathania, Jayaram, and Harshey extended these methods to derive the number of DNA crossings trapped in an unknown protein-DNA complex involving multiple DNA segments [15]. This methodology, called difference topology, was used to determine the topological structure within the Mu protein complex, which consists of three DNA segments containing five crossings. Since Mu binds DNA sequences at 3 sites, the Mu protein DNA complex can be modeled by a 3-string tangle. 3-string tangle analysis is much more complicated than 2-string tangle analysis. The experimental results in [15] were mathematically [6] and computationally [5] analyzed

*Department of Mathematics, University of Iowa, Iowa City, Iowa 52242.

C.J. Benham et al. (eds.), *Mathematics of DNA Structure, Function and Interactions*,
The IMA Volumes in Mathematics and its Applications 150,
DOI 10.1007/978-1-4419-0670-0_9,

by using a 3-string tangle model. We address a 4-string tangle model for a protein-DNA complex which binds four DNA segments.

In Section 2, we introduce basic concepts of DNA recombination. We focus on site-specific recombination since this is a very important concept for understanding difference topology. In Section 3, we introduce tangle analysis of protein-DNA complexes. In Section 4, we explain the methodology of difference topology and its application to a Mu protein-DNA complex. In Section 5, we summarize the 3-string tangle analysis of the Mu protein-DNA complex in [6]. Finally, in Section 6, we introduce a 4-string tangle model for a protein which binds four DNA segments. We conclude that a 4-string tangle (with small number of crossings) which satisfies certain experimental conditions must be R-standard.

2. DNA recombination. DNA recombination refers to a process in which DNA is rearranged within a genome. This is one of the biological processes which can change topological properties of DNA. We are interested in DNA recombination where two specific short DNA sequences are exchanged. This process is called *site-specific recombination* and the specific sequences are called *target sites.* This reaction requires specialized proteins, called *recombinases*, to recognize these sites and to catalyze the recombination reaction at these sites.

Site-specific recombination can result in either the inversion or deletion of a DNA segment. As one can see from Figure 1(a), if the orientation of target sites are opposite to one another (inverted repeat), then recombination leads to the inversion of the DNA segment between the two target sites. On the other hand, if the orientation of target sites are the same with respect to one another (directed repeat), then recombination leads to the deletion of the DNA segment between the two target sites, see Figure 1(b).

Note that the number of components is the same after inversion. But it is different after deletion, since the DNA sequence between the two target sites are deleted from the original DNA sequences. In particular, when the initial DNA is circular, inversion results in a knot and deletion results in a link as one can see from the following example.

Cre is a site-specific recombinases. The target sites of *Cre* are called *loxP*. *Cre* can catalyzes both DNA inversion and deletion. The recombination products depend on the relative orientation of the *loxP* sites, the target sites of *Cre*. When the DNA is circular, the products of DNA inversion and deletion by *Cre* are knots and catenanes, respectively (see Figure 2).

3. DNA topology and the tangle model. An *n-string tangle* is a three dimensional ball with n-strings properly embedded in it. The tangle model of a protein-DNA complex was developed by C. Ernst and D.W. Sumners [8]. This model assumes the protein is a three dimensional ball and the protein-bound DNA are strings embedded inside the ball. See Figure 3.

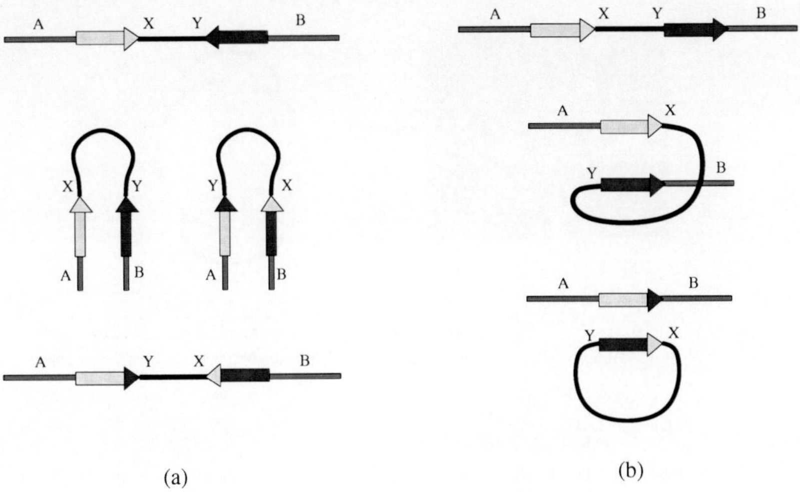

FIG. 1. *(a) Inversion. (b) Deletion. (This figure is redrawn from http://www.mun.ca/biochem/courses/3107/Lectures/Topics/Site-specific-Recomb.html.)*

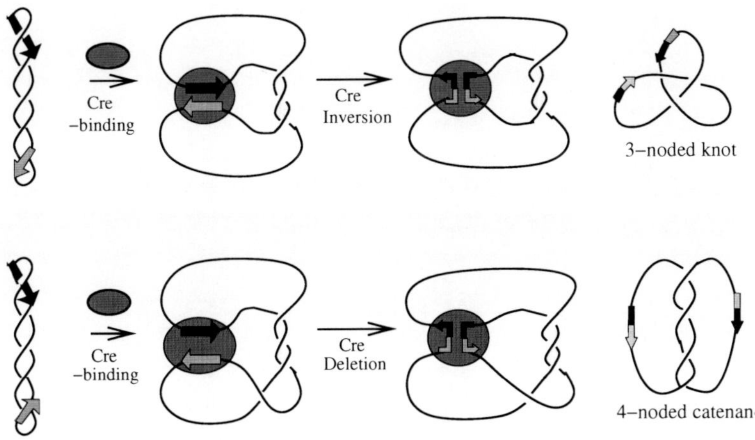

FIG. 2. *Cre recombination.*

Examples of 3-string tangles are given in Figure 4. A *rational tangle* is ambient isotopic to a tangle which has no crossings if we allow the boundary of the three ball to move. A tangle is rational if and only if its strings can be pushed to lie on the boundary of the 3D ball so that no string crosses over another string on the boundary of this ball. If the DNA wraps around

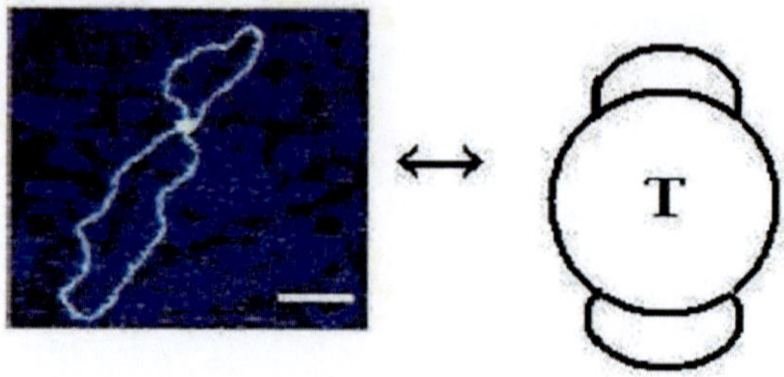

FIG. 3. *AFM image of a Cre synaptic complex formed with circular DNA (reprinted from [18] with permission from Elsevier) and a corresponding 2-string tangle model.*

the protein "ball" so that the DNA does not cross itself on the boundary of this protein ball, then the tangle modeling it is rational. Also, in nature, circular DNA is supercoiled. Protein-bound DNA is also often supercoiled. Hence rational tangles are generally believed to be the most biologically reasonable models for protein-bound DNA.

EXAMPLE 1. Figure 4 (a)–(d) give examples of 3-string tangles. Among those, (a), (c), and (d) are examples of rational 3-string tangles.

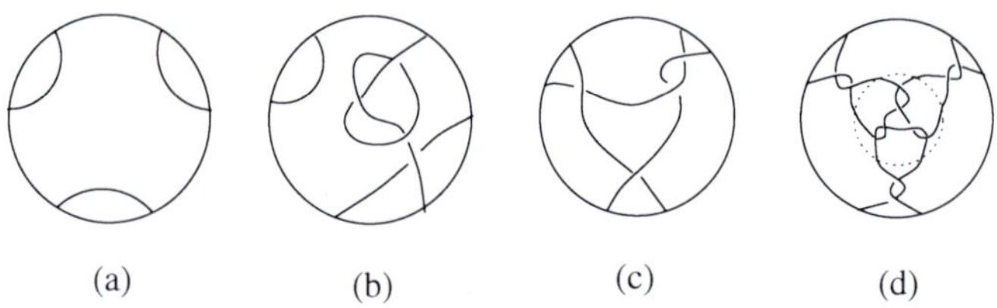

FIG. 4. *Examples of 3-string tangles.*

The original tangle model was applied to proteins which bind two segments of DNA and which will break and rejoin segments of DNA creating knotted DNA. For a review of 2-string tangle analysis, see for example [8, 9, 17, 4]. Software has been developed to solve n-string tangle equations [5]. This software was used to search through all tangles up through 8 crossings which satisfy the experimental results of [5]. But computational software which can only solve one system of equations at a time lacks the ability mathematical theories can provide for analyzing real and hypothetical experiments.

In the next section we will discuss the biological model for a Mu protein-DNA complex given in [15], while in Section 5 we will summarize the mathematical tangle analysis given in [6].

4. Difference topology and its application to Mu. DNA transposition results in the movement of a DNA segment from one location to another in a genome (http://research.utu.fi/celgenmol/molepid/savilahti.html). Bacteriophage Mu is a virus which uses transposition efficiently to replicate its DNA. During the transposition process, Mu proteins bind to 3 target sites including an enhancer sequence and two Mu ends (attL and attR) (see Figure 5). The enhancer sequence will be denoted by E, the attL site by L and the attR site by R. The protein-DNA complex consisting of Mu proteins along with these three DNA sequences is called the *transpososome*. The structure of the transpososome is very important for understanding the transposition pathway.

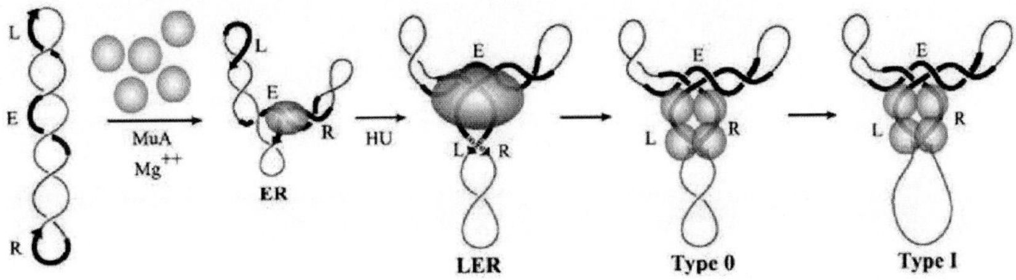

FIG. 5. *Mu transposition (reprinted from [23] with permission from Elsevier).*

The reaction pathway shown in Figure 6 was the model before the structure of the transpososome was determined in [15]. The new reaction pathway is shown in Figure 5. Note that in the older model, since there was no information available regarding the DNA shape bound by Mu, a very simple structure was assumed. The protein-bound DNA conformation in Figure 5, determined via difference topology, can be used to determine what DNA sequences are likely to be close to each other and therefore may interact [15]. Difference topology was also used to detect a new intermediate (denoted by ER in Figure 5) in the reaction pathway [14, 10, 23]. Difference topology was also used to investigate the role of supercoiling [22]. For additional applications of difference topology see [11].

Pathania, Jayaram, and Harshey used *Cre* inversion and deletion to determine the topological structure of DNA within the Mu transpososome [15]. If Cre acts on unknotted DNA not bound by any proteins except for Cre, then the main products of Cre inversion and deletion are unknots and unlinks, respectively. If, however, a protein complex such as Mu binds the DNA before Cre acts, the products can be more complicated. This difference in products was used in [15] to determine the topological conformation of the DNA bound by Mu. This methodology is called *difference topology*.

Pathania *et al.* first performed *Cre* inversion with two *loxP* sites lying on either side of E, isolating this site from L and R. In Figure 7(a), the

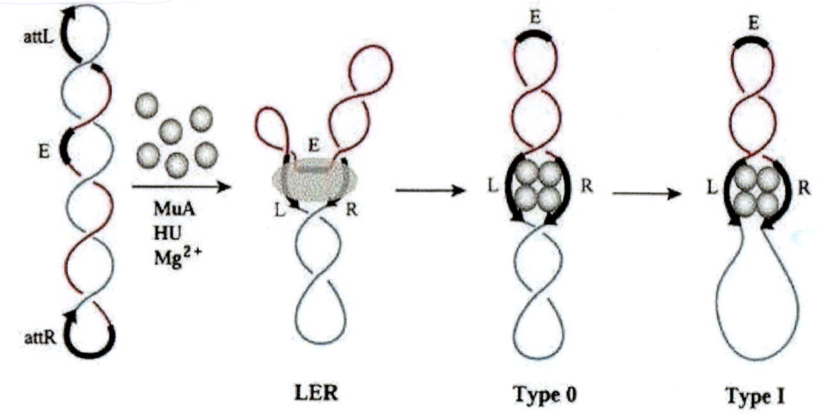

FIG. 6. *An older model of Mu transposition (reprinted from [15] with permission from Elsevier).*

loxP sites are inversely repeated. *Cre* cuts these target sites and changes the topology of the DNA before resealing it again. The product topology in this case was a three noded knot. Those three crossings resulted from E crossing R and L three times. Note that the crossings between R and L can be untwisted and thus have no affect on the topology of the product.

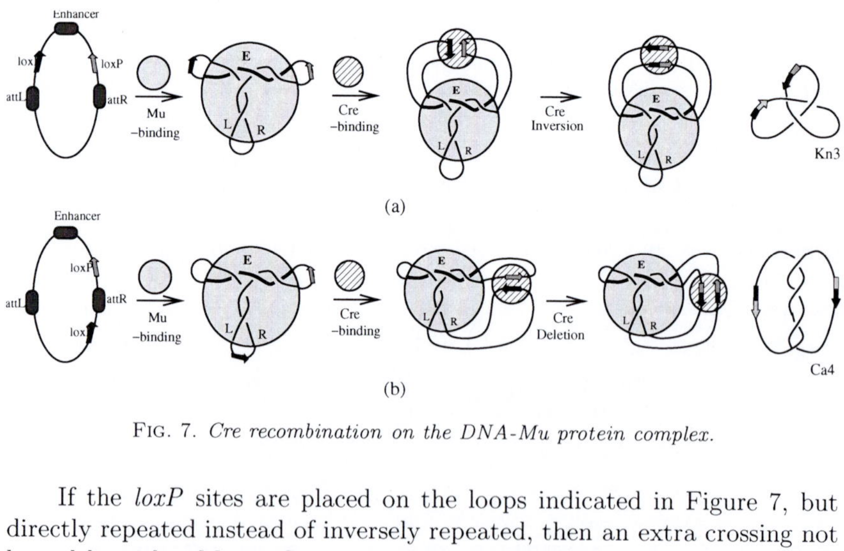

FIG. 7. *Cre recombination on the DNA-Mu protein complex.*

If the *loxP* sites are placed on the loops indicated in Figure 7, but directly repeated instead of inversely repeated, then an extra crossing not bound by either Mu or Cre is necessary to properly orient the *loxP* sites within the Cre-DNA complex. In this case the product of Cre recombination is a 4-crossing link. Note that this product has one more crossing than the product when the *loxP* sites were placed on the same pair of loops, but in inverse orientation. There are three pairs of loops on which to place the *loxP* sites. In each case the number of crossings in the product differed by one when comparing inversely repeated versus directly repeated *loxP* sites on the same pair of loops. It was assumed that the smaller crossing product corresponded to the tangle equation where no extra crossing is needed to properly orient the *loxP* sites within the Cre-DNA complex. The equations corresponding to the smaller crossing product when comparing *loxP* sites on the same pair of loops is shown in Figures 8a, 9a. In Figure 8a, the solution found in [15] is shown while Figure 9a shows the equations where the tangle corresponding to the Mu transpososome is unknown. One can prove that the solution set for T to the system of equations in Figure 9a is the same as the solution set for T if all six experiments are considered [5, 6].

To determine the number of DNA crossings within the Mu transpososome, we are interested in how many crossings are between E and R, R and L, L and E. Note that the protein-bound DNA conformation shown in Figure 8 consists of supercoiled DNA with three branches: one branch contains one crossing while the other two branches each contain two crossings. The solution found in [15] was obtained by assuming the protein-bound DNA conformation is a 3-branched supercoiled structure. Let x be the number of crossings between E and R, y the number of crossing between R and L, and z the number of crossings between L and E. If the DNA conformation bound by Mu is supercoiled with three branches, then x, y, z represent the number of crossings in each of the three branches. In this case, the equations in Figure 9a correspond to the equations $x + z = 3$, $x + y = 3$ and $y + z = 4$. Since there are three unknown variables and three linear equations, one can easily solve this linear system. The solution is that $x = 1$, $y = 2$, and $z = 2$. This implies that there is one crossing between E and R, two crossings between R and L, and two crossings between L and E. Thus if the DNA conformation bound by Mu is supercoiled with three branches, then the Mu transpososome has the five crossing configuration shown in Figure 9b as found in [15].

5. 3-String tangle analysis. Mathematically, the Mu proteins can be modeled by a three dimensional ball and the three DNA segments can be modeled by 3 strings in the ball. Pathania *et al.* found a solution to the system of equations in Figure 9a in which the DNA bound by Mu consists of supercoiled DNA with 3 branches and 5 crossings ([15], see Section 4). Pathania *et al.*'s experimental data [15] was mathematically analyzed by using 3-string tangle analysis [6] without the assumption that the tangle T represents supercoiled DNA with three branches. If a tangle T satisfies all the experimental data in [15], it can be a possible tangle model for the Mu transpososome. By using tangle theory, the following result was obtained:

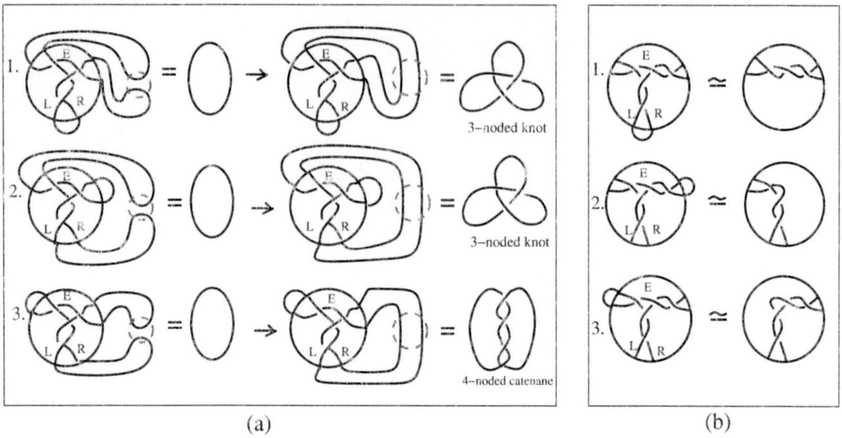

FIG. 8. *Tangle model of Mu transpososome.*

PROPOSITION 5.1. *Let T be a 3-string tangle which satisfies the system of tangle equations in Figure 9(a). If T can be freely isotopic to a projection with less than 8 crossings, then T is the tangle in the Figure 9(b).*

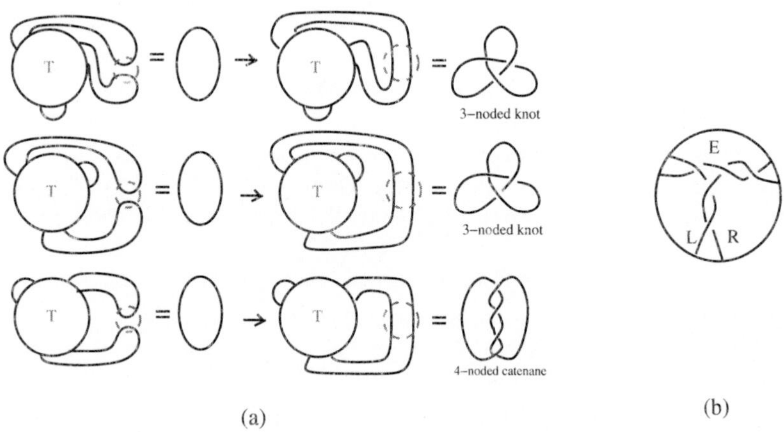

FIG. 9. *(a) Equations corresponding to the tangle model of Mu transpososome. (b) A solution to these equations.*

Two tangles are freely isotopic to each other if they are ambient isotopic allowing the boundary to move. For example, a rational tangle is freely isotopic to a tangle with no crossings. Thus Proposition 5.1 implies that the only rational tangle solution to the Figure 9(a) equations is that given in Figure 9(b).

An additional experiment not described here was used in [6] to rule out eight crossing solutions. The upper bound for the number of crossings which could be bound by Mu is unknown. However, since the solution found in [15] has five crossings, it is unlikely that a solution with more than eight crossings could be a model for the Mu transpososome. Thus the solution found in [15] is the only biologically reasonable solution.

6. 4-String tangle analysis. We do not currently have experimental data for a protein-DNA complex which binds four segments of DNA. In fact, we are not aware of such a complex. However there are a number of protein-DNA complexes, such as those involved in replicating and transcribing DNA, in which multiple proteins interact with each other and with multiple segments of DNA. Thus it is highly likely that protein-DNA complexes exist involving four or more DNA segments. We address a model for a protein complex which binds four DNA segments. Such a protein complex bound to circular DNA is modeled by a 4-string tangle with four loops outside of the tangle (Figure 10(a)).

In nature, DNA is negatively supercoiled if it is circular or if the ends are constrained (http://www.cbs.dtu.dk/staff/dave/roanoke/genetics980213a.html). There are two kinds of DNA supercoiling, plectonemic and solenoidal. Plectonemic supercoils are frequently branched [1]. Figure 10(b) shows an example of a branched supercoiled DNA-protein complex which would be a biologically reasonable model for a protein-DNA complex involving four segments of DNA. More generally, Figure 10(c) shows a biologically natural tangle model of a 4-branched supercoiled DNA-protein complex, where the n_i's are the number of left-handed half twists.

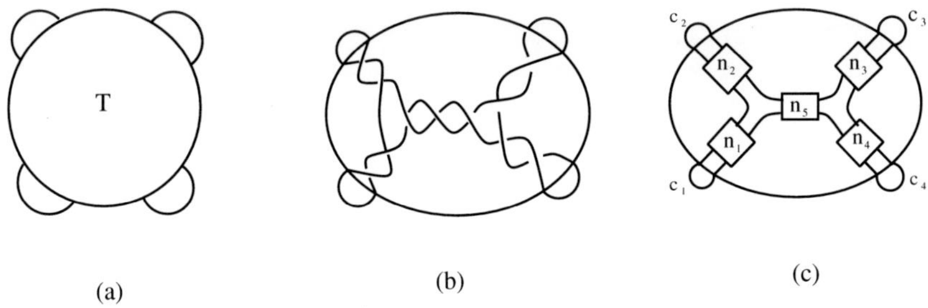

FIG. 10. *(a) A 4-string tangle model of a DNA-protein complex. (b), (c) Examples of 4-string tangle model which are biologically relevant.*

In this section, we would like to extend 3-string tangle analysis (Section 5) to 4-string tangle analysis based on difference topology. For Cre recombination, we need to put *loxP* sites on two of the outside loops. In the 3-string tangle model, there are three choices for a pair of loops on which to place Cre binding sites. On the other hand, in the 4-string tangle model, there are six different possible pairs of loops. In each case, there are two possible orientations for the Cre binding sites, directly or inversely repeated. Thus there are twelve possible Cre reactions for the 4-string tangle model (six different pairs of loops and two different orientation of *loxP* sites for each pair). By the prediction of difference topology (Section 4), the crossing number of the knotted inversion product and the catenated deletion product will differ by one when the Cre binding sites are placed on the same pair of loops but in different orientations.

As we mentioned at the beginning of this section, Figure 10(c) is a biologically relevant 4-string tangle model. Assume two *loxP* sites are located on loops c_1 and c_2 of Figure 10(c). After Cre recombination, the n_1 and n_2 crossings on two branches of the supercoiled DNA would be trapped, but the n_3, n_4, and n_5 crossings on the other three branches can be removed. The result is a $(2,\ n_1+n_2)$-torus knot if n_1+n_2 is odd or $(2,\ n_1+n_2)$-torus link if n_1+n_2 is even. For example, Cre recombination on directly repeated *loxP* sites assuming the tangle model Figure 10(b) results in the $(2,4)$-torus link. See Figure 11(a). Similarly, if two inversely repeated *loxP* sites are located on loops c_2 and c_3, Cre recombination results in the $(2,7)$-torus knot as shown in Figure 11(b). For convenience, Cre is placed on the left side and the 4-string tangle is rotated 90° counterclockwise in Figure 11(b).

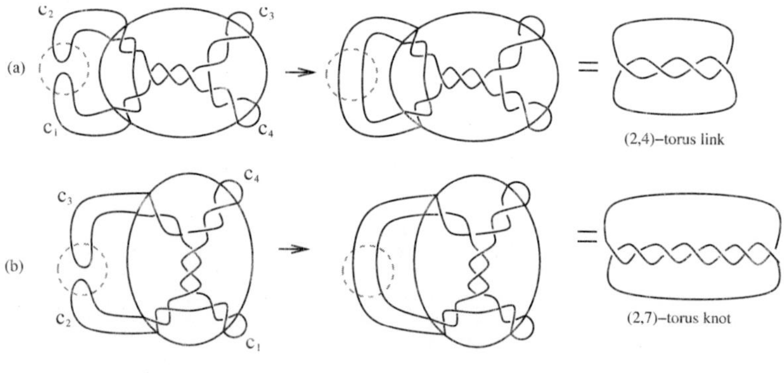

FIG. 11. *(a) If Cre acts on the loops* c_1 *and* c_2*, then Cre recombination for this protein-bound DNA conformation results in a* $(2,4)$*-torus link. (b) If the Cre binding sites are placed on loops* c_2 *and* c_3*, then Cre recombination for this protein-bound DNA conformation results in a* $(2,7)$*-torus knot.*

Hence if T is a tangle of the form shown in Figure 10(c), Cre recombination results in a $(2,p)$-torus knot when p is odd or a $(2,p)$-torus link when p is even. Note that the products of Cre recombination in the Mu/Cre experiments were $(2,p)$-torus knots and links [15]. Thus for the 4-string tangle model, we focus on equations where we assume the products are $(2, p)$-torus knots and links. This process can be modeled by Figure 12. In this figure, the tangle T represents a protein which binds to four DNA segments. The dotted circle represents Cre. For convenience, Cre is placed on the left side of T. T is rotated by 90° in (b) and (f), 180° in (c) and (e), 270° in (d) counterclockwise. We can summarize all these assumptions with Figure 12 and define a tangle satisfying these conditions a *solution tangle*.

DEFINITION 6.1. *A solution tangle is a tangle T which is a solution to the system of 12 difference topology experiments where the products are $(2, p_i)$ torus knots/links. Six of these are shown in Figure 12.*

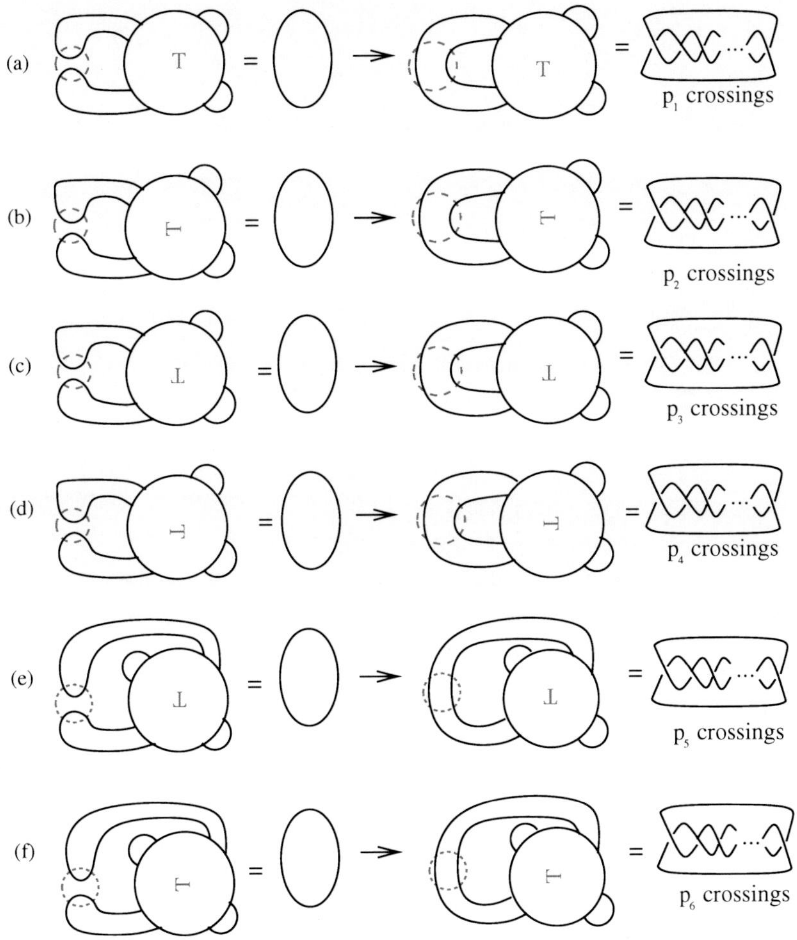

FIG. 12. *Tangle equations. In (b)–(f), T is rotated. The dotted circle represents Cre recombinase.*

We will first discuss branched supercoiled DNA solutions. A 4-string tangle model of a branched supercoiled DNA-complex can be represented by a weighted graph as shown in Figure 13. For example, the 4-string tangle in Figure 10 (b) has 2 or 3 left-handed half twists on each branch and hence $n_i = -2$ for $i = 1, ..., 4$ and $n_5 = -3$.

DEFINITION 6.2. *A tangle of the form shown in Figure 13(a) will be called standard, where n_i is the number of left-handed half twists. Note that a 4-string standard tangle T can be represented by a weighted graph G, where G is as in Figure 13(b). Call this graph G a standard graph.*

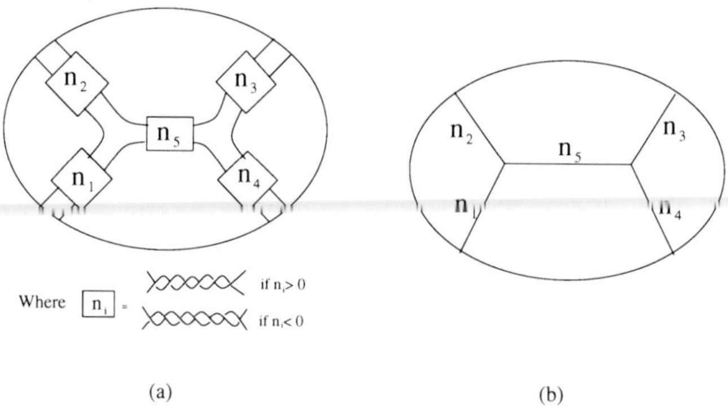

FIG. 13. *(a) Standard tangle. (b) A weighted graph G representing a 4-string standard tangle.*

We will also address the possibility that a pair of supercoiled DNA branches can be twisted. In other words, what if a tangle model is isotopic to a standard tangle allowing boundary of the corresponding graph (see Definition 6.2) to move?

DEFINITION 6.3. *A weighted graph G_R is an R-standard graph if it is isotopic to a standard graph G allowing the boundary of G to move. A tangle T is R-standard if it corresponds to an R-standard graph G_R.*

EXAMPLE 2. Examples of 4-string standard tangles are shown in Figure 14(a), (b) and an example of a 4-string R-standard tangle is shown in (c).

EXAMPLE 3. Figure 15(a) shows an example of a weighted graph G_R which represents the R-standard tangle T in Figure 15(b).

By extending 3-string tangle analysis of [6] to 4-string tangles, we determined that the biologically relevant solutions to the system of equations in Figure 12 must be R-standard:

THEOREM 6.1. *[12] Suppose T is a 4-string tangle which has less than 8 crossings up to free isotopy. If T is a solution tangle, then T is R-standard.*

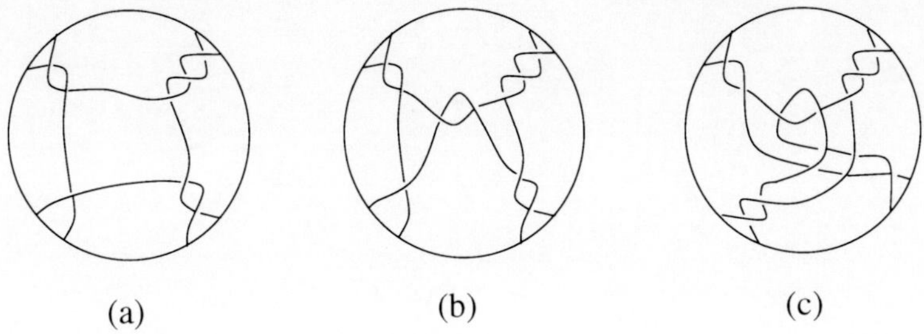

FIG. 14. *(a), (b) Examples of standard tangles; (c) Example of R-standard tangle.*

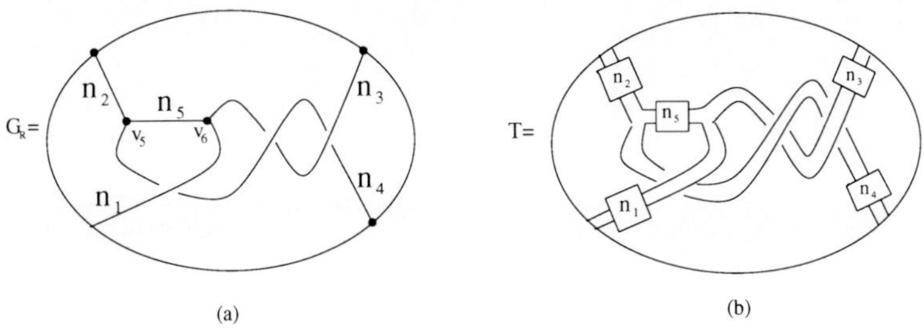

FIG. 15. *An example of a weighted graph G_R for an R-standard tangle T.*

In other words, if a 4-string tangle T satisfies all the equations of Figure 12 involving $(2, p)$-torus links and has less than 8 crossings up to free isotopy, T can be represented by an R-standard graph. Since all rational tangles are freely isotopic to a tangle which has no crossings, we can find all rational solutions.

We start with the following definition:

DEFINITION 6.4. *Let G_R be a graph which corresponds to an R-standard tangle. There are two vertices in the interior of the ball and 4 vertices on the boundary of the ball. Let $v_1 = SW$, $v_2 = NW$, $v_3 = NE$, $v_4 = SE$ be the vertices on the boundary of the ball, and v_5 and v_6 be the vertices in the interior of the ball. G_R is $(2, j)$-branched if v_5 is connected to $v_2 = NW$ and v_j for some $1 \leqslant j \leqslant 4$, $j \neq 2$.*

The vertex v_5 can only be connected with (v_1, v_2) or (v_2, v_3) or (v_2, v_4); hence there are 3 different (i, j) branchings (Figure 16). For example, the graph G_R in Figure 15(a) is $(2, 4)$-branched. Note that $n_5 = 0$ if and only if G_R is (i, j)-branched for all (i, j).

Each edge of G_R represents a branch of a branched supercoiled DNA molecule. This implies that an (i, j)-branched graph and a (k, l)-branched graph represent different geometries of a DNA molecule when $(i, j) \neq (k, l)$.

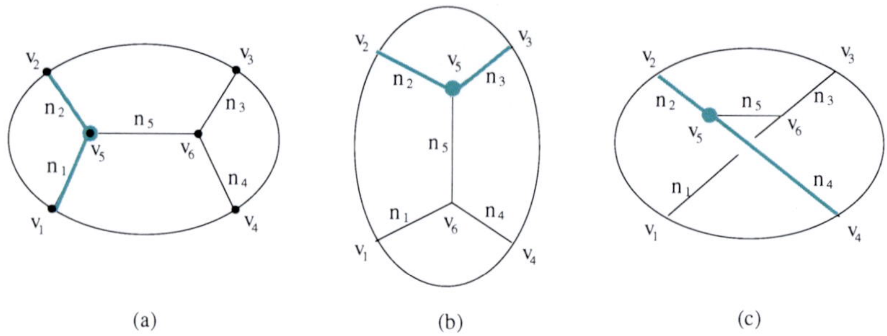

FIG. 16. *(a) (1, 2)-branched; (b) (2, 3)-branched; (c) (2, 4)-branched weighted graph for R-standard tangle.*

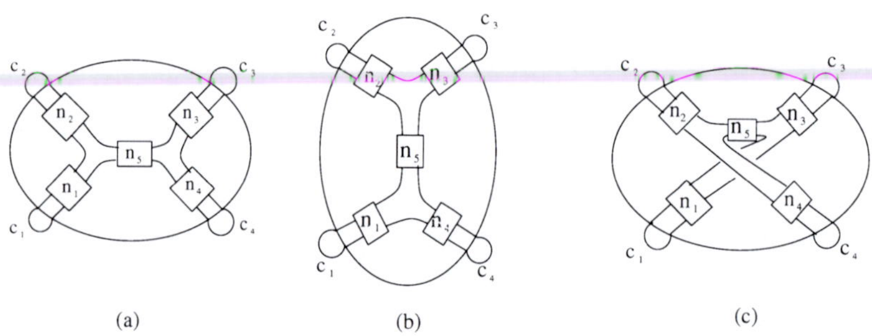

FIG. 17. *(a) Example of R-standard tangle model of a branched DNA-protein complex corresponding to a weighted graph which is (a) (1, 2)-branched; (b) (2, 3)-branched; (c) (2, 4)-branched.*

We will first focus on tangles of the form shown in Figure 17. Suppose a tangle of the form shown in Figure 17(a) is a solution to the system of equations in Figure 12. Then we have the equations in 6.1. The values $p_1, \cdots, p_6$ in Figure 12 must be determined experimentally. Our goal is to find $n_1, \cdots n_6$ in terms of the p_i's.

$$\begin{aligned} n_1 + n_2 &= p_1 \\ n_2 + n_3 + n_5 &= p_2 \\ n_3 + n_4 &= p_3 \\ n_1 + n_4 + n_5 &= p_4 \\ n_1 + n_3 + n_5 &= p_5 \\ n_2 + n_4 + n_5 &= p_6. \end{aligned} \tag{6.1}$$

The solution to the system of Equations 6.1 is the following:

$$
\begin{aligned}
n_1 &= \frac{p_1 + p_4 - p_6}{2}, & n_2 &= \frac{p_1 - p_4 + p_6}{2} \\
n_3 &= \frac{p_2 + p_3 - p_6}{2}, & n_4 &= \frac{-p_2 + p_3 - p_6}{2} \\
n_5 &= \frac{-p_1 + p_2 - p_3 + p_4}{2}, & p_2 + p_4 &= p_5 + p_6.
\end{aligned} \tag{6.2}
$$

We can solve similar equations for tangles corresponding to the graphs in Figure 17(b) and (c). The summary of the results is the following:

- The solution to the Figure 12 equations is the following if the solution is of the form Figure 17(b):

$$
\begin{aligned}
n_1 &= \frac{-p_3 + p_4 + p_5}{2}, & n_2 &= \frac{p_1 + p_2 - p_5}{2} \\
n_3 &= \frac{-p_1 + p_2 + p_5}{2}, & n_4 &= \frac{p_2 + p_3 - p_5}{2} \\
n_5 &= \frac{p_1 - p_2 + p_3 - p_4}{2}, & p_1 + p_3 &= p_5 + p_6.
\end{aligned} \tag{6.3}
$$

- The solution to the Figure 12 equations is the following if the solution is of the form Figure 17(c):

$$
\begin{aligned}
n_1 &= \frac{p_1 - p_2 + p_5}{2}, & n_2 &= \frac{p_1 - p_4 + p_6}{2} \\
n_3 &= \frac{-p_1 + p_2 + p_5}{2}, & n_4 &= \frac{-p_1 + p_4 + p_6}{2} \\
n_5 &= \frac{-p_2 + p_4 - p_5 - p_6}{2}, & p_1 + p_3 &= p_2 + p_4.
\end{aligned} \tag{6.4}
$$

Note that the n_i must be integral. To have an integer solution set $\{n_1, \cdots, n_5\}$, all numerators of Equations 6.2, 6.3, and 6.4 should be even. In fact, there are eight possible cases to have an integer solution set for Equation 6.2, shown in the following table.

	p_1	p_2	p_3	p_4	p_5	p_6
1	even	even	even	even	even	even
2	odd	odd	even	even	even	odd
3	even	odd	even	odd	odd	odd
4	odd	even	even	odd	odd	even
5	even	odd	odd	even	odd	even
6	odd	even	odd	even	odd	odd
7	even	even	odd	odd	even	odd
8	odd	odd	odd	odd	even	even

Equations 6.2, 6.3, and 6.4 have an integer solution set for the same eight cases. Thus the different ways of branching can only be distinguished by the last equation given in Equations 6.2, 6.3, and 6.4:

LEMMA 6.1. *The graph G_R corresponding to an R-standard tangle can only be branched in three different ways, (1,2), (2,3) or (2,4)-branched. The (i, j) branching of a solution of the form in Figure 17 can be determined as follows:*

- *If $p_2 + p_4 = p_5 + p_6$ holds, G_R is (1,2)-branched.*
- *If $p_1 + p_3 = p_5 + p_6$ holds, G_R is (2,3)-branched.*
- *If $p_1 + p_3 = p_2 + p_4$ holds, G_R is (2,4)-branched.*

In addition, $n_5 = 0$ if and only if G_R is (i, j)-branched for all (i, j).

We have only proved Lemma 6.1 for tangles corresponding to those shown in Figure 17. However, Lemma 6.1 can be extended to R-standard tangles as discussed in the next section.

6.1. Discussion on complicated branched solution tangles. We will now consider a more complicate branched solution tangle like that in Figure 18.

EXAMPLE 4. Let G be a graph which corresponds to the standard graph in Figure 18(a). After doing one counterclockwise half twist of v_1 and v_4 and two clockwise half twists of v_3 and v_4 moving the boundary of 3-ball, one obtains the weighted graph G_R (Figure 18(b)). Then G_R is the weighted graph which corresponds to the R-standard tangle T in Figure 18(c). Since v_5 is connected to v_2 and v_4, G_R is (2,4)-branched.

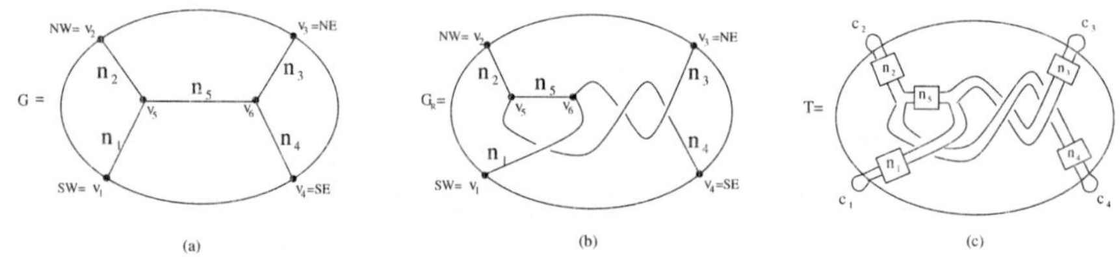

FIG. 18. *Example of R-standard tangle.*

Let's compare this example with the tangle in Figure 17(c), which we will call T'. The link obtained from Cre recombination on c_3 and c_4 of T has positive 2 writhe which can be converted to four half twists as shown in Figure 19 [2]. Hence $p_3 = n_3 + n_4 + n_5 + 4$ for T while for T', $p_3 = n_3 + n_4 + n_5$. Similarly, $p_4 = n_1 + n_4 + n_5 - 2$ for T, while $p_4 = n_1 + n_4 + n_5$ for T'. The remaining equations for T are identical to the equations for T'.

Note that a solution of the form T will satisfy the first five equations in 6.4 if and only if a solution of the form T' satisfies these equations. This is because writhe of a link diagram can be converted to an even number of

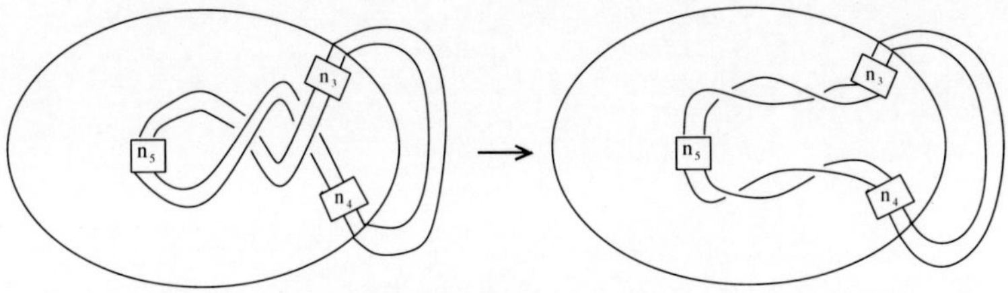

FIG. 19. *(a) A link obtained by doing Cre recombination in Figure 18 has writhe two. (b) The writhe two can be converted to 4 half twists.*

half twists. Thus, all numerators of Equations 6.2, 6.3, and 6.4 will still be even after adding or subtracting an even number. Hence the last equation in 6.4 determines if a tangle of the form T or T' can be a solution.

In conclusion, although the 4-string tangle algebra is not as simple as the 3-string case, we can determine all rational solutions and all small crossing solutions (< 8 crossings) when the experimental products are $(2, p_i)$ torus links (Figure 12). For more information see [12].

REFERENCES

[1] A.D. BATES AND A. MAXWELL. *DNA Topology.* IRL Press, Oxford, 1993.

[2] W.R. BAUER, F.H.C. CRICK, AND J.H. WHITE. Supercoiled DNA. *Scientific American*, **243**: 118–133, 1980.

[3] J.H. CONWAY. An enumeration of knots and links and some of their related properties. *Computational Problems in Abstract Algebra (John Leech, ed.) Pergamon Press, Oxford and New York*, 1969.

[4] I. DARCY. *Biological distances on DNA knots and links: Applications to Xer recombination*, Journal of Knot Theory and its Ramifications **10**: 269–294, 2001.

[5] I.K. DARCY, A. BHUTRA, J. CHANG, N. DRUIVENGA, C. MCKINNEY, R.K. MEDIKONDURI, S. MILLS, J. NAVARRA MADSEN, A. PONNUSAMY, J. SWEET, AND T. THOMPSON. Coloring the Mu transpososome. *BMC Bioinformatics*, **7**(435), 2006.

[6] I.K. DARCY, J. LUECKE, AND M. VAZQUEZ. Tangle analysis of difference topology experiments: Applications to a Mu protein-DNA complex. Preprint.

[7] F.B. DEAN, A. STASIAK, T. KOLLER, AND N.R. COZZARELLI. Duplex DNA knots produced by Escherichia Coli topoisomerase I. *I. Biol. Chem.*, **260**: 4795–4983, 1985.

[8] C. ERNST AND D. W. SUMNERS. A calculus for rational tangles: Applications to DNA recombination. *Math. Proc. Camb. Phil. Soc.*, **108**: 489–515, 1990.

[9] C. ERNST AND D. W. SUMNERS. Solving tangle equations arising in a DNA recombination model. *Math. Proc. Camb. Phil. Soc.*, **126**: 23–36, 1999.

[10] R. HARSHEY AND M. JAYARAM. The Mu transpososome through a topological lens. *Crit. Rev. Biochem. Mol. Biol.*, **41**(6): 387–405, 2006.

[11] M. JAYARAM AND R. HARSHEY. Difference topology: Analysis of high-order DNA-protein assemblies. Included in this volume, Mathematics of DNA Structure, Function, and Interactions edited by Craig John Benham, Stephen Harvey, Wilma K. Olson, De Witt L. Sumners, and David Swigon,Springer Science + Business Media, LLC, New York, 2009. Preprint.

[12] S. Kim and I.K. Darcy. A 4-string tangle analysis of DNA-protein complexes based on difference topology. Preprint.

[13] K. Kimura, V.V Rybenkov, N.J. Crisona, T. Hirano, and N.R. Cozzarelli. 13s condensin actively reconfigures DNA by introducing global positive writhe: Implications for chromosome condensation. *Cell*, **98**(2): 239–248, 1999.

[14] S. Pathania, M. Jayaram, and R. Harshey. A unique right end-enhancer complex precedes synapsis of Mu ends: The enhancer is sequestered within the transpososome throughout transposition. *EMBO J.*, **22**(14): 3725–36, 2003.

[15] S. Pathania, M. Jayaram, and R.M. Harshey. Path of DNA within the Mu transpososome. transposase interactions bridging two Mu ends and the enhancer trap five DNA supercoils. *Cell*, **109**(4): 425–436, 2002.

[16] S.J. Spengler, A. Stasiak, and N.R. Cozzarelli. The stereostructure of knots and catenanes produced by phage λ integrative recombination: Implications for mechanism and DNA structure. *Cell*, **42**: 325–334, 1985.

[17] D.W. Sumners, C. Ernst, N.R. Cozzarelli, and S.J. Spengler *Mathematical analysis of the mechanisms of DNA recombination using tangles*, Quarterly Reviews of Biophysics **28**, 1995.

[18] A.A. Vetcher, A.Y. Lushnikov, J. Navarra-Madsen, R.G. Scharein, Y.L. Lyubchenko, I.K. Darcy, and S.D. Levene. DNA topology and geometry in Flp and Cre recombination. *J. Mol. Biol.*, **357**: 1089–1104, 2006.

[19] S.A. Wasserman and N.R. Cozzarelli. Determination of the stereostructure of the product of Tn3 resolvase by a general method. *Proc. Nat. Acad. Sci. U.S.A.*, **82**: 1079–1083, 1985.

[20] S.A. Wasserman and N.R. Cozzarelli. Biochemical topology: Applications to DNA recombination and replication. *Science*, **232**: 951–960, 1986.

[21] S.A. Wasserman, J.M. Dungan, and N.R. Cozzarelli. Discovery of a predicted DNA knot substantiates a model for site-specific recombination. *Science*, **229**: 171–174, 1985.

[22] Z. Yin, M. Jayaram, S. Pathania, and R. Harshey. The Mu transposase interwraps distant DNA sites within a functional transpososome in the absence of DNA supercoiling. *J. Biol. Chem.*, **280**(7): 6149–56, 2005.

[23] Z. Yin, A. Suzuki, Z. Lou, M. Jayaram, and R.M. Harshey. Interactions of phage Mu enhancer and termini that specify the assembly of a topologically unique interwrapped transpososome. *J. Mol. Biol.*, **372**: 382–396, 2007.

CLOSING THE LOOP ON PROTEIN-DNA INTERACTIONS: INTERPLAY BETWEEN SHAPE AND FLEXIBILITY IN NUCLEOPROTEIN ASSEMBLIES HAVING IMPLICATIONS FOR BIOLOGICAL REGULATION

STEPHEN D. LEVENE* AND YONGLI ZHANG†

Abstract. The formation of DNA loops by proteins bound at distant sites along a single molecule is an essential mechanistic aspect of many biological processes including gene regulation, DNA replication, and recombination. The biological importance of DNA loop formation is underscored by an abundance of architectural proteins in cells such as HU, IHF, and HMGs, which facilitate looping by bending the intervening DNA between cognate protein-binding sites. We have developed a rigorous theory for DNA loop formation that connects the global mechanical and geometric properties of both DNA and protein, including previously neglected phenomena such as the conformational flexibility of protein domains. The theory is applied to the problem of loop-mediated gene repression *in vivo* by *lac* repressor.

Key words. DNA looping, wormlike chain, J factor, gene regulation, *lac* repressor.

AMS(MOS) subject classifications. Primary 92C05, 92C40, 82D60.

1. Introduction.

Large macromolecular assemblies involving multiple proteins and genomic DNA are recognized as an important theme in biological mechanism and regulation in all organisms. We have recently focused on the general problem of multiple protein-DNA interactions involving two or more specific recognition sites on a single DNA molecule. Such nucleoprotein structures adopt a looped conformation for the intervening DNA between successive protein-binding sites with an average loop geometry dictated by the interplay of multiple influences. These include the local DNA geometry required for protein binding, elastic/mechanical properties of looped DNA segments and the protein assembly, and additional interactions between the DNA loop and other proteins, such as accessory factors. The importance of DNA-loop formation is such that all cells maintain high levels of architectural proteins such as HU, IHF, and HMGs, which facilitate looping by promoting bending of DNA [1]. In recognition of the fundamental importance of this topic, there has been rapidly growing interest in the problem of DNA looping on the part of both nucleic-acid experimentalists and theoreticians [2–11].

The absence of an explicit theory for DNA looping motivated us to develop a semi-analytical solution for the J factor, the generalization of the Jacobsen-Stockmayer factor, J. J expresses the effective concentration of loop ends and is directly related to the free energy of DNA looping

* Departments of Molecular and Cell Biology and Physics, University of Texas at Dallas, Richardson, TX 75080. This work supported by NIH grant GM67242.

†Department of Physiology and Biophysics, Albert Einstein College of Medicine, Bronx, NY 10461.

C.J. Benham et al. (eds.), *Mathematics of DNA Structure, Function and Interactions*, The IMA Volumes in Mathematics and its Applications 150,
DOI 10.1007/978-1-4419-0670-0_10,

[12, 13]. The approach is based on the statistical mechanics of a semiflexible polymer chain subject to particular constraints and fully takes into account bending and twisting elasticity of DNA and proteins, thermal fluctuations, and sequence-dependent inhomogeneities in DNA structure. A major advantage of the method is that solutions for the J factor are computed orders of magnitude more rapidly than by Monte Carlo simulation, making this new approach especially suitable for fitting experimental data using nonlinear-regression techniques. In this chapter we review a number of general principles regarding DNA looping that emerge from the analysis and its application to lac-repressor-mediated gene regulation by DNA looping *in vivo*.

2. Statistical mechanics of DNA looping. Our knowledge about the roles of DNA bending, twist, and their respective energetic contributions to DNA looping has come almost exclusively from analyses of DNA cyclization [14–16]. Shore and Baldwin first showed that cyclization-kinetics measurements of the J factor for DNA molecules between 237 and 254 bp in size oscillate with DNA length and therefore helical phasing [17]. In this context the J factor can be defined as the free DNA-end concentration whose bimolecular ligation efficiency equals that of the two DNA ends on the same molecule. Hence, the free energy of DNA cyclization is given by $\Delta G_{\text{cyc}} = -k_B T \ln J$ [18]. For short DNA fragments J factors are limited by the significant bending and twisting energies required to form closed circles, whereas for long DNA, the chain entropy loss during circularization exceeds the elastic-energy decrease and reduces the J factor. Because of this competition between bending and twisting energetics and entropy, there is an optimal DNA length for cyclization near 500 bp [19]. Analogous behavior has been expected for DNA looping, especially with respect to the helical dependence.

Two emerging needs in structural and molecular biology motivated a rigorous solution to the problem of DNA looping: (i.) the need for new structural approaches to characterize the properties of large nucleoprotein assemblies in solution and *in vivo*; (ii.) the introduction of novel single-molecule experiments to characterize DNA looping that require a proper theoretical framework for their interpretation. The probability of loop formation can, in principle, be computed by Monte Carlo simulation for a given set of end constraints. However, that approach is not practical for fitting experimental data in cases where the values of geometric variables that describe the end constraints of a particular looped conformation are being sought.

The statistical-mechanical approach combines computation of the mechanical equilibrium conformation of nucleoprotein complexes with subsequent evaluation of thermodynamic quantities using a harmonic approximation [12, 19]. DNA conformations are described by parameters defined at base-pair steps, i.e., tilt, roll, and twist, [20] which allows straightforward

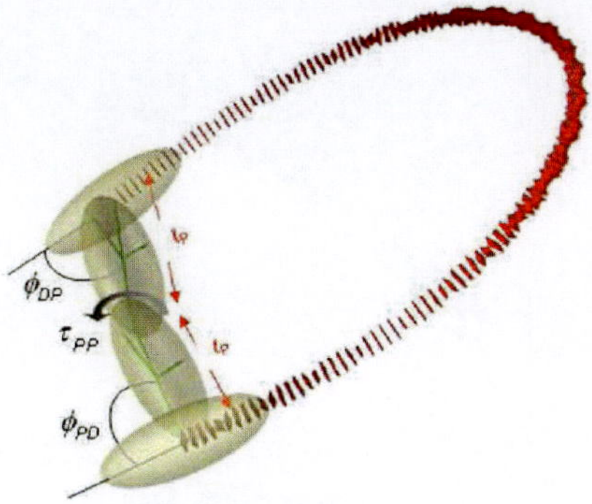

FIG. 1. *Minimum elastic-energy conformation of a 137-bp DNA loop computed using the statistical-mechanical approach. DNA base pairs are represented by rectangular slabs (red). Two sets of coordinate axes (green) represent the local coordinate frames embedded in the protein subunits (gold) that mediate DNA looping. Lengths of the virtual bonds that connect the loop ends are identical and equal to ℓ_P. The coupling of protein and DNA geometry is characterized by tilt, roll, and twist values for the DNA-protein, protein-protein, and protein-DNA interfaces. Three of these variables are shown here: the DNA-protein roll angle, ϕ_{DP}; the protein-protein twist angle, τ_{PP}; and the protein-DNA roll angle, ϕ_{PD}.*

incorporation of intrinsic or protein-induced DNA bending at the base-pair level. Similar to Shimada and Yamakawa's approach to DNA cyclization, [21] the theory takes advantage of fluctuations around one stable mechanical configuration defined by particular constraints, which is found with an iterative energy-minimization algorithm. Unlike the Shimada-Yamakawa theory, our method is suitable for a ring polymer of *arbitrary* intrinsic geometry, a critical advance required in order to solve the J-factor problem for a looped DNA structure having arbitrary boundary conditions.

The extension from cyclization to DNA looping was accomplished by treating the protein subunits as a pair of elastically coupled rigid bodies, assigning non-canonical segment lengths and base-step parameters to two consecutive segments of a closed chain (Figure 1). Two sets of base-step geometry and flexibility parameters model the interactions of protein subunits with the DNA, whereas one set of parameters describe the interactions between protein subunits. Unlike previous treatments, this method accomodates intrinsic flexibility among and within protein domains that mediate the loop, a feature that is essential for a fully quantitative theory of DNA looping [12, 13].

3. DNA looping and cyclization: a distinction with a difference. Using the new approach, we systematically examined the DNA-length (loop-size) dependence of J for small DNA loops ($\lesssim$ 300 bp) as a function of separation between loop ends and the twist angle between protein subunits, τ_{PP}. Figure 2 shows that the amplitude of the helical-

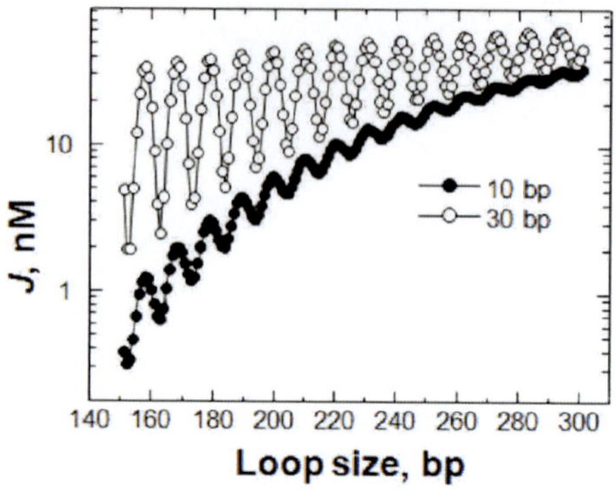

FIG. 2. *Helical dependence of DNA looping for loops with anti-parallel end orientations and end-to-end separations of 10 and 30 bp.*

phase dependence of J for a loop with anti-parallel ends increases over a range of end-to-end separation values that are typical for many looped complexes. Moreover, the phase-dependent amplitudes for loops near 300 bp are strongly attenuated relative to those found for cyclization: the periodicity for loops with 10-bp end-to-end separation is essentially negligible, whereas the peak-to-peak amplitude of loops with 30-bp end-to-end distance is only about four-fold. Cyclization J factors vary by more than 10-fold in this range and significant periodicity persists for circle lengths beyond the maximum in the phase-independent average of J near 500 bp. The decreased helical-phase dependence of J for looping versus cyclization can be mainly explained in terms of the major role that writhe plays in DNA looping. In contrast to small circles, for which writhe is strongly suppressed, [17, 18, 22–24] significant levels of superhelical stress in small loops can be partitioned into writhe, reflecting a generally greater degree of DNA bending flexibility in the looped state.

The J-factor curves shown in Figure 2 are sums of J values for two sets of loop topoisomers, both of which are elastic-energy minima for a loop specified in terms of size and end constraints [12] (because J is effectively a concentration term, the observed J value is simply the sum of J values for each topoisomer). This situation also stands in contrast to that with small DNA circles. Cyclization of DNA fragments with sizes below 500 bp normally yields only a single topoisomer species [22, 25]. In the case of DNA loops in this size range, however, two topoisomers are thermally accessible at room (or physiological) temperature. Figure 3 shows pairs of topoisomer solutions for anti-parallel loops having 10-bp end separation. One set of conformations corresponds to the case of a local J-factor maximum, 210 bp, whereas the other gives an example of solutions for a J-factor minimum, 215 bp. One of the 210-bp loops is clearly planar; the second topoisomer solution contains a single DNA crossing in projection and has a J factor

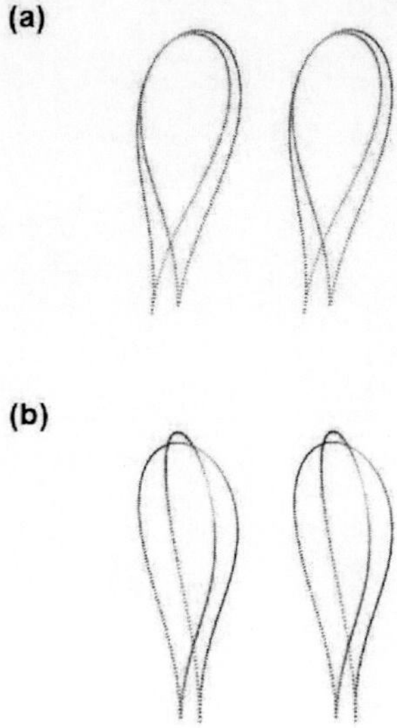

FIG. 3. *Stereo models of the two mechanical-equilibrium conformations for (a) 210-bp and (b) 215-bp anti-parallel DNA loops with end-to-end separation equal to 10 bp. Pairs of 210- and 215-bp loop topoisomers respectively contribute to J factor values of an adjacent peak and valley of the curve in Figure 2.*

that is about one order of magnitude lower than that of the planar solution. In contrast, both 215-bp solutions are non-planar and have similar levels of writhe, but of opposite sign. The J values for each of these solutions are also similar. Note that the interconversion of loop topoisomers can occur only upon disruption/re-formation of one or both protein-DNA interfaces.

In addition to a strong dependence on end separation, the J factor is also expected to be sensitive to the crossing angle between loop ends. This quantity is equal to the twist angle between protein subunits in our model, τ_{PP}. Figure 4 shows how the length dependence of DNA looping depends on the crossing, or axial, angle. On increasing τ_{PP} from 0 to 60°, there is a phase shift in the J-factor curve, which indicates that the maxima in DNA looping no longer correspond to DNA lengths that are integral multiples of the helical repeat. Increasing τ_{PP} further to 120° causes a more substantial phase shift in addition to a dramatic overall reduction in J values. At $\tau_{PP} = 120°$ there is more than a than 10-fold decrease in the phase-independent loop-size-averaged J value due to the greatly increased elastic-energy cost of loop formation.

The sensitivity of J to τ_{PP} has important implications in cases where a single protein conformation can mediate multiple looped geometries. An example is shown in Figure 5(a), where protein-binding DNA sequences flank a symmetrical motif; this is the case with certain type-II restriction

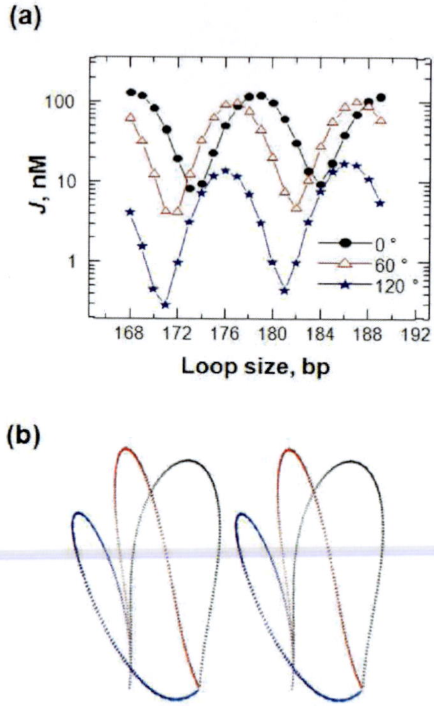

FIG. 4. *(a) DNA-length dependence of J for protein-protein twist angles, τ_{PP}, of 0°, 60°, and 120° with the end-to-end separation set equal to 40 bp. Note that the positions of the extrema shift to the left with increasing values of the axial angle. (b) Stereo models of minimum elastic-energy conformations of 179-bp loops color coded in accord with the corresponding axial-angle values in (a).*

endonucleases [26] and site-specific recombinases [27–29]. A single crossed geometry of the sites is consistent with two possible loops that differ in the connectivity of the loop ends and have correspondingly different values of τ_{PP}. As in the previous case where the loop-closure probability involves contributions from multiple topoisomers, J is simply the sum of J values for each of the conformations (loop topoisomers having already been taken into account). We consider two such cases in Figure 5: an included angle of 90°, which generates two geometrically equivalent looped structures with $\tau_{PP} = \pm 90°$, and an included angle of 45°, which results in distinct loops having $\tau_{PP} = 45°$ and $\tau_{PP} = -135°$. The loops formed in the case where $\tau_{PP} = \pm 90°$ have J factors that are nearly opposite in phase and therefore their sum has nearly negligible amplitude. In contrast, the sum of J factors for loops with $\tau_{PP} = 45°$ and $\tau_{PP} = -135°$ is almost identical to J for the $\tau_{PP} = 45°$ loop alone. This is expected because the structure with the smaller included angle should be the dominant conformation in this case.

Thus, failure to observe helical phasing effects in an experiment should not be interpreted as evidence for the absence of DNA looping.

Significant differences in basic physical behavior therefore militate against interpreting DNA-loop-mediated processes in terms of cyclization theory, especially for small DNA loops. These can be summarized as follows: (i.) the geometry of a protein complex involved in DNA looping can impose a non-planar structure on the looped DNA, implying a degree of shape complimentarity between protein and loop geometry; (ii.) J-factor maxima for loops having non-planar optimum geometries will generally correspond to non-integral numbers of helical turns; (iii.) looping of small DNA segments, unlike cyclization of small fragments, generally involves contributions from two or more topoisomers; (iv.) a single protein conformation may mediate multiple classes of DNA loops; (v.) the superposition of multiple loop solutions (as in (iv.)) can dramatically attenuate the amplitude of the J-factor dependence on loop size and thus lack of a helical phase dependence does not necessarily imply the absence of DNA looping.

4. Analysis of *in-vivo* lac-repressor-mediated gene repression based on the statistical mechanics of DNA looping. The interaction of *E.coli lac* repressor (LacR) with its cognate operator sequences in the wild-type *lac* operon is a major paradigm for gene regulation via DNA looping. A large body of evidence supports the conclusion that expression of structural genes regulated by the *lac* operator in *E. coli* is under thermodynamic control *in vivo* [30–32]. The abundance of *in-vivo* data from different laboratories on the helical-phase dependence of loop-mediated repression in the *lac* system [33–35] suggested that LacR-mediated gene repression would provide an excellent model system for quantitatively evaluating our statistical-mechanical treatment of DNA looping [13].

The biologically relevant structure of LacR is itself controversial: in crystal structures of LacR and the LacR-operator complex, the repressor is present as a "v-shaped" tetramer formed from two dimers via a four-helix bundle involving a tetramerization domain (Figure 6(a)) [36, 37]. However, electron microscopy [38] and solution studies [39, 40] support an extended conformation for the repressor ($\sim 180°$ between the two arms, Figure 6(b)). In principle, a conformational equilibrium between these forms can exist, leading to a bimodal distribution of looped conformations. Such a model was recently considered by Swigon *et al.* for LacR-mediated looping *in vitro* [6].

We first addressed the question of determining the relative thermodynamic stability of loops formed by each of the canonical repressor conformations. Although only a single class of loop geometries exists for the extended form of the repressor, several classes of loops are possible in the case of the v-shaped LacR tetramer (Figure 7). Following Mehta and Kahn [39], we denote these WT ("wrapping toward") and WA ("wrapping away"), whereas the class of loops formed by the extended LacR conformation is designated

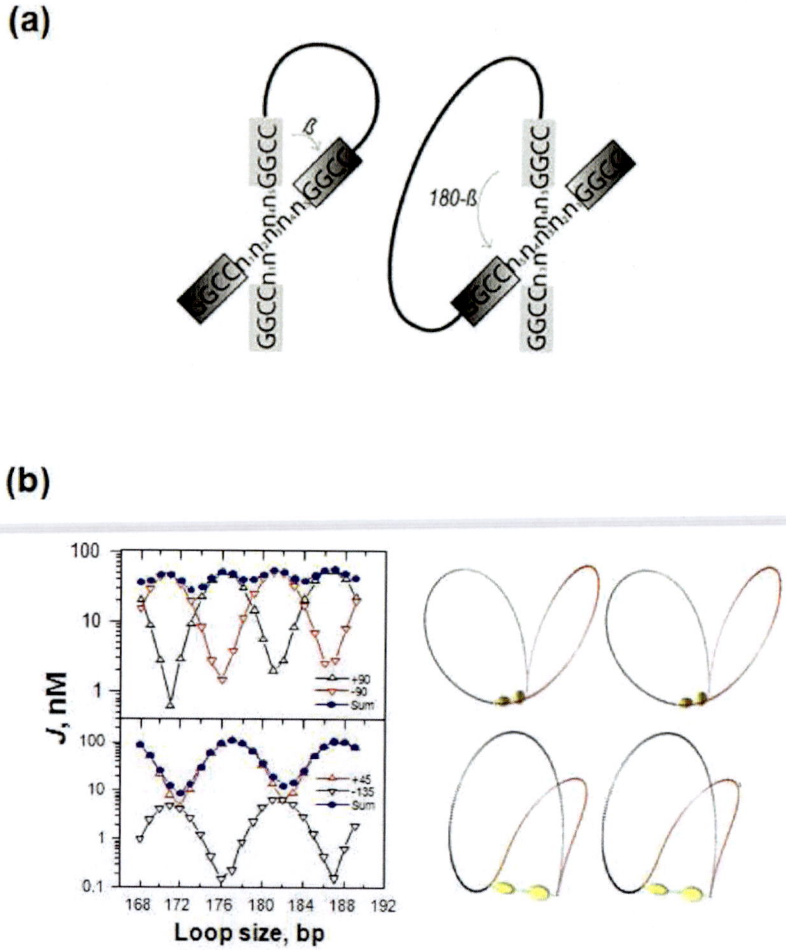

FIG. 5. *(a) Formation of alternative loop geometries by proteins with two-fold symmetric DNA-binding domains or cognate sites having two-fold symmetry. The example given here is the Sfi*I *recognition sequence, GGCCnnnn*$^{\downarrow}$*nGGCC, where n is any base and* $^{\downarrow}$ *indicates the location of the cleavage site. The two geometries are related by reversing the intrinsic DNA direction at the binding site if the protein is rigid, forming a negative (left) or positive (right) crossing according to the right-hand rule. Due to the two-fold symmetry of the protein dimer, this reversal does not affect protein-DNA interactions. Given the angle shown in the figure, the protein twist* τ_{PP} *used to model the geometry is* $-\beta$ *and* $180-\beta$ *for the left and right configurations, respectively.*
(b) Individual J factors computed for the two alternative looping geometries of a synaptic complex with two-fold symmetry and their sum. The τ_{PP} *values for these conformations are* $\pm 90°$ *(upper panel) and* $+45/-135°$ *(lower panel).*
(c) Stereo models of the equilibrium loop geometries color coded to correspond to the J-factor values in (b).

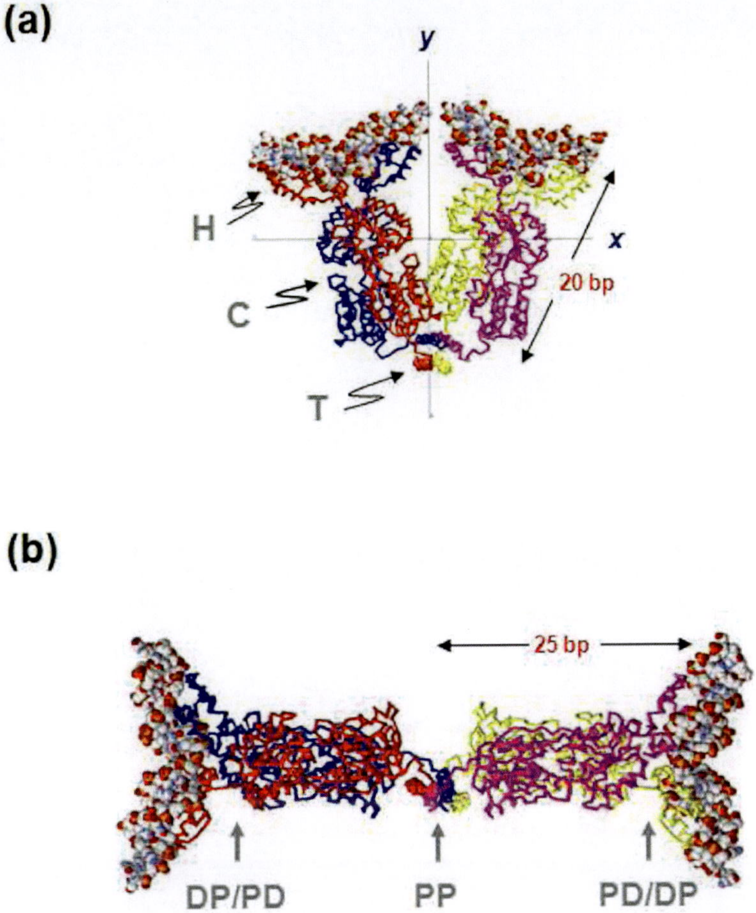

FIG. 6. *(a) Crystallographic structure of the "v"-shaped lac repressor complexed with a symmetric operator sequence (PDB accession number 1LBG) [37], shown along the z axis. The α-carbon trace of each LacR monomer is rendered in a separate color with the tetramerization domain indicated by "T".*
(b) A hypothetical structure of lac repressor in its extended conformation, generated from the "v"-shaped structure shown in (a) by increasing the interior angle from 40° to $\sim 180°$.

SL ("simple loop"). In addition, we investigated a novel hybrid of the v-shaped WT and WA loop classes that we named LB ("loop beside"), also shown in Figure 7.

Figure 7(b) compares the helical-phase dependence of the J factor for loops formed by the extended SL repressor conformation with that of the

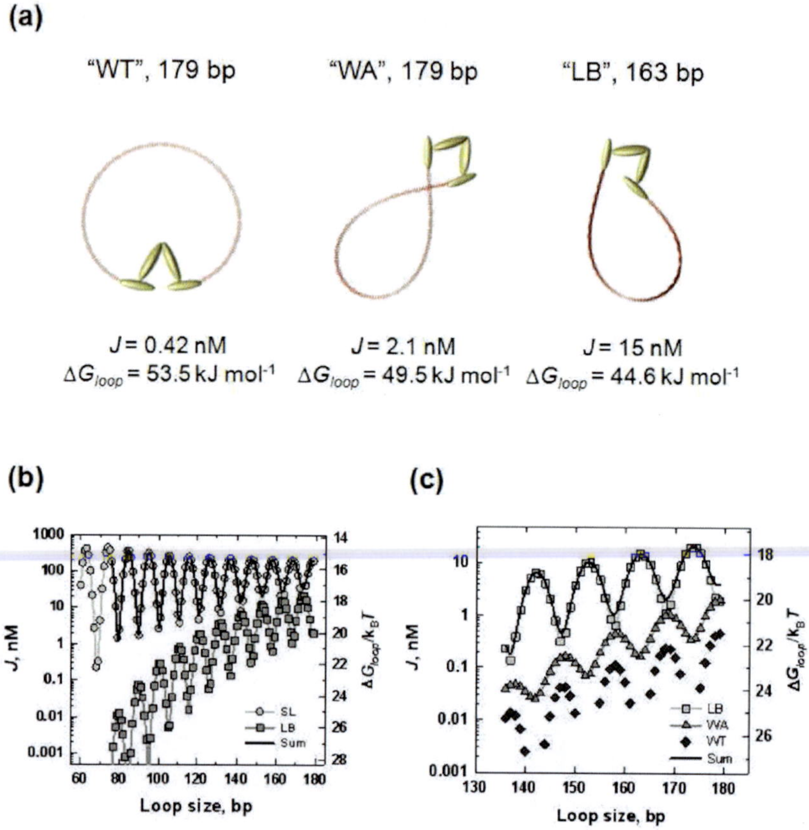

FIG. 7. *(a) Examples of looped DNA conformations mediated by the v-shaped LacR tetramer structure. The three v-shaped conformations that are solutions of the DNA-looping problem are "wrap toward" ("WT"), "wrap away" ("WA"), "loop beside" ("LB") [39]. Loop geometries for the most-probable topoisomer are shown along with their respective J factors and ΔG_{loop} values.*
(b) Length dependence of J factors and ΔG_{loop} for the extended (SL) and LB tetramer conformations.
(c) Comparison of J factors and ΔG_{loop} for the three classes of loop conformations mediated by the v-shaped lac repressor. The sum of J values for extended and LB loops in (b) is nearly indistinguishable from the J-factor dependence for SL loops alone, indicating that the extended LacR conformation dominates all of the v-shaped forms for loops smaller than 180 bp.

most thermodynamically favorable of the v-shaped forms, LB. The results clearly show that SL loops are thermodynamically favored over the LB loop class by at least 5 kJ mol^{-1} (2 RT) for DNA lengths below 100 bp. Other v-shaped conformations fare much worse; Figure 7(c) shows a trend in which the WA and WT conformations are 2–4 RT greater in free energy

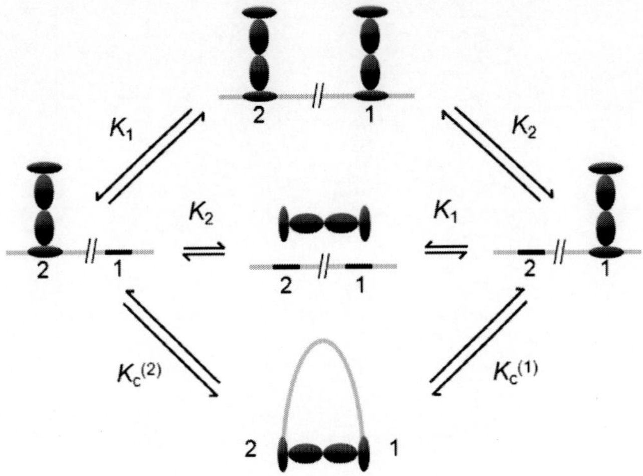

FIG. 8. *Coupled equilbria involving different LacR-DNA complexes. A LacR tetramer can bind to two operator sites with different affinities, for which K_1 and K_2 are apparent dissociation constants. Assuming site "1" is the primary site near the promoter of the lac operon, its occupancy by LacR prevents RNA polymerase from binding to the promoter, thereby blocking transcription of genes under its control. Here $K_c^{(1)}$ and $K_c^{(2)}$ denote the unimolecular association constants associated with DNA looping and are related to K_1, K_2, λ, and J [13].*

than LB loops. It is noteworthy that the J factor for LB loops is about 180 degrees out-of-phase with respect to the other loop types. Given the significantly higher free energy of all of the v-shaped conformations for loops smaller than 100 bp, which is the range of greatest experimental interest, we chose to simplify our analysis of loop-mediated repression by neglecting the contribution to the overall value of J from v-shaped LacR complexes.

The model for LacR-mediated gene repression is based on strict thermodynamic regulation and shown in Figure 8. We used this model in conjunction with the statistical-mechanical computation of the J factor to analyze data from two independent *in-vivo* studies of LacR-dependent repression by the groups of Müller-Hill [33] and Maher [34]. The latter case is especially interesting, because repression was measured both in the case of wild-type *E. coli* strains having normal levels of the architectural DNA-bending protein HU and in a mutant strain with HU deleted. *In-vivo* repression, R, was calculated as a ratio of reporter-gene expression levels in the respective absence and presence of a LacR operator sequence upstream of a promoter-proximal operator [12]. Considering the statistical weight of each of the unlooped and looped states of repressor bound to a dual-operator construct leads to a simple expression for gene repression,

$$R = \frac{E_{\text{noloop}}}{E_{\text{loop}}} = 1 + \frac{\lambda J P_t}{(K_1 + P_t)(K_2 + P_t)} \equiv 1 + \Gamma J \qquad (1)$$

where E_{noloop} and E_{loop} denote rates of gene expression in the absence of DNA looping (*i.e.*, deletion of the auxiliary operator indicated by site 2 in Figure 8), and that in its presence, respectively. In Eq. 1 P_t is the LacR-tetramer concentration in the cell and K_1 and K_2 are equilibrium dissociation constants of LacR for the primary and auxiliary operators, respectively. The dimensionless parameter λ mainly accounts for possible allosteric effects when one LacR tetramer associates with two DNA sites, and was set equal to unity because formation of the bidentate LacR-operator complex is non-cooperative [41]. The factor Γ therefore contains all information concerning protein-DNA association exclusive of the looping contribution. The J factor is an implicit function of DNA and protein elasto-mechanical parameters that affect the stability of looped conformations. Fitting experimental repression data as a function of operator spacing (Figure 9) with our model yields values for DNA bending and torsional rigidities, DNA helical repeat, and a composite value for bending and torsional rigidity of the protein assembly.

Best-fit values of DNA- and protein-flexibility parameters (Table 1) were identical for the two wild-type data sets within experimental uncertainty and show that DNA bending rigidity (*i.e.*, persistence length) is reduced about 40% *in vivo* relative to its canonical solution value. When we compared the fitted values for wild-type cells with those for repression in cells deleted for the genes encoding *E. coli* HU protein, DNA bending rigidity rises to the value measured for DNA free in solution at high ionic strength [42]. To our knowledge, these results provide the first quantitative estimate of HUs effect on DNA flexibility *in vivo*.

5. Implications for biological regulation. The most frequently discussed biological role for DNA looping is to raise the local concentration of a regulatory protein in the vicinity of a specific site [14, 43]. Results shown in Figure 10 make this argument quantitative: DNA looping in HU-containing wild-type cells boosts the local LacR concentration (J value) at the primary operator (O_1) from its bulk value of 0.017 μM to between 0.28 and 2.6 μM. This effect raises the occupancy of the primary operator, the fraction of primary operator sequences bound by LacR, from 0.79 to a narrow range between 0.985 and 0.998 with virtually no dependence on helical phasing (Figure 10, upper panel). Such pronounced enhancement of operator occupancy has the consequence of decreasing the expression rate of β-galactosidase (molecules per hour per cell) [33, 44] from 1,300 to a range of 12 to 90, in excellent agreement with direct *in-vivo* measurements [45, 46].

For a two-operator system, occupancy of the primary operator involves a looped state and two unlooped states (Figure 8). To relate the enhanced operator occupancy and gene repression to DNA looping, we calculated

TABLE 1

Best-fit Values of Adjustable Parameters for LacR-loop-mediated Repression Data.

Data Set	No. of data points, N_d	$\mathbf{\Gamma}^c$, x 10^2	Fitting error [d]	Best-fit values			
				Persistence length, bp	Torsional rigidity, 10^{-19} erg·cm	Helical repeat, bp·turn^{-1}	Protein flexibility, deg.
Müller et al. [33]	51	1.17[a]	1.0	95 (±1)	1.1 (±0.1)	11.60 (±0.01)	20.7 (±0.5)
Becker et al. [34] WT	26	4.66[b]	1.1	95 (±3)	0.7 (±0.1)	11.08 (±0.04)	19.0 (±1)
Becker et al. [34] Δ HU	25	4.66[b]	1.0	128 (±2)	0.8 (±0.1)	10.95 (±0.03)	16.0 (±1)

[a,b] *An identical auxillary-operator sequence, O_{id}, was used in the experiments of references [33] and [34] with LacR-dissociation constant $K_2 = 0.036$ nM. However, the two sets of experiments differ significantly with respect to primary operators and in-vivo LacR concentration. In [33][a]: $K_1 = 0.36$ nM (O_1) and $P_t = 85$ nM, whereas in [34][b], $K_1 = 4.4$ nM (O^2) and $P_t = 17$ nM. The five-fold difference in LacR levels between Müller et al.'s experiments and those of Becker et al. are due to the former's use of a LacR expression system that has a stronger promoter than that of the wild-type LacR gene.*

[c] $\Gamma = \frac{\lambda P_t}{(K_1+P_t)(K_2+P_t)}$, *with* $\lambda = 1$ *for LacR.*

[d] *The fitting error is calculated according to* $1/N_d\{\sum_{i=1}^{N_d}[(\log R_{comp,i} - \log R_{exp,i})/\delta \log R_{exp,i}]^2\}^{1/2}$ *where* $R_{comp,i}$, $R_{exp,i}$ *denote the computed and experimental enhanced-repression values, respectively, and* $\delta \log R_{exp,i} = \delta_i / R_{exp,i}$ *with* δ_i *the reported experimental error for the i-th data point.*

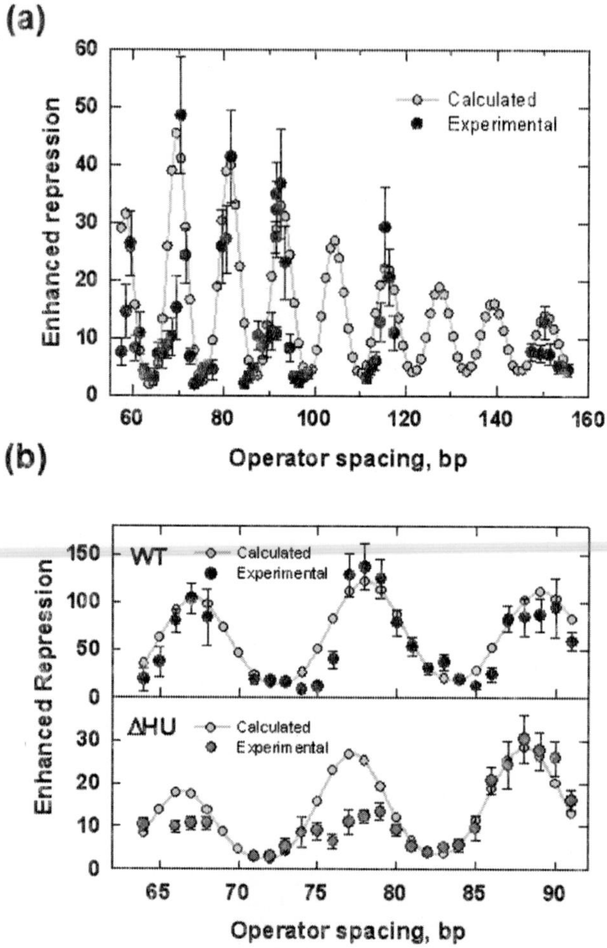

FIG. 9. *(a) Analysis of experimental in-vivo LacR repression data of Müller et al. [33] using a thermodynamic model that accounts for DNA looping (see Figure 8). Optimized fit to the data is shown applying the loop-closure theory to the extended tetramer conformation exclusively. The best-fit value for the DNA persistence length is* 95 ± 1 *bp.*
(b) Experimental data of Becker et al. [34] for LacR-mediated repression in wild-type E. coli and a mutant E .coli strain lacking HU protein (ΔHU). The best-fit DNA persistence length is 95 ± 3 *bp for the wild-type strain and* 128 ± 2 *bp for the ΔHU background.*

the loop yield (Figure 10, bottom panel), which is the proportion of looped states among all possible states. The loop yield directly correlates with the J factor, operator occupancy, and enhanced gene repression as demonstrated by their identical dependence on DNA helical phase. Furthermore,

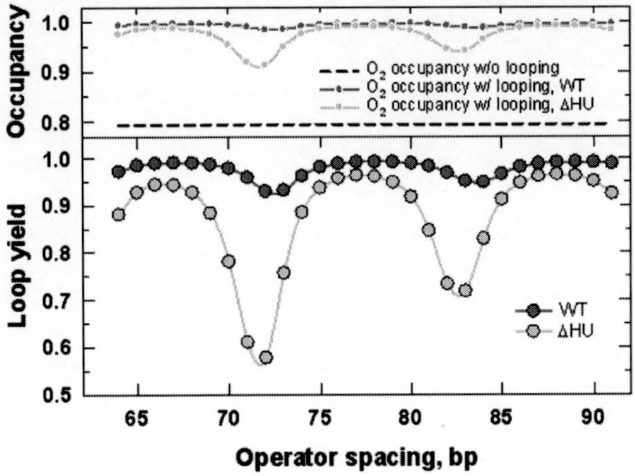

FIG. 10. *Predicted repressor occupancy of the primary operator and loop yield as a function of operator spacing corresponding to the data of Becker et al. [34] Upper panel: the proportion of the primary operator site (O_2) bound by LacR for wild-type and HU-deletion strains computed from their respective best-fit parameters using the formula* $1 - E_{loop}$. *The occupancy of the primary operator in the absence of the looping contribution is shown for reference (dashed line). Lower panel: the loop yield or fraction of DNA molecules in the looped state, determined for both wild-type and HU-deletion strains.*

the high loop yield (0.93–0.99) confirms that enhanced gene repression is almost exclusively attributable to DNA looping.

In the absence of HU, an increase primarily in effective DNA bending rigidity reduces measured gene repression by up to twelve-fold, depending on operator spacing, with an average reduction of 5.6-fold. The effect of HU is also apparent from decreases in J factor, operator occupancy, and loop yield for the HU-deficient E. coli strain compared with wild-type. These comparisons quantitatively confirm HUs putative role in facilitating the formation of small DNA loops *in vivo*.

6. Summary and concluding remarks. We describe here a quantitative analysis of DNA looping that connects the global mechanical properties of DNA and DNA-binding proteins through a wormlike chain model. Our method computes the minimum elastic-energy geometry of a DNA loop with specified end constraints and evaluates the statistical-mechanical partition function of the loop based on a harmonic approximation. Computing the J factor with this method is more than 10^4 times faster than Monte Carlo simulation; thus, J-factor calculations and fitting of experimental data, in particular, can be carried out efficiently.

As one of the most thoroughly studied examples of DNA looping both *in vitro* and *in vivo*, the *lac*-repressor gene regulatory system provides an excellent test case for the theory. We analyzed two independent sets of *in-vivo* data for gene repression as a function of helical spacing between two *lac* operator sites. Our results suggest that over the experimental range of loop sizes, DNA looping is well accounted for by the extended form of the repressor tetramer, exclusively. Moreover, the presence of the architectural DNA-bending protein HU in wild-type *E. coli* cells facilitates loop-mediated regulation mainly through an effective 40-percent reduction in DNA bending stiffness relative to its canonical solution value. Loop-mediated repression is strongly attenuated in the case of HU-deletion strains and our theoretical analysis gives a corresponding best-fit value of the DNA persistence length in the absence of HU that is close to *in-vitro* values measured in 1 M Na^+. Thus, HU appears to play a crucial role, along with other DNA-bending proteins, in regulating DNA-loop-dependent mechanisms in *E. coli.*

We dedicate this paper to the memory of Nick Cozzarelli, whose work in the last fifteen years of his career was increasingly devoted to connecting the *in-vitro* and *in-vivo* behavior of biological systems. Nick was virtually unique among his peers in keenly applying rigorous mathematical modeling to macromolecular systems while simultaneously keeping the biological aspects of a problem in sharp focus. His intellectual approach and scientific style provided strong inspiration for the work described here. Those of us who choose to follow the path that Nick blazed owe much to his legacy.

REFERENCES

[1] E.D. Ross, P.R. Hardwidge, and L.J. Maher, HMG proteins and DNA flexibility in transcription activation. Mol. Cell. Biol. **21**: 6598–6605, 2001.

[2] J. Yan, R. Kawamura, and J.F. Marko, Statistics of loop formation along double helix DNAs. Phys. Rev. E. Stat. Nonlin. Soft. Matter. Phys. **71**: 061905, 2005.

[3] G.J. Gemmen, R. Millin, and D.E. Smith, DNA looping by two-site restriction endonucleases: heterogeneous probability distributions for loop size and unbinding force. Nucleic Acids Res. **34**: 2864–2877, 2006.

[4] G.J. Gemmen, R. Millin, and D.E. Smith, Dynamics of single DNA looping and cleavage by Sau3AI and effect of tension applied to the DNA. Biophys. J. **91**: 4154–4165, 2006.

[5] L.E. Catto, S. Ganguly, S.E. milsom, A.J. Welsh, and S.E. Halford, Protein assembly and DNA looping by the FokI restriction endonuclease. Nucleic Acids Res. **34**: 1711–1720, 2006.

[6] D. Swigon, B.D. Coleman, and W.K. Olson, Modeling the Lac repressor-operator assembly: The influence of DNA looping on Lac repressor conformation. Proc. Natl. Acad. Sci. USA **103**: 9879–9884, 2006.

[7] F. Vanzi, C. Broggio, L. Sacconi, and F.S. Pavone, Lac repressor hinge flexibility and DNA looping: Single molecule kinetics by tethered particle motion. Nucleic Acids Res. **34**: 3409–3420, 2006.

[8] B. van den Broek, F. Vanzi, D. Normanno, F.S. Pavone, and G.J. Wuite, Real-time observation of DNA looping dynamics of Type IIE restriction enzymes NaeI and NarI. Nucleic Acids Res. **34**: 167–174, 2006.

[9] L.E. Catto, S.R. Bellamy, S.E. Retter, and S.E. Halford, Dynamics and consequences of DNA looping by the FokI restriction endonuclease. Nucleic Acids Res. **36**: 2073–2081, 2008.

[10] N.J. Agrawal, R. Radhakrishnan, and P.K. Purohit, Geometry of mediating protein affects the probability of loop formation in DNA. Biophys. J. **94**: 3150–3158, 2008.

[11] D. Normanno, F. Vanzi, and F.S. Pavone, Single-molecule manipulation reveals supercoiling-dependent modulation of lac repressor-mediated DNA looping. Nucleic Acids Res., 2008.

[12] Y. Zhang, A.E. McEwen, D.M. Crothers, and S.D. Levene, Statistical-mechanical theory of DNA looping. Biophys. J. **90**: 1903–1912, 2006.

[13] Y. Zhang, A.E. McEwen, D.M. Crothers, and S.D. Levene, Analysis of in-vivo LacR-mediated gene repression based on the mechanics of DNA looping. PLoS ONE **1**: e136, 2006.

[14] R. Schleif, DNA looping. Annu. Rev. Biochem. **61**: 199–223, 1992.

[15] K. Rippe, P.H. von Hippel, and J. Langowski, Action at a distance - DNA-looping and initiation of transcription. Trends Biochem. Sci. **20**: 500–506, 1995.

[16] K. Rippe, Making contacts on a nucleic acid polymer. Trends Biochem. Sci. **26**: 733–740, 2001.

[17] D. Shore and R.L. Baldwin, Energetics of DNA twisting. I. Relation between twist and cyclization probability. J. Mol. Biol. **170**: 957–981, 1983.

[18] D.M. Crothers, J. Drak, J.D. Kahn, and S.D. Levene, DNA bending, flexibility, and helical repeat by cyclization kinetics. Methods Enzymol **212**: 3–29, 1992.

[19] Y. Zhang, and D.M. Crothers, Statistical mechanics of sequence-dependent circular DNA and its application for DNA cyclization. Biophys. J. **84**: 136–153, 2003.

[20] V.A. Bloomfield, D.M. Crothers, and I.J. Tinoco, 2000. Nucleic Acids: Structures, Properties and Functions. University Science Books, Herndon, VA. 800.

[21] J. Shimada, and H. Yamakawa, Ring closure probabilities for twisted wormlike chains: Applications to DNA. Macromolecules **17**: 689–698, 1984.

[22] D. Shore and R.L. Baldwin, Energetics of DNA twisting. II. Topoisomer analysis. J. Mol. Biol. **170**: 983–1007, 1983.

[23] S.D. Levene and D.M. Crothers, Topological distributions and the torsional rigidity of DNA. A Monte Carlo study of DNA circles. J. Mol. Biol. **189**: 73–83, 1986.

[24] K.V. Klenin, A.V. Vologodskii, V.V. Anshelevich, V. Klishko, A.M. Dykhne, and M.D. Frank-Kamenetskii, Variance of writhe for wormlike DNA rings with excluded volume. J. Biomol. Struct. Dyn. **6**: 707–714, 1989.

[25] D.S. Horowitz and J.C. Wang, Torsional rigidity of DNA and length dependence of the free energy of DNA supercoiling. J. Mol. Biol. **173**: 75–91, 1984.

[26] M.A. Watson, D.M. Gowers, and S.E. Halford, Alternative geometries of DNA looping: an analysis using the SfiI endonuclease. J. Mol. Biol. **298**: 461–475, 2000.

[27] H. Benjamin and N. Cozzarelli, DNA-directed synapsis in recombination: Slithering and random collision of sites. The Robert A. Welch Foundation Conferences on Chemical Research **29**: 107–126.

[28] H. Tsen and S.D. Levene, Supercoiling-dependent flexibility of adenosine-tract-containing DNA detected by a topological method. Proc. Natl. Acad. Sci. USA **94**: 2817–2822, 1997.

[29] G.D. van Duyne, A structural view of Cre-loxP site-specific recombination. Annu. Rev. Biophys. Biomol. Struct. **30**: 87–104, 2001.

[30] S.M. Law, G.R. Bellomy, P.J. Schlax, and M.T. Record, In-vivo thermodynamic analysis of repression with and without looping in lac constructs - estimates of free and local lac repressor concentrations and of physical proper-

ties of a region of supercoiled plasmid DNA *in vivo*. J. Mol. Biol. **230**: 161–173, 1993.

[31] L. Bintu, N.E. Buchler, H.G. Garcia, U. Gerland, T. Hwa, J. Kondev, and R. Phillips, Transcriptional regulation by the numbers: models. Curr. Opin. Genet. Dev. **15**: 116–124, 2005.

[32] L. Bintu, N.E. Buchler, H.G. Garcia, U. Gerland, T. Hwa, J. Kondev, T. Kuhlman, and R. Phillips, Transcriptional regulation by the numbers: applications. Curr. Opin. Genet. Dev. **15**: 125–135, 2005.

[33] J. Müller, S. Oehler, and B. Müller-Hill, Repression of lac promoter as a function of distance, phase and quality of an auxiliary lac operator. J. Mol. Biol. **257**: 21–29, 1996.

[34] N.A. Becker, J.D. Kahn, and L.J. Maher, 3RD, Bacterial repression loops require enhanced DNA flexibility. J. Mol. Biol. **349**: 716–730, 2005.

[35] N.A. Becker, J.D. Kahn, and L.J. Maher, 3RD, Effects of nucleoid proteins on DNA repression loop formation in Escherichia coli. Nucleic Acids Res. **35**: 3988–4000, 2007.

[36] A.M. Friedman, T.O. Fischmann, and T.A. Steitz, Crystal structure of lac repressor core tetramer and its implications for DNA looping. Science **268**: 1721–1727, 1995.

[37] M. Lewis, G. Chang, N.C. Horton, M.A. Kercher, H.C. Pace, M.A. Schumacher, R.G. Brennan, and P. Lu, Crystal structure of the lactose operon repressor and its complexes with DNA and inducer. Science **271**: 1247–1254, 1996.

[38] G.C. Ruben and T.B. Roos, Conformation of lac repressor tetramer in solution, bound and unbound to operator DNA. Microsc. Res. Tech. **36**: 400–416, 1997.

[39] R.A. Mehta and J.D. Kahn, Designed hyperstable lac repressor·DNA loop topologies suggest alternative loop geometries. J. Mol. Biol. **294**: 67–77, 1999.

[40] L.M. Edelman, R. Cheong, and J.D. Kahn, Fluorescence resonance energy transfer over 130 basepairs in hyperstable lac repressor-DNA loops. Biophys. J. **84**: 1131–1145, 2003.

[41] P.H. von Hippel, On the molecular bases of the specificity of interaction of transcriptional proteins with genome DNA. In Biological Regulation and Development, R. F. Goldberger, editor. Plenum, New York, pp. 279–347, 1979.

[42] D. Pörschke, Persistence length and bending dynamics of DNA from electrooptical measurements at high salt concentrations. Biophys. Chem. **40**: 169–179, 1991.

[43] M.C. Mossing and M.T. Record, Jr., Upstream operators enhance repression of the lac promoter. Science **233**: 889–892, 1986.

[44] J.M. Vilar and L. Saiz, DNA looping in gene regulation: from the assembly of macromolecular complexes to the control of transcriptional noise. Curr. Opin. Genet. Dev. **15**: 136–144, 2005.

[45] S. Sasse-Dwight and J.D. Gralla, Probing co-operative DNA-binding in vivo. The lac O1:O3 interaction. J. Mol. Biol. **202**: 107–119, 1988.

[46] J. Yu, J. Xiao, X. Ren, K. Lao, and X.S. Xie, Probing gene expression in live cells, one protein molecule at a time. Science **311**: 1600–1603, 2006.

FOUR-WAY HELICAL JUNCTIONS IN DNA MOLECULES

DAVID M.J. LILLEY*

Four-way junctions in DNA. Four-way (Holliday) junctions [1] are branchpoints in DNA where four helices are interconnected by the mutual exchange of strands [2] (Figure 1). They are important intermediates in the rearrangement of DNA, and may be created by strand invasion in recombination, double-strand break repair and fork reversal during replication. Additionally, four-way junctions are the central intermediate of the tyrosine recombinases such as Cre [3]. Four-way junctions (and other branched structures) are also important structural and functional elements in RNA, but this aspect will not be considered further here. Nevertheless, many of the conformational principles deduced for the DNA junction apply equally to RNA [4]. This short review focuses on recent developments in understanding the structure and dynamics of DNA four-way junctions.

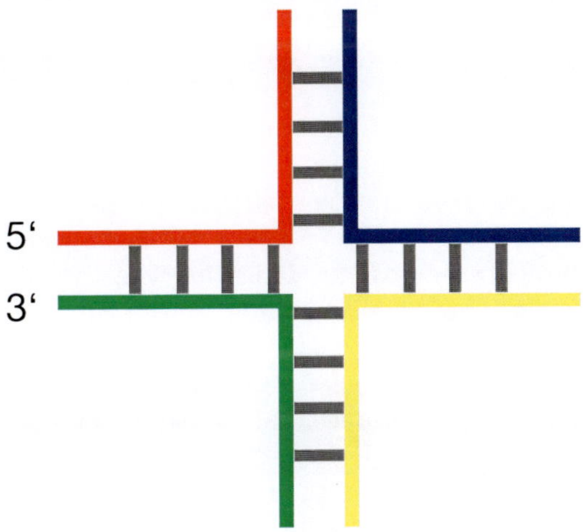

FIG. 1. *Four-way helical junctions. Four-way junctions are the points of connection between four double-stranded helices, connected by the exchange of strands. For DNA, these are often called Holliday junctions. They are conventionally drawn with the strand polarity indicated, presenting the major groove side of the junction.*

The stacked X-structure of the four-way DNA junction. The structure of the four-way DNA junction has been studied for almost 25 years [5–10]. In the absence of added metal ions, the four-way junction adopts a (probably rather flexible) structure where the arms are directed

*Cancer Research UK Nucleic Acid Structure Research Group, MSI/WTB Complex, The University of Dundee, Dundee DD1 5EH, U.K. (d.m.j.lilley@dundee.ac.uk).

C.J. Benham et al. (eds.), *Mathematics of DNA Structure, Function and Interactions*,
The IMA Volumes in Mathematics and its Applications 150,
DOI 10.1007/978-1-4419-0670-0_11,

to the corners of a square [8, 11]. This is discussed further below. Upon addition of divalent metal ions junctions fold by the pairwise coaxial stacking of helical arms [2], termed the stacked X-structure (Figure 2). This was first demonstrated by electrophoretic [8] and fluorescence resonance energy transfer (FRET) experiments [9], but the detailed structure has been solved by X-ray crystallography more recently [12–14]. The structure can be considered to be formed by coaxial stacking of helices together with a rotation of the axes by 40–60° in a right-handed sense.

When the original structure of the stacked X-structure was determined [8, 9] the data new indicated that it must be antiparallel, in contrast to earlier assumptions and the virtually universal parallel depiction in text books. However the antiparallel structure was later confirmed by the crystallographic structures [12–14]. More recent single-molecule studies have shown that the probability of parallel forms of the junction with a lifetime greater than 1 ms is less than 2×10^{-5} [15].

The stacked X-structure has two-fold symmetry (compared to the four-fold symmetry of the open, extended form), dividing the component strands into two continuous strands each with a single axis, and two exchanging strands that pass between axes (Figure 3).

Stacking conformers of the four-way junction. There are two ways that a given junction can be folded into the stacked X-structure, depending on the choice of stacking partner (Figure 3). The two structures are stereochemically equivalent, but will usually differ in terms of the stacking interactions across the junction. Thus they will not be energetically equivalent, and hence most junctions exhibit a distinct conformational bias towards one conformer. The bias has never been found to be total however, and a significant population of the minor conformer has been found in all junctions studied. Miick et al. [16] studied the distribution of stacking conformers in three different four-way junctions using the chemical shift of ^{15}N-labelled thymine and distance distributions calculated from time-resolved FRET experiments. These experiments showed that the distribution of stacking conformers varied considerably with junction sequence. On the other hand, at least one sequence (termed junction 7) was identified where the populations of the two conformers were very close to being equal [17]. In principle it should be possible to calculate conformer bias given a sufficiently good quantitative understanding of the interactions around the branchpoint, but this has not been achieved to date. It was shown that sequence changes two or three nucleotides distant from the point of strand exchange could influence conformer bias, so the problem is quite complex. Crystallographic studies have revealed sequence-specific hydrogen-bonding contacts can form across the junction [13], resulting in increased stability for particular sequences. The potential for this has been further demonstrated by the observation of halogen bonds stabilizing junctions incorporating halogen-substituted nucleotides [18].

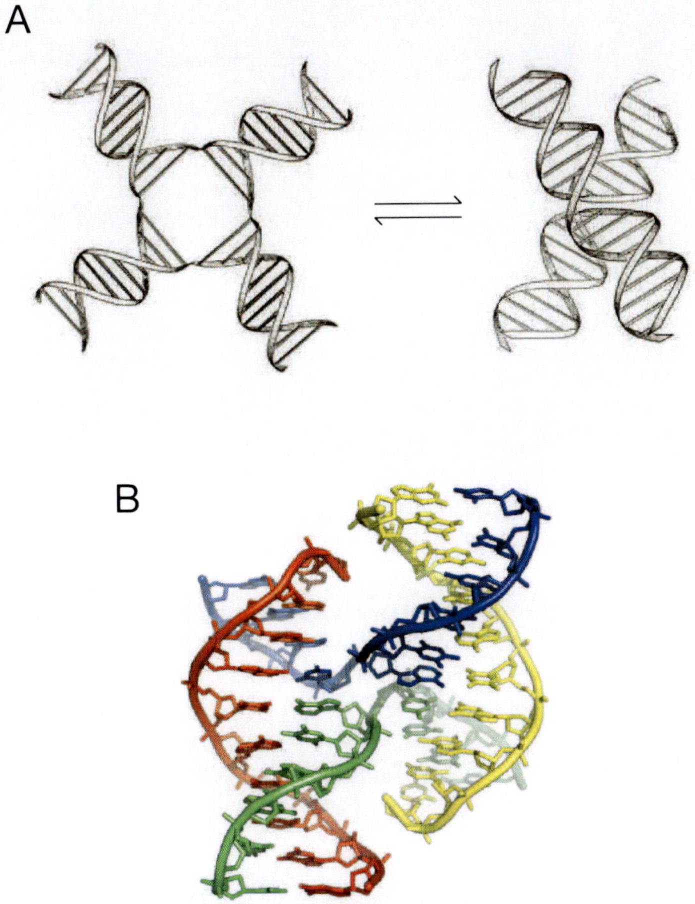

FIG. 2. *The stacked X-structure of the DNA junction.*

A. The structure of the four-way DNA junction is dependent upon the presence of metal ions. In the absence of ions, electrostatic repulsion results in a square structure with an open center (left), in which the four helical arms point to the four corners of a square. On lowering the electrostatic interactions between charged phosphate groups by addition of metal ions, the junction can undergo a folding process by the pairwise coaxial of helical arms to form the stacked X-structure (right).

B. The structure of the four-way DNA junction in the crystal [13]. The junction adopts a right-handed stacked X-structure. The red and yellow strands are continuous strands, while the blue and green strands exchange at the centre. The continuous strands run antiparallel. The view shown is the major groove side of the junction.

Conformer exchange. The existence of two populations of stacking conformers suggested that the two forms might interconvert. Early studies using *Mbo*II cleavage of junctions confirmed the existence of conformer exchange that was rapid relative to the timescale of gel electrophoresis [17, 19]. This was also demonstrated by NMR, showing that a minor con-

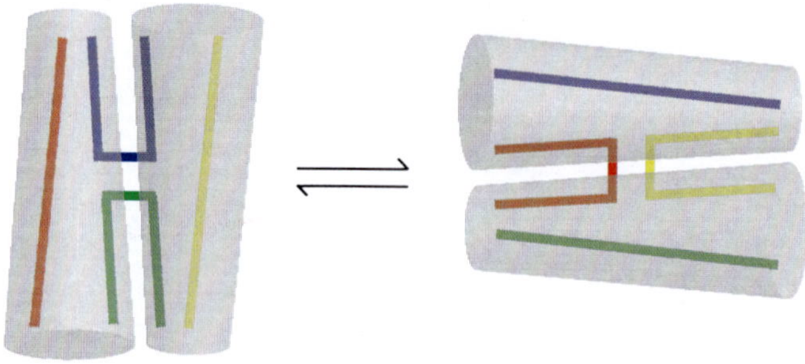

FIG. 3. *Conformer exchange in a four-way junction. There are two alternative stacking conformers for a four way junction, that differ in the choice of stacking partners. The nature of the strands differs in the two conformers – e.g. red and yellow strands are continuous in the conformer shown on the left, but are exchanging in that on the right.*

former existed for one junction that was in rapid exchange on the NMR timescale [20].

Since there is no way to synchronize conformer exchange, it was not possible to confirm this process directly before the development of single-molecule methods. However, FRET studies of single junction molecules showed the exchange between junction conformers [21], occurring at a surprisingly slow rate in the presence of 50 mM Mg^{2+} ions (Figure 4). These experiments further showed that the rate of conformer exchange became significantly faster as the ionic strength of the solution was lowered [15]. This indicates that the transition state is destabilized by metal ions, suggesting that it resembles the open form of the junction that exists in the absence of metal ions. This of course makes structural sense, since it is hard to imagine how the required exchange of stacking partners could occur without substantial opening of the junction. The lifetime of the open species was estimated to be < 1 ms.

Very recently, conformer exchange has been studied by the application of stretching force [22]. By measuring the gradient of the rate of conformer exchange as a function of applied force it was possible to determine the position of the transition state along the reaction coordinate. Using relatively low forces (< 4 pN) it was determined that the open structure is a shallow intermediate, flanked by transition states on either side. The application of stretching force tilts the entire energy landscape, thereby altering the current rate determining step.

Branch migration. In homologous junctions there is no fixed position for the branchpoint, which can therefore change position due to the exchange of basepairing. Branchpoints can migrate over long distances, and the process occurs as a step-wise random walk. A single step of branch migration requires the breakage of two basepairs at the point of strand exchange located on diametrically opposite arms, rotation, and pairing to

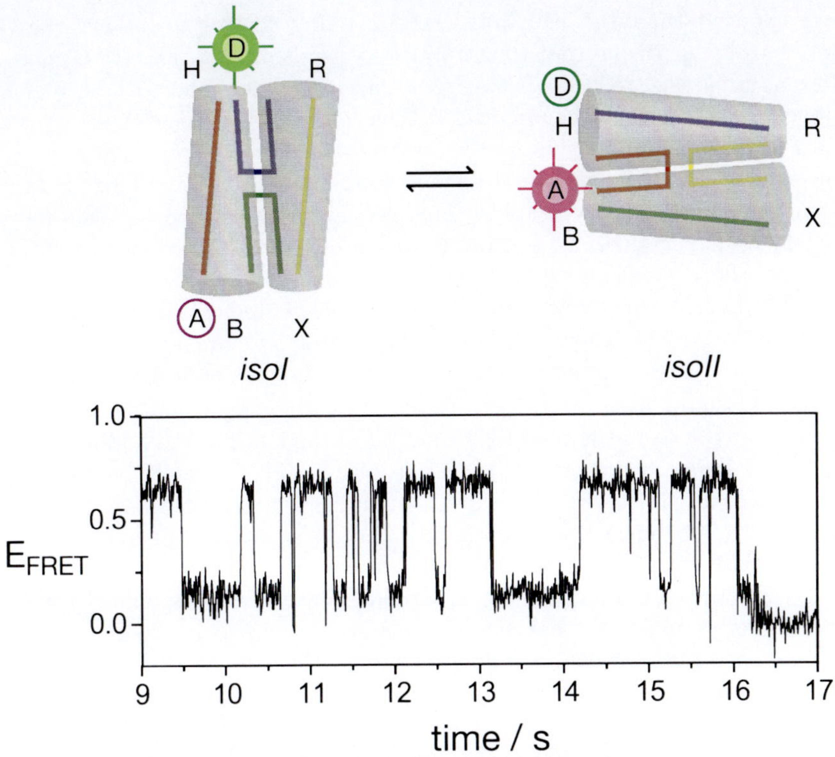

FIG. 4. *Conformer exchange observe in single junction molecules by FRET. Donor and acceptor fluorophores have been attached to the 5′-termini of two arms, so that there is low FRET efficiency in the conformer on the left, but high FRET efficiency in the conformer on the right. A record of FRET efficiency (E_{FRET}) as a function of time for a single junction molecule shows repeated interconversion between the low efficiency ($E_{FRET} = 0.1$) and the high efficiency ($E_{FRET} = 0.6$) states [21].*

form the terminal basepairs on the other two arms. This would be expected to require substantial opening of the junction, much as envisaged for conformer exchange. In ensemble experiments, Panyutin and Hsieh (1994) observed that the rate of branch migration is extremely sensitive to the presence of divalent metal ions, being greatly accelerated as the ionic strength was reduced. This is consistent with a requirement for an opening of the structure.

The process was studied in single junction molecules that could undergo a single step of branch migration [23]. Individual steps of branch migration could be observed, together with the conformer exchange occurring in each form of the junction. It was concluded that conformer exchange and branch migration share a common intermediate that is likely to resemble the open state deduced for free junctions in the absence of divalent metal ions.

Electrostatic interactions. DNA is a polyelectrolyte and it is therefore hardly surprising that the structure depends on the presence of metal ions. Simple electrostatic calculations show that there should be a region of very high electrostatic potential at the center of the stacked X-structure, and metal ions will be required to neutralize the charge repulsion. In the absence of this, the repulsive energy dominates and the extended open structure will be more stable. The overall shape of the low-salt form would be expected to be square as this provides the maximal extension around the centre, and would therefore probably minimize electrostatic repulsion. But this would not necessarily be planar; given that the two sides of the extended structure are inequivalent (there are major and minor groove faces) this is indeed unlikely, and the structure is probably to some degree pyramidal.

Steady-state ensemble FRET analysis of the ion-induced folding of the four-way junction indicates that the process is well described by a simple two-state model [24, 25]. The experimental data can be fitted with a K_A^{app} for ion divalent metal ion binding of $\sim 10^5$ M^{-1}, and a Hill coefficient of $n = 1$. This is consistent with a model in which folding is induced by the non-cooperative binding of one metal ion. However this is unlikely to be site bound as an inner-sphere complex for two reasons. First, folding can be induced using monovalent metal ions alone, and second, no metal ion bound at the point of strand exchange has been identified by crystallography [13, 14].

Nevertheless, it is likely that there is a high occupancy of rapidly exchanging metal ions at the center of the junction, as outer-sphere complexes. Uranyl-induced photocleavage experiments indicated the presence of an ion binding site near the point of strand exchange in the folded junction [26]. The two phosphates that define the point of strand exchange on the exchanging strands are only separated by 6.5 Å, and there is a box of six phosphates comprising these two plus those 5′ and 3′ to them on each strand (Figure 5). Replacement of any of these phosphates by the electrically-neutral methyl phosphonate groups increases the stability of the folded form of the junction at lower ionic concentrations [27]. However, these effects can be rather more subtle. When methyl phosphonate groups were substituted for the central phosphate on one of the continuous strands of a junction (in its major conformer) it was found that two stacking conformers co-existed in solution without interconversion [28]. Upon deeper analysis these proved to contain the alternative diastereomers of the substituted phosphate group. It is therefore probable that the location of the methyl group determines the relative stability of the alternative stacking conformers due to differential interactions with metal ions at the point of strand exchange. Substitution of the *pro*-R oxygen atom stabilizes the folded junction, and therefore represents the simple electrostatic effect. But substitution of the *pro*-S oxygen atom destabilizes the stacked X-structure, probably arising from steric interference with ionic interactions.

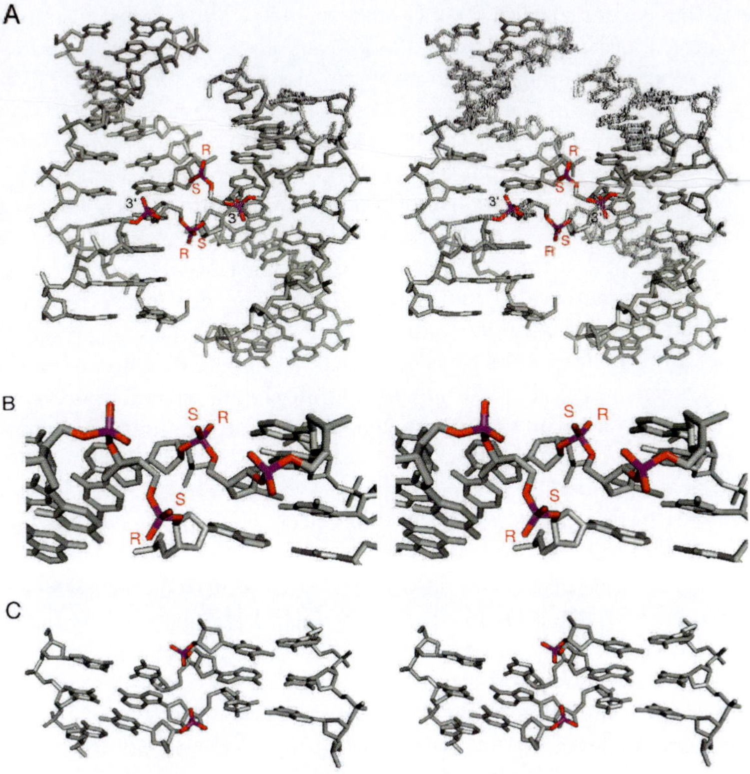

FIG. 5. *Parallel-eye stereo images of the disposition of phosphate groups around the point of strand exchange in the four-way DNA junction.*

A. The minor-groove face of a crystal structure [13] of a DNA junction, with the central and 3′ phosphate groups highlighted. The pro-R and pro-S oxygen atoms of the central phosphate groups are labelled R and S respectively.

B. Expanded view of the central region of the minor-groove face.

C. The major groove face, with the phosphate groups 5′ to the point of strand exchange highlighted.

Recognition by proteins. The structure of four-way DNA junctions is recognized by a number of proteins, especially the group of nucleases that introduce specific pairs of cleavages in order to resolve the junction into nicked duplex species [29]. Junction-resolving enzymes have been isolated from many organisms, from bacteria and their phage, yeast, archaea, through to mammalian cells and their viruses [30, 31]. The enzymes bind four-way DNA junctions as dimers, with affinities close to $K_d = 1$ nM. Binding is highly structure selective; the complex of a dimer of a given junction-resolving enzyme and a four-way DNA junction is normally not disrupted by a 1000-fold excess of duplex DNA of the same sequence. Yet despite recognizing junction structure, all the junction-resolving enzymes distort the very structure that they recognize. Many substantially open the structure of the junction to different degrees — the most extreme case is Cce1, where the junction adopts an open-square structure [32] like that of the free junction in the absence of metal ions. In fact, it is likely that the binding of Cce1 results in a significant disruption of base pairing adjacent to the point of strand exchange in addition to the global change [33].

The structural distortion of the DNA junctions induced by enzyme binding may be related to the their function, helping to assure a well-ordered resolution of the junction. Detailed kinetic analysis of the cleavage of cruciform structures (a four-way junction in a different guise [34]) revealed that there is an acceleration of the second strand cleavage (by a factor of 100 in the case of RuvC), so that it immediately follows cleavage of the first strand [35–37]. We suggested that the intact junction is strained in the complex containing the intact junction, and the first cleavage event is followed by a relaxation that accelerates the second cleavage.

For a long time the fundamental process by which these enzymes recognize the structure of branched DNA has been unclear. Crystal structures have been determined for most of the proteins in isolation [38–46], but diffracting crystals of the complexes with DNA junctions have proved elusive. However, that has changed recently with the solution of crystal structures for complexes of four-way DNA junctions with the phage enzymes T7 endonuclease I [48] and T4 endonuclease VII [47]. These recognize the structure of the junction in rather different ways.

T4 endonuclease VII [47] distorts the junction into an almost planar open X-shape. The DNA is bound via its minor groove face to a fairly flat surface of the enzyme, making contacts with all four helical arms. By contrast, when a junction is bound by T7 endonuclease I [48] the helical arms remain coaxial (though the axes are rotated by 130°), conferring a much more three-dimensional aspect to its shape. The protein dimer presents two 30 Å-long hemi-cylindrical clefts that are mutually perpendicular (Figure 6). The DNA arms are bound along the length of these electropositive channels, and thus the enzyme is selective for a DNA structure that can adopt the near-perpendicular geometry of the junction [48, 49]. The active sites of the enzyme [50] are also located in the grooves, where two metal ions [51] provide the hydrolytic water molecules for nucleophilic attack on the scissile phosphate groups of the continuous strands in a mechanism [52] that is strikingly similar to that of many restriction enzymes.

Clearly these two enzymes have evolved rather different strategies for recognizing and binding four-way junctions, but both require substantial distortion of the DNA structure. This raises a question of whether the change in DNA structure is induced by the protein, or whether the protein binds to altered conformations that form transiently in the free junction. The force experiments have shown that the open structure is an intermediate with a finite lifetime [22], and perhaps this might represent the entry point for the protein. This is certainly easy to imagine for T4 endonucle-

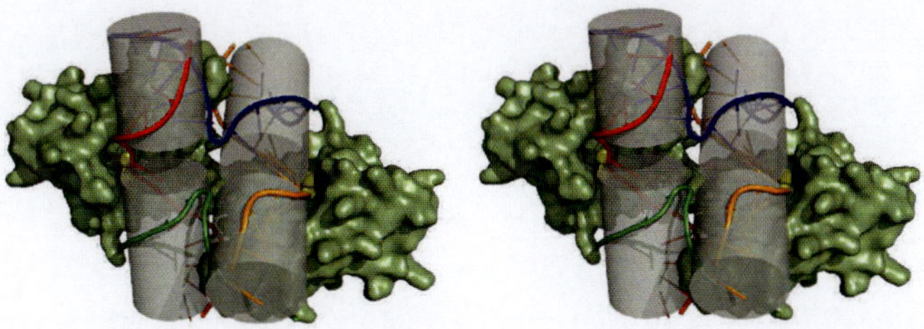

FIG. 6. *The crystal structure of a complex between the resolving enzyme T7 endonuclease I and a DNA junction [48], shown as a parallel-eye stereo pair. The positions of the helical arms of the junction are highlighted by the transparent cylinders. In this complex, pairwise coaxial alignment of the helices is preserved, although the stacking between them is not (the basepairs at the center are separated by $\sim$ 10 Å). The axes of the junction are almost perpendicular. The protein is shown as a molecular surface (green), and the two calcium ions of the active sites are shown yellow. Each pair of helices is accommodated in a 30 Å-long cleft formed by the dimeric protein.*

ase VII, but harder for T7 endonuclease I. Perhaps future single-molecule experiments will cast more light on these questions.

Conclusion. We have learned a lot about the structural and dynamic properties of the four-way DNA junction, and are now beginning to understand its interactions with enzymes too. The Holliday junction is important in so many facets of DNA rearrangement. It is important to uncover all the properties of this central intermediate, as the interactions with the resolving enzymes have taught us that all aspects of the process are intimately connected.

Nick Cozzarelli. This review is dedicated to the memory of Nick Cozzarelli. I will forever remember Nick as a towering intellect, a wicked sense of humor and a wonderful friend. I knew Nick for about 25 years. When he passed away it left a big hole, and I felt I lost a very special friend. But when I attended his memorial symposium in Berkeley in 2006 I came to realize that many people felt the same way. He was unique.

REFERENCES

[1] HOLLIDAY R. (1964), *A mechanism for gene conversion in fungi.* Genet. Res, **5**:282–304.

[2] LILLEY D.M.J. (2000), *Structures of helical junctions in nucleic acids.* Quart. Rev. Biophys., **33**:109–159.

[3] GOPAUL D.N., GUO F., AND VAN DUYNE G.D. (1998), *Structure of the Holliday junction intermediate in Cre-loxP site-specific recombination.* EMBO J., **17**:175–4187.

[4] DUCKETT D.R., MURCHIE A.I.H., AND LILLEY D.M.J. (1995), *The global folding of four-way helical junctions in RNA, including that in U1 snRNA.* Cell, **83**:1027–1036.

[5] KALLENBACH N.R., MA R.-I., AND SEEMAN N.C. (1983), *An immobile nucleic acid junction constructed from oligonucleotides.* Nature, **305**:829–831.

[6] Cooper J.P. and Hagerman P.J. (1987), *Gel electrophoretic analysis of the geometry of a DNA four-way junction.* J. Molec. Biol., **198**:711–719.

[7] Churchill M.E., Tullius T.D., Kallenbach N.R., and Seeman N.C. (1988), *A Holliday recombination intermediate is twofold symmetric.* Proc. Natl. Acad. Sci. USA, **85**:4653–4656.

[8] Duckett D.R., Murchie A.I.H., Diekmann S., von Kitzing E., Kemper B., and Lilley D.M.J. (1988), *The structure of the Holliday junction and its resolution.* Cell, **55**:79–89.

[9] Murchie A.I.H., Clegg R.M., von Kitzing E., Duckett D.R., Diekmann S., and Lilley D.M.J. (1989), *Fluorescence energy transfer shows that the four-way DNA junction is a right-handed cross of antiparallel molecules.* Nature, **341**:763–766.

[10] Cooper J.P. and Hagerman P.J. (1989), *Geometry of a branched DNA structure in solution.* Proc. Natl. Acad. Sci. USA, **86**:7336–7340.

[11] Clegg R.M., Murchie A.I.H., Zechel A., and Lilley D.M.J. (1994), *The solution structure of the four-way DNA junction at low salt concentration; a fluorescence resonance energy transfer analysis.* Biophys. J., **66**:99–109.

[12] Ortiz-Lombarda M., González A., Erijta R., Aymamí, J., Azorín F., and Coll M. (1999), *Crystal structure of a DNA Holliday junction.* Nature Struct. Biol., **6**:913–917.

[13] Eichman B.F., Vargason J.M., Mooers B.H.M., and Ho P.S. (2000), *The Holliday junction in an inverted repeat DNA sequence: Sequence effects on the structure of four-way junctions.* Proc. Natl. Acad. Sci. USA, **97**:3971–3976.

[14] Thorpe J.H., Gale B.C., Teixeira S.C., and Cardin C.J. (2003), *Conformational and hydration effects of site-selective sodium, calcium and strontium ion binding to the DNA Holliday junction structure d(TCGGTACCGA)4.* J. Molec. Biol., **327**:97–109.

[15] Joo C., McKinney S.A., Lilley D.M.J., and Ha T. (2004), *Exploring rare conformational species and ionic effects in DNA Holliday junctions using single-molecule spectroscopy.* J. Molec. Biol., **341**:739–751.

[16] Miick S.M., Fee R.S., Millar D.P., and Chazin W.J. (1997), *Crossover isomer bias is the primary sequence-dependent property of immobilized Holliday junctions.* Proc. Natl. Acad. Sci. USA, **94**:9080–9084.

[17] Grainger R.J., Murchie A.I.H., and Lilley D.M.J. (1998), *Exchange between stacking conformers in a four-way DNA junction.* Biochemistry, **37**:23–32.

[18] Voth A.R., Hays F.A., and Ho P.S. (2007), *Directing macromolecular conformation through halogen bonds.* Proc. Natl. Acad. Sci. USA, **104**:6188–6193.

[19] Murchie A.I.H., Portugal J., and Lilley D.M.J. (1991), *Cleavage of a four-way DNA junction by a restriction enzyme spanning the point of strand exchange.* EMBO J., **10**:713–718.

[20] Carlström G. and Chazin W.J. (1996), *Sequence dependence and direct measurement of crossover isomer distribution in model Holliday junctions using NMR spectroscopy.* Biochemistry, **35**:3534–3544.

[21] McKinney S.A., Déclais A.-C., Lilley D.M.J., and Ha T. (2003), *Structural dynamics of individual Holliday junctions.* Nature Struct. Biol., **10**:93–97.

[22] Hohng S., Zhou R., Nahas M.K., Yu J., Schulten K., Lilley D.M.J., and Ha T. (2007), *Fluorescence-force spectroscopy maps two-dimensional reaction landscape of the Holliday junction.* Science, In the press.

[23] McKinney S.A., Freeman A.D., Lilley D.M.J., and Ha T. (2005), *Observing spontaneous branch migration of Holliday junctions one step at a time.* Proc. Natl. Acad. Sci. USA, **102**:5715–5720.

[24] Clegg R.M., Murchie A.I.H., Zechel A., Carlberg C., Diekmann S., and Lilley D.M.J. (1992), *Fluorescence resonance energy transfer analysis of the structure of the four-way DNA junction.* Biochemistry, **31**:4846–4856.

[25] Fogg J.M., Kvaratskhelia M., White M.F., and Lilley D.M.J. (2001), *Distortion of DNA junctions imposed by the binding of resolving enzymes: A fluorescence study.* J. Molec. Biol., **313**:751–764.

[26] Møllegaard N.E., Murchie A.I.H., Lilley D.M.J., and Nielsen P.E. (1994), *Uranyl photoprobing of a four-way DNA junction: Evidence for specific metal ion binding.* EMBO J., **13**:1508–1513.

[27] Liu J., Déclais A.-C., and Lilley D.M.J. (2004), *Electrostatic interactions and the folding of the four-way DNA junction: Analysis by selective methyl phosphonate substitution.* J. Molec. Biol., **343**:851864.

[28] Liu J., Déclais A.-C., McKinney S.A., Ha T., Norman D.G., and Lilley D.M.J. (2005), *Stereospecific effects determine the structure of a four-way DNA junction.* Chem. Biol., **12**:217–228.

[29] Déclais A.C. and Lilley D.M.J. (2007), *New insight into the recognition of branched DNA structure by junction-resolving enzymes.* Curr. Opin. Struct. Biol., In the press.

[30] Lilley D.M.J. and White M.F. (2001), *The junction-resolving enzymes.* Nature Rev. Molec. Cell Biol., **2**:433–443.

[31] West S.C. (2003), *Molecular views of recombination proteins and their control.* Nature Rev. Mol. Cell Biol., **4**:435–445.

[32] White M.F. and Lilley D.M.J. (1997), *The resolving enzyme CCE1 of yeast opens the structure of the four-way DNA junction.* J. Molec. Biol., **266**: 122–134.

[33] Déclais A.-C. and Lilley D.M.J. (2000), *Extensive central disruption of a four-way junction on binding CCE1 resolving enzyme.* J. Molec. Biol., **296**:421–433.

[34] Lilley D.M.J. (1980), *The inverted repeat as a recognisable structural feature in supercoiled DNA molecules.* Proc. Natl. Acad. Sci. USA, **77**:6468–6472.

[35] Giraud-Panis M.-J.E. and Lilley D.M.J. (1997), *Near-simultaneous DNA cleavage by the subunits of the junction-resolving enzyme T4 endonuclease VII.* EMBO J., **16**:2528 – 2534.

[36] Fogg J.M., Schofield M.J., Déclais A.-C., and Lilley D.M.J. (2000), *The yeast resolving enzyme CCE1 makes sequential cleavages in DNA junctions within the lifetime of the complex.* Biochemistry, **39**:4082–4089.

[37] Fogg J.M. and Lilley D.M.J. (2001), *Ensuring productive resolution by the junction-resolving enzyme RuvC: Large enhancement of second-strand cleavage rate.* Biochemistry, **39**:16125–16134.

[38] Ariyoshi M., Vassylyev D.G., Iwasaki H., Nakamura H., Shinagawa H., and Morikawa K. (1994), *Atomic structure of the RuvC resolvase: A Holliday junction-specific endonuclease from E.coli.* Cell, **78**:1063–1072.

[39] Raaijmakers H., Vix O., Toro I., Golz S., Kemper B., and Suck D. (1999), *X-ray structure of T4 endonuclease VII: a DNA junction resolvase with a novel fold and unusual domain-swapped dimer architecture.* EMBO J., **18**: 1447–1458.

[40] Bond C.S., Kvaratskhelia M., Richard D., White M.F., and Hunter W.N. (2001), *Structure of Hjc, a Holliday junction resolvase, from Sulfolobus solfataricus.* Proc. Natl. Acad. Sci. USA, **98**:5509–5514.

[41] Nishino T., Komori K., Tsuchiya D., Ishino Y., and Morikawa K. (2001), *Crystal structure of the archaeal Holliday junction resolvase Hjc and implications for DNA recognition.* Structure, **9**:197–204.

[42] McGregor N., Ayora S., Sedelnikova S., Carrasco B., Alonso J.C., Thaw P., and Rafferty J. (2005), *The structure of Bacillus subtilis RecU Holliday junction resolvase and its role in substrate selection and sequence-specific cleavage.* Structure, **13**:1341–1351.

[43] Hadden J.M., Convery M.A., Déclais A.-C., Lilley D.M.J., and Phillips S.E.V. (2001), *Crystal structure of the Holliday junction-resolving enzyme T7 endonuclease I at 2.1 Å resolution.* Nature Struct. Biol., **8**:62–67.

[44] Ceschini S., Keeley A., McAlister M.S.B., Oram M., Phelan J., Pearl L.H., Tsaneva I.R., and Barrett T.E. (2001), *Crystal structure of the fission yeast mitochondrial Holliday junction resolvase Ydc2.* EMBO J., **20**:6601–6611.

[45] Middleton C.L., Parker J.L., Richard D.J., White M.F., and Bond C.S. (2004), *Substrate recognition and catalysis by the Holliday junction resolving enzyme Hje.* Nucleic Acids Res., **32**:5442–5451.

[46] Kelly S.J., Li J., Setlow P., and Jedrzejas M.J. (2007), *Structure, flexibility, and mechanism of the Bacillus stearothermophilus RecU Holliday junction resolvase.* Proteins, **68**:961–971.

[47] Biertümpfel C., Yang W., and Suck D. (2007), *Crystal structure of T4 endonuclease VII resolving a Holliday junction.* Nature, **449**:616–620.

[48] Hadden J.M., Déclais A.-C., Carr S., Lilley D.M.J., and Phillips S.E.V. (2007), *The structural basis of Holliday junction resolution by T7 endonuclease I.* Nature, **449**:621–624.

[49] Déclais A.C., Liu J., Freeman A.D.J., and Lilley D.M.J. (2006), *Structural recognition between a four-way DNA junction and a resolving enzyme.* J. Molec. Biol., **359**:1261–1276.

[50] Déclais A.-C., Hadden J.M., Phillips S.E.V., and Lilley D.M.J. (2001), *The active site of the junction-resolving enzyme T7 endonuclease I.* J. Molec. Biol., **307**:1145–1158.

[51] Hadden J.M., Déclais A.-C., Phillips S.E.V., and Lilley D.M.J. (2002), *Metal ions bound at the active site of the junction-resolving enzyme T7 endonuclease I.* EMBO J., **21**:3505–3515.

[52] Liu J., Déclais A.C., and Lilley D.M.J. (2006), *Mechanistic aspects of the DNA junction-resolving enzyme T7 endonuclease I.* Biochemistry, **45**:3934–3942.

MICROMECHANICS OF SINGLE SUPERCOILED DNA MOLECULES

JOHN F. MARKO*

Abstract. The theory of the mechanical response of single DNA molecules under stretching and twisting stresses is reviewed. Using established results for the the semiflexible polymer including the effect of torsional stress, and for the free energy of plectonemic supercoils, a theory of coexisting plectonemic and extended DNA is constructed and shown to produce phenomena observed experimentally. Analytical results for DNA extension and torque are presented, and effects of anharmonicities in the plectonemic free energy are described. An application of the theory to the problem of torsional-stress-induced cruciform extrusion is also discussed.

Key words. DNA, molecular biology, statistical mechanics, polymer physics.

AMS(MOS) subject classifications. 82D60, 92C05, 92C40.

1. Introduction. Single-molecule stretching is a powerful tool not just for the study of the DNA double helix, but also for the study of proteins which interact with it. A wide variety of such experiments have followed from seminal DNA-stretching work of Smith *et al.* [1]. The semiflexible polymer model provides a quantitative starting point for theories describing these types of experiments [2–5]. This model, which describes DNA bending fluctuations in terms of a single "persistence length", is useful thanks to the separation of the double-helix persistence length (50 nm) from both the base-pair scale (0.34 nm) and the total molecular length scale ($>$ 1000 nm for $>$ 3 kb molecules; 1 nm $= 10^{-9}$ m). The availability of a simple but quantitative theoretical framework for DNA elasticity has greatly facilitated analysis of a wide variety of single-DNA-based experiments.

DNA's double-helix structure gives it a twist modulus. Both this twisting elasticity, and the topology of wrapping of the DNA strands in the double helix are of paramount biological importance, with whole families of enzymes known to play various roles in control of and response to DNA linking number *in vivo*. Many of these enzymes, including site-specific recombinases, topoisomerases, nucleic acid polymerases, and the mysterious "structural maintenance of chromosome" cohesin and condensin protein complexes, had their functions dissected by Nick Cozzarelli, to whom this volume is dedicated.

Elegant single-DNA manipulation techniques have been developed to control the linking numbers of individual DNA molecules, and quantitative studies have been made of the elasticity of twisted DNA molecules [6–9]. A dominant feature of the elastic behavior of twisted DNA is that

*Department of Physics and Astronomy and Department of Biochemistry, Molecular Biology and Cell Biology, Northwestern University, Evanston, Illinois 60208-3500.

C.J. Benham et al. (eds.), *Mathematics of DNA Structure, Function and Interactions*,
The IMA Volumes in Mathematics and its Applications 150,
DOI 10.1007/978-1-4419-0670-0_12,

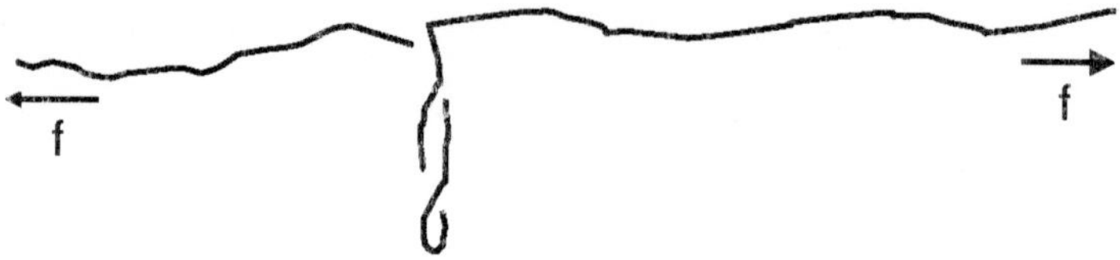

FIG. 1. *Sketch of a DNA molecule under tension f, and with linking number fixed so as to put the double helix under torsional stress. Over a range of applied tension, the molecule breaks up into "domains" of extended and plectonemically supercoiled DNA.*

at fixed force, introduction of sufficient linking number to a DNA causes it to start to wrap around itself or *supercoil* in the manner of a twisted wire. This buckling is driven by the reduction of twist elastic energy occurring when linking number is transferred to chiral bending, or "writhe". Further introduction of linking number causes folding up of the molecule into a plectonemic supercoiled DNA, and a gradual reduction in molecule extension. This folding can be discussed in terms of coexisting domains of extended and supercoiled DNA [10–16], as sketched in Fig. 1.

An important feature of this domain, or phase-coexistence-like behavior is that as linking number is converting DNA between plectonemic and extended form at a constant force, the torque in the molecule is constant [8, 9]. In such experiments, the torques applied to single DNAs are typically on the order of a few k_BT or pN·nm (1 pN = 1 piconewton = 10^{-12} newton; $k_BT = 4.1$ pN·nm at $T = 300$ K).

Precise measurements of elasticity of twisted DNAs have been used as the basis of studies of enzymes whose interactions with DNA depend on that twisting, notably topoisomerases [17–23]. Nick Cozzarelli played a key role in many of these experiments, as well as in single-molecule experiments examining basic twisting elasticity of DNA [24, 25].

This paper presents a brief review of the theoretical description of elastic properties of twisted DNA molecules. Of particular interest is the question of exactly what the torque is in a twisted stretched DNA and how it is related to other measurable properties of the molecule, since torque drives linking number relaxation in the experiments mentioned above. New optical tweezer techniques using rotationally polarized light promise to bring direct and precise torque measurements [26].

Sec. 2 reviews results of the statistical mechanical semiflexible polymer theory used to describe DNA stretching. Then, Sec. 3 discusses how the theory can be modified to describe twisted DNA molecules, using a mixed-state treatment of plectonemic and extended-twisted molecule domains. Sec. 3.4 reviews recently published results for that theory [27], while Sec. 4 presents an extension to that model which should permit precise description of experimental data. Sec. 5 discusses an application of the same general type of theory to the problem of extrusion of a cruciform structure from DNA containing a palindromic sequence. Finally the Conclusion discusses prospects for further application and development of the theory.

2. The DNA double helix is a semiflexible polymer with twist rigidity. In this section the basic polymer physics of the DNA double helix are reviewed (for a pedagogical introduction, see Ref. [28]). Thanks to its not-too-large bending stiffness, few-kilobase-long DNAs have conformational fluctuations that are well described by a single bending stiffness, or "bending persistence length". The double helical structure of the molecule gives the double helix the additional feature of a twist rigidity.

2.1. Persistence length and polymer stretching elasticity. Molecules of double-helix DNA under physiological solution conditions (aqueous solution with pH buffered to be near 7.5, salt concentration in the 10 mM to 500 mM range, temperature between 20 and 37°C) behave as semiflexible polymers, with a well-defined bending persistence length of $A \approx 50$ nm [29]. Recalling that there are just about 3 base pairs (bp) per nm of contour length, this indicates that an otherwise unconstrained DNA molecule in solution undergoes thermally excited bending causing its local tangent to reorient every ≈ 150 bp. However, if tension is applied to the molecule, the tangent vector will align with that applied tension. Here, molecules which are many persistence lengths long are of interest, *i.e.* in the multi-kilobase range, the typical size of circular plasmid molecules or chromosome loop domains in bacteria.

For forces well below k_BT/A a DNA molecule will be only slightly stretched; for forces well in excess of k_BT/A the molecule stretches out. Given that $k_BT \approx 4.1$ pN·nm, this threshold force for stretching a double-helix DNA is $k_BT/A = 0.08$ pN. This is a rather low force by molecular biological standards: most molecular motors use chemical energy of a few k_BT to take steps of a few nm in length, so that the scale for molecular motor forces is on the scale of a few pN. The low value of k_BT/A is due to the long persistence length of the double helix ($A = 50$ nm). In contrast single-stranded nucleic acid molecules have a much shorter persistence length (≈ 1 nm) and consequently do require few-pN forces for their full extension [1]. Thus, double-helix DNAs can be easily stretched out to their full extension by molecular motors in the cell; single-stranded DNA (or RNA, or denatured proteins) require larger forces to be stretched out.

The semiflexible polymer model has been shown to give a good account of the stretching properties of multi-kilobase DNAs from forces of below 0.1 pN up to roughly 20 pN [2, 3, 5]. For higher forces the secondary structure of the double helix starts to be deformed [4, 5], eventually leading to an abrupt first-order-like "overstretching" transition whereby the double helix contour length increases by about 70% [30, 31].

The statistical mechanics of the semiflexible polymer model can be solved to arbitrary accuracy numerically; the asymptotic high- and low-force limits are understood analytically [5]. The global variation of the force f as a function of extension X is captured well by the expression

$$f = \frac{k_B T}{A}\left[\frac{X}{L} + \frac{1}{4(1-X/L)^2} - \frac{1}{4}\right] \tag{2.1}$$

where L is the total molecule contour length (recall that there are 0.34 nm of contour length per base pair, so that a 10 kb molecule is about 3000 nm = 3 μm in contour length). Eq. 2.1 has the correct low-extension ($X/L << 1$) linear elasticity response. For high extensions ($X/L \to 1$) the force increases drastically due to an increasing entropic cost of quenching thermal bending fluctuations with progressively smaller wavelengths. An even more accurate approximate representation of the exact solution is obtained by adding a term $-(3/4)(X/L)^2$ inside the square brackets [32], eliminating any low-extension X^2 dependence which should be absent via symmetry considerations [11].

The integral of force over extension, or $W(X) = \int_0^X dX' f(X')$, gives the work done stretching a DNA at constant temperature or the free energy for the semiflexible polymer as a function of extension [11]. In what follows the free energy at fixed force is needed, which is obtained via the Legendre transformation $g(f) = fX - W(X)$; the extension X is eliminated using the inverse of (2.1) [11]. The leading behavior of $g(f)$ and the average extension $X = \partial g/\partial f$ behaves asymptotically for high force as

$$\begin{aligned} g(f) &= f - \sqrt{k_B T f/A} + \cdots \\ X(f) &= 1 - \sqrt{\frac{k_B T}{4Af}} + \cdots \end{aligned} \tag{2.2}$$

For the DNA double helix, these leading terms are sufficient for situations where forces are between 0.2 pN and 10 pN, which applies over much of the force range for single-DNA stretching experiments. To describe slightly higher forces up to 40 pN, linear contour length stretching elasticity may be added [4, 11].

2.2. Twisting stiffness of the double helix. If one could twist a DNA molecule without bending it, one could measure the work required to introduce a total twist by an angle Θ, or the twisting free energy. In the absence of bending, Θ would correspond to the change in linking number of the double helix (number of wraps of one strand around the other), via $\Delta\text{Lk} = \Theta/2\pi$. Molecular biologists often describe linking number changes normalized by the linking number of the relaxed double helix, $\text{Lk}_0 = L/h$ where the helix repeat is $h = 3.6$ nm = 10.5 bp: $\sigma = \Delta\text{Lk}/\text{Lk}_0$. In the bacterium *E. coli*, it is known that circular plasmid and chromosomal DNAs have $\sigma \approx -0.05$, *i.e.*, their DNAs have about 5% fewer links than would be expected from their contour length.

For small pure twists, the energy per contour length is expected by symmetry [10] to be that of a twisted elastic rod [33]:

$$\frac{E_{\mathrm{twist}}(\sigma)}{L} = \frac{k_B T C}{2} \left(\frac{2\pi \Delta \mathrm{Lk}}{L} \right)^2 = \frac{k_B T C \omega_0{}^2}{2} \sigma^2 \tag{2.3}$$

where C is the twist persistence length of DNA (so that $k_B T C$ is the twist rigidity elastic constant [33]), and where $\omega_0 = 2\pi/h = 1.85\ \mathrm{nm}^{-1}$ is the spatial rate of circulation of the relaxed double helix.

This form of simple twist energy has been shown to apply to DNA for small linking number changes in direct micromechanical measurements as well as in biochemical studies. However, precise measurement of the twist rigidity C of the double helix has proven to be problematic, with different types of experiments yielding different results [34]. Even among single-molecule mechanical experiments there have been disagreements between different analyses of experimental data [35].

A cause of difficulty in determining C precisely has been that experiments where DNA is twisted are always subject to chiral bending effects. First, even for small amounts of twisting, chiral bending fluctuations cause a shift (or "renormalization") of the apparent twist rigidity; the energetic cost of twisting the double helix can be reduced by chiral bending fluctuations [36, 37]. A second complication is that for larger amounts of twisting, plectonemic supercoiling of DNA generates a large amount of writhe which further reduces the twist energy cost of adding linking number [11].

A third problem is that DNA tends to be more easily denatured by untwisting than by overtwisting due to its right-handed helix chirality, and also that twisting is directly coupled to changes in DNA helix contour length [38–41]. This effect, which becomes more important as force is increased, further complicates comparison of experiments done with opposite signs of σ, as well as fits to microscopic elastic theories which often do not account for chiral asymmetry and denaturation effects. Despite all these challenges, there is a rough consensus between single-DNA experiments in the few-pN ranges that the "bare" twist rigidity of DNA at forces below a few pN is in the range $C = 95 \pm 20$ nm [36, 37, 24, 35, 16].

3. Supercoiling and torque in stretched twisted DNA. A general framework is now described, aimed at describing micromechanical experiments on single DNAs, taking into account effects of "coexistence" of domains of DNA in different conformations. The main application discussed in this section is to coexistence of plectonemic supercoils and extended DNA. The calculations here are similar to those discussed in more detail in a recent paper [27]. Here, new results will be presented for the effect of anharmonicity in the plectonemic supercoiling free energy on DNA extension obtained as a function of force and linking number.

3.1. Coexistence of DNA states. If a DNA molecule is stretched by force f, its free energy and structure change as its linking number density

σ is changed. Here, the DNA molecule is treated as being divided into domains of two "pure" states, each of which is described by a free energy per molecular length, dependent on applied force f and the linking number density σ. As the forces on the two domains are equal, the focus will be on the σ dependences of the two states, which in most of this paper will describe "stretched" and "plectonemic" DNA [10, 11, 14, 13, 15, 16].

The free energies per length of the pure states are taken to be $\mathcal{S}(\sigma)$ for stretched and $\mathcal{P}(\sigma)$ for plectonemic DNA. From a statistical-mechanical perspective, these free energies are $-k_BT$ times the logarithm of the partition function for the molecule at fixed force and linking number, divided by the relaxed double helix contour length; note that all the free energies per length can be converted to free energies per base pair by multiplying by 0.34 nm/bp. For these pure states, the rate that work is done injecting linking number is proportional to torque:

$$\tau = \frac{1}{\omega_0}\frac{\partial \mathcal{S}(\sigma)}{\partial \sigma}. \tag{3.1}$$

The prefactor converts the σ derivative to one with respect to rotation angle; a similar equation holds for the $\mathcal{P}$ state.

Along a molecule which is a fraction $x_{\rm s}$ of state $\mathcal{S}$ and fraction $x_{\rm p} = 1 - x_{\rm s}$ of state $\mathcal{P}$, the free energy per base pair of the mixed phase is

$$\mathcal{F}(\sigma) = x_{\rm s}\mathcal{S}(\sigma_{\rm s}) + x_{\rm p}\mathcal{P}(\sigma_{\rm p}). \tag{3.2}$$

The equilibrium length fraction $x_{\rm s}$ and the free energy is determined by minimization of this free energy subject to the constraint $\sigma = x\sigma_{\rm s} + x_{\rm p}\sigma_{\rm p}$.

This last equation indicates that linking number is being *partitioned* between the two states. This must be justified, since while twist is locally defined, in general writhe cannot be considered as a local variable that can be partitioned in this way. At the workshop, Craig Benham emphasized this point.

However, in the case of interest here, the plectonemic regions are essentially closed loops (see Fig. 1). By "pinching" of those loops off to form circular plectonemic supercoils separated from the extended DNA, the calculation of writhe can be decoupled into separate writhes for extended and plectonemic regions. The "cross terms" ignored by this decoupling correspond to the Gauss linking invariant of the plectonemic regions with one another and with the extended DNA, all of which are zero. The entire chain is an unknot, guaranteeing that this pinching-off decoupling can always be done, but this also indicates that the physically relevant case of unknotted and self-avoiding DNA must be considered. For a "phantom DNA" model lacking the unknot topology constraint, nontrivial knotting of the chain could cause this partitioning to break down [12]. At the workshop de Witt L. Sumners described a sketch of a proof of the additivity of linking numbers of extended and plectonemic regions in this experimentally relevant unknot case.

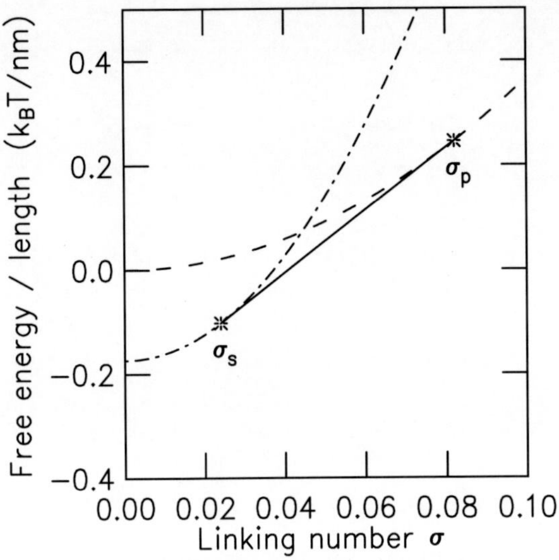

FIG. 2. *Free energies of extended (dot-dashed curve, $\mathcal{S}(\sigma)$) and plectonemic supercoil (dashed curve, $\mathcal{P}(\sigma)$) DNA states as a function of linking number σ. For $\sigma < \sigma_s$, the $\mathcal{S}$ state is lower in free energy than either $\mathcal{P}$ or any mixture of the two. Similarly, for $\sigma > \sigma_p$, pure $\mathcal{P}$ is the lowest-free energy configuration. However, for σ between σ_s and σ_p the tangent construction shown (solid line segment between tangent points indicated by stars), representing coexisting domains of $\mathcal{S}(\sigma_s)$ and $\mathcal{P}(\sigma_p)$, is the lowest free energy state. Note that the gap between the two states at $\sigma = 0$ is the free energy difference between random coil DNA [$\mathcal{S}(0)$] and stretched unsupercoiled DNA [$\mathcal{P}(0)$]; this difference grows with applied force.*

If the two pure state free energy densities plotted as a function of linking number density never cross, then one pure state or the other will be the equilibrium state, *i.e.*, one of the two extreme cases $x_s = 0$ or $x_s = 1$ will always minimize Eq. 3.2. However, if the two free energy densities cross, then there will be a range of σ over which there will be coexisting domains of the two states. Fig. 2 shows this situation, sketched to correspond to the case of main interest here, where at low values of σ the stretched state is stable (lower in free energy) relative to the plectoneme state, but where at large σ the stability reverses due to "screening" of the twist energy by the plectonemic state's writhe [10, 11, 14, 15].

Minimization of Eq. 3.2 is accomplished by a double-tangent construction familiar from other examples of phase coexistence (e.g., liquid-gas); in this case the conserved density is that of linking number (Fig. 2). The two coexisting states of linking number densities σ_s and σ_p satisfy $\partial\mathcal{S}(\sigma_s)/\partial\sigma_s = \partial\mathcal{P}(\sigma_s)/\partial\sigma_p$, *i.e.*, they have equal torques. They mix in proportions x_s and x_p, so the free energy in the coexistence region is

$$\mathcal{F}(\sigma) = \mathcal{S}(\sigma) + \frac{\partial \mathcal{S}(\sigma_s)}{\partial \sigma_s}(\sigma - \sigma_s) = \mathcal{P}(\sigma) + \frac{\partial \mathcal{P}(\sigma_p)}{\partial \sigma_p}(\sigma - \sigma_p). \tag{3.3}$$

In the coexistence region, the fractions of the two states in the mixed state depend linearly on σ, as

$$x_s = \frac{\sigma_p - \sigma}{\sigma_p - \sigma_s} \qquad x_p = \frac{\sigma - \sigma_s}{\sigma_p - \sigma_s}. \tag{3.4}$$

The coexistence construction guarantees that the free energy is a convex function of linking number, and therefore that the torque is a monotonic function of linking number, as required for mechanical stability. In the coexistence region (σ between the limits σ_s and σ_p) the torques in the two types of domains are equal and σ-independent; *i.e.*, the σ-derivative of Eq. 3.3 is constant.

In the coexistence region Eq. 3.4 indicates that the rate of change of the length fractions with σ is constant; $\partial x_s/\partial\sigma = -1/(\sigma_p - \sigma_s)$. This generates the linear dependence of molecule extension on linking number observed experimentally once the threshold for generating plectonemic DNA is reached, as can be seen by computing the molecule extension (as a fraction of relaxed double helix contour length L):

$$\frac{z}{L} = -\frac{\partial \mathcal{F}}{\partial f} = -x_s \frac{\partial \mathcal{S}(\sigma_s)}{\partial f} - x_p \frac{\partial \mathcal{P}(\sigma_p)}{\partial f}. \tag{3.5}$$

In the coexistence region, the only σ dependence is the linear variation of x_s and x_p, making the dependence of extension on σ entirely linear.

In the main case of interest here where $\mathcal{P}$ is the plectonemic supercoil state, its zero length eliminates its contribution to Eq. 3.5 (*i.e.*, $\partial\mathcal{P}/\partial f = 0$), yielding

$$\frac{z}{L} = -x_s \frac{\partial \mathcal{S}(\sigma_s)}{\partial f} = \frac{\sigma_p - \sigma}{\sigma_p - \sigma_s} \frac{z(\sigma_s)}{L} \tag{3.6}$$

where the final extension per length factor is the extension per length of the extended DNA state.

Experimentally, σ_s and σ_p may be measured from the beginning and the end of the linear coexistence regime of extension as a function of σ. Likewise, $z(\sigma_s)/L$ is the extension per length of the molecule at the onset of the linear regime. Thus Eq. 3.6 can be used to determine the coexisting state linking number values, the extension of the stretched DNA state as a function of force and linking number, and via integration the free energy of the stretched state. Then through use of the tangent construction (Fig. 2), the free energy of the plectonemic state can be measured. Note that when the molecule is entirely converted to plectoneme ($x_s = 0$, $x_p = 0$) the extension reaches zero. The point $\sigma = \sigma_p$ where this occurs can be estimated experimentally from extrapolation of extension data to zero.

3.2. Free energy of extended twisted DNA. The previous subsection outlines the basic scheme of calculation of equilibrium domain coexistence along a stretched and supercoiled DNA, but to calculate a experimentally relevant result, explicit forms for the extended and plectonemic state free energies are needed. Fortunately, suitable formulae based on statistical-mechanical treatments of microscopic models are available.

For a DNA under torsional stress there occur chiral fluctuations even when the chain is fully extended [10, 11, 36, 37]. The free energy of the extended DNA at a fixed force f can be written as an expansion in σ [36, 37, 15]:

$$\mathcal{S}(\sigma) = -g + \frac{c_{\rm s}}{2}\sigma^2 + \cdots \tag{3.7}$$

The leading constant g is the free energy of stretched torsionally unconstrained DNA of Eq. 2.2. The parameter $c_{\rm s}$ is the twist stiffness of the extended DNA state and can be measured from the curvature of the extension as a function of σ near its maximum. Note that $c_{\rm s}$ has been defined to have dimensions of energy per length, the same as that of g and force f.

The first two terms of the expansion of Eq. 3.7 are sufficient to illustrate the basic properties of state coexistence in a quantitative way. As for the plectonemic state, higher-order terms can be added (*e.g.* σ^3 to generate positive-negative twisting asymmetry). The most general expansion would replace σ with $\sigma - \sigma_0$ in Eq. 3.7 where this additional force-dependent σ_0 parameter sets the linking number at which the double helix is relaxed. Variation of σ_0 with force takes into account the stretch-twist coupling. However, for the relatively low forces of interest here (a few pN) this coupling has only weak effects and may be neglected (*i.e.*, $\sigma_0 \approx 0$). In any case, σ_0 for a given force is determined by finding the σ value at which a local maximum of extension is obtained. Here, $\sigma_0 = 0$ is assumed for the $\mathcal{S}$ state.

Although $c_{\rm s}$ could be determined directly from experimental data, a theoretical formula does exist based on a large-force perturbative calculation for the semiflexible polymer with harmonic twist rigidity by Moroz and Nelson [36, 37]:

$$c_{\rm s} = k_B T \omega_0^2 C \left[1 - \frac{C}{4A} \left(\frac{k_B T}{Af} \right)^{1/2} \right]. \tag{3.8}$$

The main effect introduced by this theoretical formula for $c_{\rm s}$ is a force-dependence of the twisting rigidity. Reducing force reduces $c_{\rm s}$ since more linking number can be absorbed into writhe fluctuations at lower forces [36, 37].

3.3. Free energy of plectonemically supercoiled DNA. A plectonemic supercoil has essentially no extension between its ends, and therefore its free energy has no force dependence (a nonzero force derivative would indicate that it has finite extension). In addition, the bare twist energy (2.3) is heavily screened by the large writhing generated by the supercoil [10, 11]. Given its minimum at $\sigma = 0$, the free energy for the plectoneme must have the form

$$\mathcal{P}(\sigma) = \frac{p}{2}\sigma^2 + \cdots \tag{3.9}$$

The parameter p describes the twist stiffness of the plectonemic state, and like c_s and g has dimensions of a force. The stiffness p can be converted to a persistence-like quantity via $p = k_BTP\omega_0^2$; P represents the effective twist persistence length of the plectonemic state, accounting for the screening of the bare twist energy (2.3) by the strong writhe of the plectoneme. This screening effect indicates that $P < C$; available data indicate $P \approx 25$ nm [42] for $\approx$ 100 mM univalent salt in pH 7.5 buffer solution (alternately $p \equiv k_BTP\omega_0^2$).

The stiffness P varies with salt concentration, since that in turn adjusts the effective diameter of the double helix; for higher salt, P is reduced, while for lower salt P increases. This effect arises because the bending energy in a plectonemic superhelix is reduced when the effective diameter is reduced. The torque of the $\mathcal{P}$ state is, by Eq. 3.1, $\tau = (p/\omega_0)\sigma = k_BTP\omega_0\sigma$.

In Sec. 3.4 only the first, quadratic term of (3.9) is used, which is a reasonable model for $|\sigma| < 0.05$, as it allows analytical calculation of all the domain coexistence properties, as well as providing a semi-quantitative description of experiment. However, biochemically it has long been known that anharmonic corrections are present [43]; for large $|\sigma|$ where plectonemic interwinding becomes tight, one can expect appreciable deviations from quadratic behavior. Sec. 4 will show how anharmonicity in (3.9) can modify the extension *vs.* linking number behavior.

3.4. Analytical results for the harmonic-σ state free energies. With the $\mathcal{O}(\sigma^2)$ expressions for plectonemic and extended DNA free energies, the state coexistence behavior at constant force f can be computed following the procedure outlined in Sec. 3.1. The mixed-state free energy is

$$\mathcal{F} = x_{\rm s}\left(-g + \frac{c_{\rm s}\sigma_{\rm s}^2}{2}\right) + x_{\rm p}\frac{p\sigma_{\rm p}^2}{2}. \tag{3.10}$$

Details of the calculation of the equilibrium mixed state can be found in Ref. [27]. This amounts to elimination of σ_p and x_p from (3.10) using $x_{\rm s} + x_{\rm p} = 1$ and $\sigma = x_{\rm s}\sigma_{\rm s} + x_{\rm p}\sigma_{\rm p}$, followed by minimization to determine the remaining free parameters σ_s and x_s.

The coexisting state linking number densities for this harmonic model work out to be:

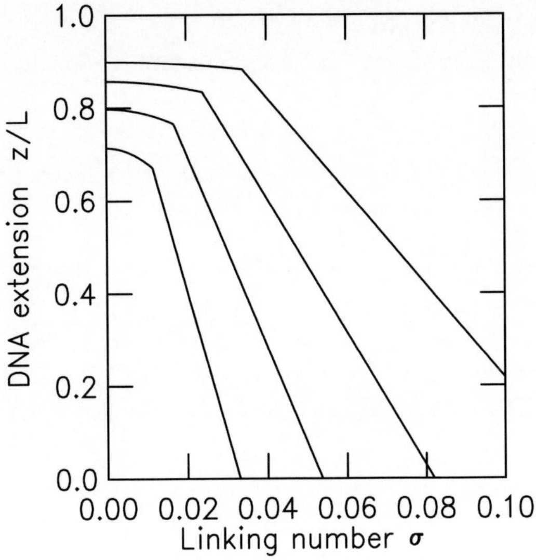

FIG. 3. *Extension versus linking number, for forces 0.25 pN (lowest curve), 0.5 pN, 1.0 pN and 2.0 pN (highest curve), for positive linking number change ($\sigma > 0$). As force is increased, the extension increases, and the effect of torsional stress (linking number) is reduced. Parameter values are $C = 95$ nm, $A = 50$ nm, $P = 24$ nm. The parabolic peak of each extension curve occurs when the DNA is pure extended state; extended and plectonemic DNA are in coexistence on the linear part of each extension curve. The beginning of the linear segments indicates σ_s, and their σ-intercepts indicate σ_p.*

$$|\sigma_s| = \frac{1}{c_s}\left(\frac{2pg}{1-p/c_s}\right)^{1/2}$$

$$|\sigma_p| = \frac{1}{p}\left(\frac{2pg}{1-p/c_s}\right)^{1/2} . \tag{3.11}$$

Since σ_s can be determined from experimental data by determining where the linear (coexistence) regime begins, and since the slope $dx_s/d\sigma$ is similarly determined from experimental data, and finally since g is independently well known, it is possible to determine the stiffnesses c_s and p. This would amount to a measurement of the free energy of extended and plectonemic DNA in a way relatively independent of details of specific microscopic theories.

Fig. 3 shows extension versus linking number for this model via (3.5) and (3.6). The curves are computed for $f = 0.25$ pN (lowest curve), 0.5 pN, 1.0 pN and 2.0 pN (highest curve). The stiffnesses used were $A = 50$ nm, $C = 95$ nm and $P = 24$ nm. The parabolic peak at the top of each curve is the region of pure extended DNA. This joins to a linear coexistence segment which stretches between σ_s and σ_p, the latter being the point where extension reaches zero. The curves have the characteristic "hat" shape seen experimentally [9].

The equilibrium free energy is easily computed as a function of force and linking number:

$$\mathcal{F} = \begin{cases} -g + \frac{1}{2}c_s\sigma^2 & |\sigma| < |\sigma_s| \\ -g/(1-p/c_s) + [2pg/(1-p/c_s)]^{1/2}|\sigma| & |\sigma_s| < |\sigma| < |\sigma_p| \\ \frac{1}{2}p\sigma^2 & |\sigma| > |\sigma_p| \end{cases} . \quad (3.12)$$

In the coexistence region, the linear dependence of the free energy on σ indicates that torque ($\tau = \omega_0^{-1}\partial\mathcal{F}/\partial\sigma$) is a constant. The molecule extension ($X/L = \partial\mathcal{F}/\partial f$) varies linearly with σ in the coexistence region, as anticipated in (3.5).

3.5. DNA torque and its force dependence for the harmonic model. Eq. 3.11 gives the torque, via Eq. 3.1 and either of the pure state free energies Eq. 3.9 or Eq. 3.7:

$$\tau = \begin{cases} (c_s/\omega_0)\sigma & |\sigma| < |\sigma_s| \\ (2pg/[1-p/c_s])^{1/2}/\omega_0 & |\sigma_s| < |\sigma| < |\sigma_p| \\ (p/\omega_0)\sigma & |\sigma| > |\sigma_p| \end{cases} . \quad (3.13)$$

For constant force, as $|\sigma|$ is increased from zero, the equilibrium state is first pure extended DNA, with torque growing linearly with $|\sigma|$. Then at $\sigma = \sigma_s$, the coexistence point is reached, and until $\sigma = \sigma_p$ the torque is constant. Finally for $|\sigma| > |\sigma_p|$, the entire DNA is plectonemic supercoil, and again the torque changes with $|\sigma|$ but at a reduced rate (recall $P < C_s$) due to the efficient removal of twist by the large plectonemic writhe.

The middle line of Eq. 3.13 is a main result of this calculation, as it gives the dependence of the coexisting state torque on force:

$$\tau = \sqrt{\frac{2k_BTPg}{1-P/C_s}}. \quad (3.14)$$

This formula is written in terms of the twist persistence length of the plectoneme (P) and of the extended state $C_s = C[1-(C/4A)(k_BT/Af)^{1/2}]$ (note $P/C_s = p/c_s$).

The force-dependence of the torque enters mainly through the extended state free energy $g = f - (k_BTf/A)^{1/2} \approx f$ in the numerator. However, note that the twist persistence length of the extended state C_s increases with increasing force.

The simplest application of Eq. 3.14 is estimating torques in situations where one is in the coexistence range of σ and where one knows the force.

This is commonly the case in magnetic tweezer experiments where force and linking number are both fixed by the position of a macroscopic magnet [9]. An elegant example of use of this is the constant-torque driving of rotational relaxation of DNA [21]. One must be in the coexistence range of σ in order to have this constant torque; below σ_s, the torque will drop below the coexistence value, and above σ_p the torque will increase with $|\sigma|$.

It is useful to consider the situation of fixed σ and varied force. This is experimentally accessible using various micromanipulation schemes, including magnetic tweezers as long as the force is not reduced to so low a value that linking number can leak away by having the DNA hop over the bead.

Imagine holding σ fixed at a value sufficient to form plectonemic supercoils at a low force, and then slowly increase f. At low forces, the entire molecule will remain supercoiled (with constant torque $\tau_p = k_B T P \omega_0 \sigma$ until the force reaches a threshold value f_p, *i.e.*, until the work done by the external force can overcome the plectoneme's "length binding energy" [10, 11]. This effect can be analyzed by solving for the force f_p where $x_s = 0$, which leads to

$$g(f_p) = \frac{p[1 - p/c_s]}{2}\sigma^2. \tag{3.15}$$

The point $g(f_p)$ is the minimum extended state free energy (essentially force) needed to start extending the plectonemic DNA. The f_p that solves this equation is the tension inside a plectonemically supercoiled DNA, and for physiological levels of supercoiling ($\sigma = -0.05$) this force is ≈ 0.5 pN.

As force increases further the torque will increase as linking number is shifted increasingly from plectoneme to extended DNA, and the torque will be described by Eq. 3.14. In the coexistence range the torque will be independent of σ.

Finally, when sufficient force f_s is applied, the plectonemic domains will be destroyed; f_s can be found by solving for when $x_s = 1$:

$$g(f_s) = \frac{c_s - p}{2}\sigma^2. \tag{3.16}$$

The torque at this point will be $\tau_s = (c_s/\omega_0)\sigma$. As force is increased further, the small amount of linking number remaining will be forced into DNA twist, gradually forcing the molecule torque towards its limit of $\tau = k_B T C \omega_0 \sigma$.

Fig. 4 plots torque versus force for $\sigma = 0.03$, 0.04, 0.05 and 0.06 using the same stiffnesses as used in Fig. 3 ($C = 95$ nm, $A = 50$ nm, $P = 24$ nm). The torque starts at a constant (horizontal segments to left; lowest corresponds to $\sigma = 0.03$; highest to $\sigma = 0.06$). Then, the torque starts to increase when extended DNA starts to be created at f_p. In the coexistence regime, the torque does not depend on σ, so all the curves overlap. When f_s is reached, the torque curves separate again,

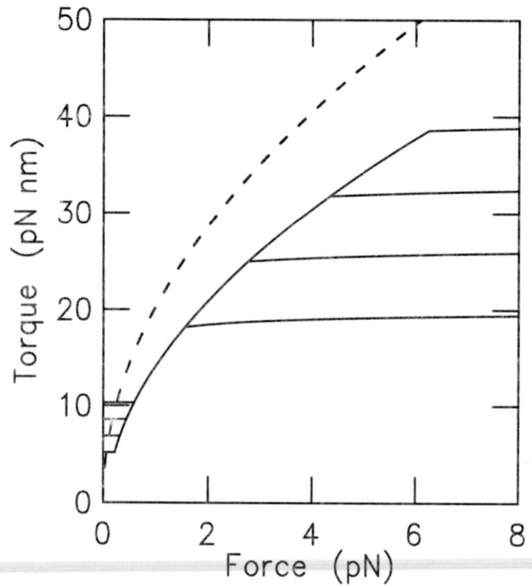

FIG. 4. *Torque versus force curve for coexisting state densities corresponding to fixed* $\sigma = +0.03$, *0.04, 0.05 and 0.06, with other parameters as in Fig. 2. For forces below the lower critical forces* $f_{\rm P}$ *where the extended DNA state disappears (horizontal line segments at left of graph), torque is constant. Above the upper critical force* $f_{\rm s}$ *where the plectonemic DNA disappears, the torque slowly approaches its maximum value* $(c/\omega_0)\sigma$ *(nearly flat regions to right). Between* $f_{\rm P}$ *and* $f_{\rm s}$*, the torque follows the same curve for each* σ *value (concave part of curve). Dashed curve shows* $(2k_BTAf)^{1/2}$ *for comparison.*

and then torque only slowly increases with further force increase (nearly horizontal curves to right). These results have been shown to be in good agreement with a numerical calculation of torque in stretched twisted DNA [27] computed using a previously developed Monte Carlo computation [12].

It is worth noting that the torque in the coexistence range of σ (3.14) is not equal to $\tau = (2k_BTAf)^{1/2}$ where A is the bending persistence length, as has been suggested in Refs. [8, 9, 21]. This formula is reminiscent of that for the critical torque for the linear instability of buckling of a rod under tension f from classical elasticity theory, $(4k_BTAf)^{1/2}$ [46, 36, 37, 15], but with the factor of 4 replaced by a 2. The derivation of the formula with the 2 uses an approximate calculation [9] based on mechanical energy without including effects of thermal fluctuations.

The formula $\tau = (2k_BTAf)^{1/2}$ (dashed curve in Fig. 4) overestimates the torque calculated above, Eq. 3.14, by roughly 25%. Given the accuracy of single-DNA micromechanical measurements plus the consequences of this overestimate for other measurements *e.g.*, energy landscape parameters for topoisomerases [21], this discrepancy is significant. Experimentally,

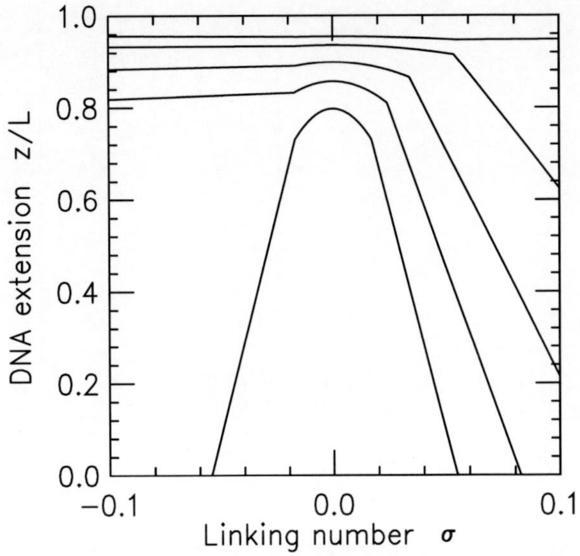

FIG. 5. *Extension vs. linking number σ for DNA at constant forces, including denaturation effects. Results are shown for forces of 0.5 pN (lowest curve), 1 pN, 2 pN, 5 pN and 10 pN (highest curve), using $A = 50$ nm, $C = 95$ nm, $P = 24$ nm, plus description of denaturation as described in Ref. [27]. For 0.5 pN the extension is symmetric under sign change of σ, due to coexistence occurring between extended and plectonemically supercoiled DNA states, which have this symmetry. However, for forces of 1 pN, 2 pN and 5 pN, coexistence occurs between extended and plectonemic DNA for $\sigma > 0$, but between extended and denatured DNA for $\sigma < 0$, exhibiting a strong breaking of $\sigma \to -\sigma$ symmetry. Finally for 10 pN, denaturation occurs for both positive and negative σ.*

the DNA torque, being dependent on the free energies of the coexisting phases, can be expected to depend on factors which shift around those free energies, notably solution ionic conditions. The torque could be estimated in a model-independent way, using c_s and g extracted directly from experimental data.

3.6. Negative supercoiling and DNA denaturation. The results given in this section are most applicable to positively supercoiled DNA ($\sigma > 0$) where DNA remains in double-helix form for forces up to ≈ 10 pN. For $\sigma < 0$ the situation is quite different; negative supercoiling of drives denaturation rather readily, leading to ranges with $\sigma < 0$ where extension is nearly constant. Denaturation also occurs for positive supercoiling, but only for more extreme values of σ [49]. This behavior is straightforward to describe via the framework presented above, by introducing additional pure states corresponding to denatured (strand-separated) DNA [47].

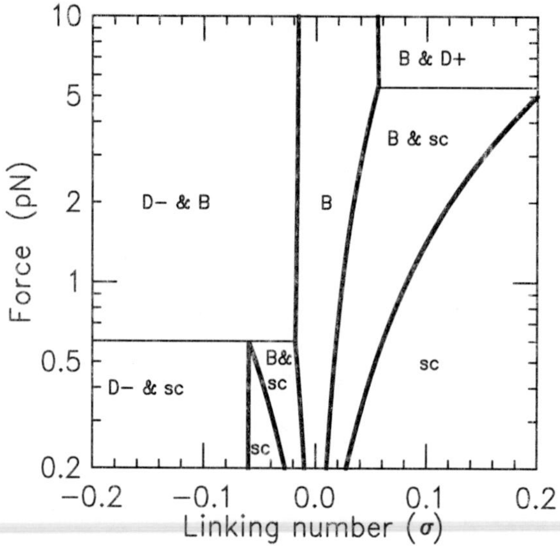

FIG. 6. *Force-linking number phase diagram showing transitions between extended double helix (B), plectonemically supercoiled B-DNA (sc), underwound denatured DNA (D-) and overwound denatured DNA (D+). Regions labeled by two pure phases (e.g., "B & sc") represent two-phase coexistence regions; thick curves indicate transitions between pure-state and mixed-state regions. The horizontal thin lines are lines of three-state coexistence.*

DNA which has been denatured by unwinding has a short persistence length, and has a large negative value of linking number corresponding to the $\sigma \approx -1$ necessary to unwind the double helix. In addition, denatured DNA has a free energy shifted up relative to that of the base-paired DNA double helix, by a sequence-averaged amount of about $2.5k_BT$/bp [48]. These effects are discussed in some detail in Ref. [27], including showing comparisons with experimental extension data.

Fig. 5 shows extension versus linking number combining the extended and plectonemic states with denatured states. For $\sigma > 0$, coexistence of extended and plectonemically supercoiled DNA occurs as before, but for $\sigma < 0$, denatured underwound-melted states preempt the plectonemic state, leading to nearly constant extension with linking number. The nearly constant variation of extension with σ once denaturation begins to occur is mainly a consequence of the small amount of the large change of linking number associated with denaturation (complete separation of the DNA strands corresponds $\sigma = -1$) combined with the high extensibility of denatured DNA (the DNA backbones can be extended about a factor of two by unwinding them).

By tabulating the linking numbers at which the various transitions occur, one can construct a force-linking number "phase diagram" (Fig. 6). Since linking number is partitioned between coexisting states, this phase diagram contains "pure state" and "mixed state" regions. In the two-state coexistence regions, torque is constant, and extension varies linearly with σ. Fig. 6 also contains lines on which three states coexist; the most experimentally accessible of these is the boundary where regions of coexisting extended B-DNA and plectonemic supercoiled DNA ("B & sc" in Fig. 6) and regions of coexisting B-DNA and denatured underwound DNA ("B & D-") meet. Notably, there are also regions of denatured overwound DNA, corresponding to the "P-DNA" state observed by Allemand *et al.* [49]. Ref. [27] presented the corresponding phase diagram plotted in the force-torque plane.

4. Effect of plectonemic free energy anharmonicity on extension of twisted DNA. The harmonic (quadratic-σ) model of Sec. 3.4 has the virtues of being simple and analytically solvable, while still describing experimental data reasonably well [27]. It does suffer from limitations, for example a lack of any accounting for effects of DNA sequence, known to profoundly affect stress-driven DNA denaturation [50]. Also, the basic scheme introduced in Sec. 3.1 whereby two coexisting domains are considered additively could be generalized to include a thermally excited spectrum of "droplets" of the two pure states, along the lines of calculations presented in Ref. [51]. These particular improvements will not be discussed further here.

A limitation of the theory of Sec. 3.4 that will be examined here is the effect of truncation of the pure state free energies at order σ^2. Truncation of the extended state free energy (3.7) at this order does not introduce large errors, since the extended state linking number (for $\sigma > 0$) is less than σ_s, the point where plectoneme-extended coexistence begins, and σ_s remains relatively small. Consistently, experimental data in the pure-extended regime ($\sigma < \sigma_s$) is well-fit by the extension function resulting from truncation at σ^2, for forces > 0.2 pN. For larger forces (> 0.5 pN) bending fluctuations in the extended state are suppressed, making the σ^2 term even more dominant over higher-order terms [36] (see Eq. 3.8).

The situation is exactly the opposite for the plectonemic state: the coexistence model indicates that the linking number in the plectonemic domains is never *less* than σ_p (for $\sigma > 0$). Consequently σ_p is often large: for moderate forces (*e.g.* 2 pN) $\sigma_p \approx 0.1$ (the boundary between "sc & B" and "sc" in Fig. 6). While the free energy of plectonemic DNA is well described by the initial σ^2 dependence for $\sigma < 0.05$ [43, 42, 44, 45], for large σ one expects tight interwinding to generate a strong upturn in free energy [11]. It is important to consider the effects of addition of higher-order-σ contributions to (3.9).

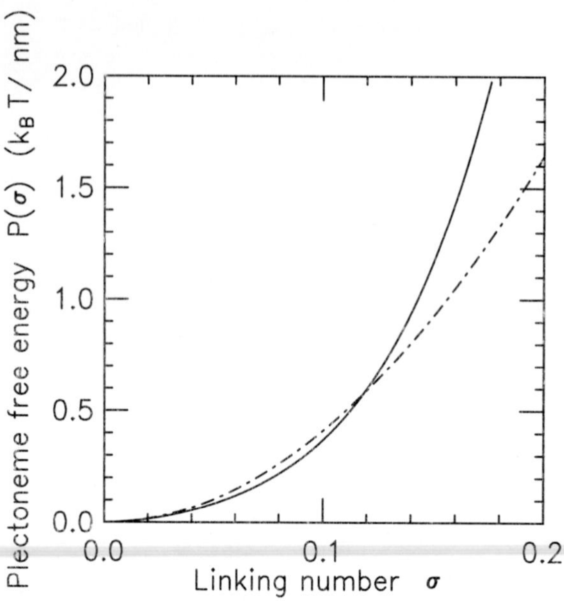

FIG. 7. *Harmonic and anharmonic plectonemic free energies. Dashed curve shows purely harmonic plectonemic free energy with $P = 24$ nm. Solid curve shows anharmonic plectoneme free energy with the same quadratic stiffness $P = 24$ nm (recall $p = k_BTP\omega_0^2$) but with nonlinear terms $t = -800k_BT/nm$, and $q = 9000k_BT/nm$. The anharmonic free energy has the same initial curvature, but then is softer than the harmonic free energy for intermediate values of σ. For large σ the quartic term makes the anharmonic free energy become greater than the harmonic free energy, providing a model for the breakdown of writhe increase for tight wrapping.*

Here anharmonic cubic- and quartic-σ terms are added to (3.9):

$$\mathcal{P}(\sigma) = \frac{p}{2}\sigma^2 + \frac{t}{3}\sigma^3 + \frac{q}{4}\sigma^4. \tag{4.1}$$

This general expansion permits one to have an initial quadratic behavior as required by symmetry, to have a softening of that initial stiffness ($t < 0$) as expected theoretically [42, 11], and then to have an upturn in free energy for $\sigma \approx 0.1$ from tight wrapping (Fig. 7).

The free energy model with this addition is

$$\mathcal{F} = x_{\rm s}\left(-g + \frac{c_{\rm s}\sigma_{\rm s}^2}{2}\right) + x_{\rm p}\left(\frac{p}{2}\sigma_{\rm p}^2 + \frac{t}{3}\sigma_{\rm p}^3 + \frac{q}{4}\sigma_{\rm p}^4\right). \tag{4.2}$$

This is just (3.10) with cubic and quartic terms added to $\mathcal{P}(\sigma_{\rm p})$. Minimization of this free energy proceeds exactly as in Sec. 3.1 with the change that the calculation cannot be completed in closed form, but can easily be done numerically.

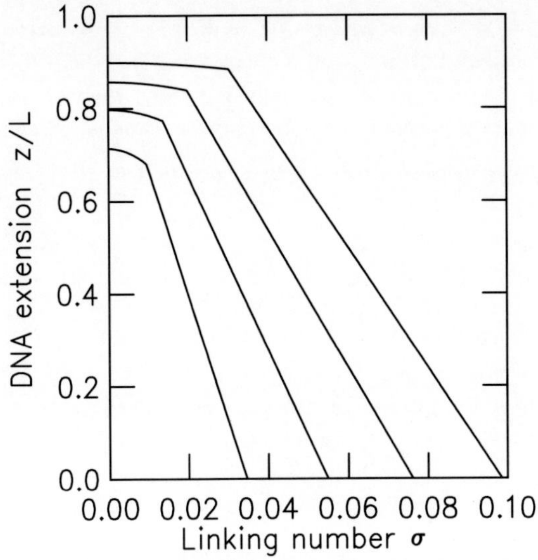

FIG. 8. *Extension vs. linking number curves for model of the text, with cubic and quartic term included in the plectoneme free energy. Parameters used are* $A = 50$ *nm,* $C = 95$ *nm,* $P = 24$ *nm,* $t = -800$ k_BT/nm*, and* $q = 9000$ k_BT/nm *(plectoneme parameters are as in the solid curve of Fig. 7). Curves are plotted for forces of 0.25 (lowest curve), 0.5, 1 and 2 pN (highest curve). These curves should be compared with those of Fig. 3 (see text).*

Fig. 8 shows the results of the anharmonic-$\mathcal{P}$ model for extension *vs.* linking number, for the same series of forces ($f = 0.25$, 0.5, 1 and 2 pN) and extended-state parameters ($A = 50$ nm, $C = 95$ nm) used in Fig. 3. The only difference between the calculation leading to Fig. 3 and that leading to Fig. 8 is that for the latter, cubic and quartic terms ($t = -800$ k_BT/nm, $q = 9000$ k_BT/nm) corresponding to the solid curve of Fig. 7 have been included. The large numerical values of t and q reflect the fact that the scale for σ is ≈ 0.05.

Comparing Fig. 3 and Fig. 8, the effect of the anharmonic terms in $\mathcal{P}$ is clear. The cubic softening of the plectoneme free energy leads to an increase in the value of σ_{p} for smaller forces, while the quartic stiffening at larger σ retards the increase of σ_{p} for larger forces. The result is that the successive extended-plectoneme coexistence curves of Fig. 8 are compressed together, and are more parallel than the rather more splayed ones of Fig. 3. At the same time, the behavior of $\sigma_s(f)$ is only mildly affected. Preliminary consideration of available experimental data (see, *e.g.*, left panel of Fig. 3 in Ref. [16]) indicates that this effect will be important to making the theory quantitatively describe experiment.

Fits of experimental data sets to the theory are still in preparation at the time of this writing, but will be published elsewhere in the near future. It appears likely that this model will be able to arrive at not just values of DNA elastic constants, but will provide insight into the variation of plectonemic supercoil free energy over a wider range of linking number than is accessible using conventional biochemical measurements.

5. Extrusion of a cruciform from twisted and stretched DNA. During the workshop, Vincent Croquette discussed experiments whereby special-sequence DNA molecules containing long palindromic repeats were subjected to twisting and pulling [52]. These molecules are able to respond to their underwinding in a way which is not usual for random-sequence DNA, by forming "cruciform" structures, where a region of the palindromic sequence strand-separates, and then where each strand is "extruded" as a hairpin-like double-helical arm. Croquette described how this arrangement may be used to measure the linking number of relaxed DNA, by measuring the length change of the extended region of the DNA per turn introduced, after the cruciform region starts to form. Since a region of DNA must be strand-separated to form the cruciform, this is driven by DNA underwinding ($\sigma < 0$). Croquette and co-workers have shown how this type of molecule may be used to carry out a novel measurement of the helix repeat of relaxed DNA in solution [52].

This situation can also be attacked using the methods presented above, allowing corrections for thermal fluctuations to be calculated. Suppose that a fraction $x_{\rm s}$ of our molecule is in extended (stretched and twisted) form, while the remaining fraction $x_{\rm c} = 1 - x_{\rm s}$ forms the two hairpin arms. Since the extruded arms are not under any torque or tension, and since they are reconstructed into stretches of double helix, their free energy can be taken to be that of unperturbed double helix, or zero. Thus the free energy of the whole molecule is just that of the extended region:

$$\mathcal{F}(\sigma) = x_{\rm s}\mathcal{S}(\sigma_{\rm s}). \tag{5.1}$$

Since the cruciform arms are entirely unlinked from one another, the linking number of the cruciform region is $\sigma_{\rm c} = -1$, and since $x_{\rm s}\sigma_{\rm s} + x_{\rm c}\sigma_{\rm c} = \sigma$, $\sigma_{\rm s} = (1 + \sigma - x_{\rm s})/x_{\rm s}$. Plugging this into the extended state free energy (3.7) gives

$$\mathcal{F}(\sigma) = -x_{\rm s} g + \frac{c_{\rm s}}{2}\frac{(\sigma + 1 - x_{\rm s})^2}{x_{\rm s}}. \tag{5.2}$$

The equilibrium value of $x_{\rm s}$ is obtained by minimization of (5.2):

$$x_{\rm s} = \begin{cases} 1 & \sigma > \sigma_{\rm c} \\ (1+\sigma)/\sqrt{1 - 2g/c_{\rm s}} & \sigma < \sigma_{\rm c} \end{cases} \tag{5.3}$$

where the critical linking number for cruciform extrusion is $\sigma_{\rm c} = -1 + \sqrt{1 - 2g/c_{\rm s}}$. As expected, the critical linking number is negative.

Recall that g is the free energy per length of untwisted stretched DNA, while c_s is the effective twisting stiffness of extended twisted DNA, and that both are essentially known functions of force. For most cases of interest, $g/c_s << 1$; for typical forces in the piconewton range, $g \approx k_BT/\text{nm}$, while $c_s > 100k_BT/\text{nm}$. Expansions in g/c_s are therefore useful to simplify some of the algebra, as in $\sqrt{1-2g/c_s} \approx g/c_s$. For example, in this approximation $\sigma_c \approx -g/c_s$.

Putting the result (5.3) into (5.2) allows us to calculate the total free energy of the extended-cruciform structure ($\sigma < \sigma_c$):

$$\mathcal{F}(\sigma) = (1+\sigma)c_s\left[\sqrt{1-2g/c_s}-1\right] \approx -(1+\sigma)g \tag{5.4}$$

where the final, approximate term is the leading order in expansion in g/c_s. The linearity in σ indicates that the torque during cruciform formation is constant:

$$\tau = \frac{1}{\omega_0}\frac{\partial \mathcal{F}}{\partial \sigma} = \frac{c_s}{\omega_0}\left[\sqrt{1-2g/c_s}-1\right] \approx -\frac{g}{\omega_0}. \tag{5.5}$$

Notably, the effect of the twist modulus cancels out of the $\mathcal{O}(g/c_s)$ result, leaving a dependence of the critical torque for cruciform formation on only the stretching free energy, and on the relaxed B-DNA helix parameter ω_0. This could be used to provide a torque standard in the k_BT range since both g and ω_0 are rather well known ($g \approx f \approx k_BT/\text{nm}$, $\omega_0 \approx 1.85\ \text{nm}^{-1}$). A series of measurements at different forces could therefore be used for a rather precise measurement of $c_s(f)$ in the extended state free energy (3.7).

The extension follows as

$$\frac{X}{L} = -\frac{\partial \mathcal{F}}{\partial f} \approx (1+\sigma)\frac{\partial g}{\partial f} \tag{5.6}$$

where only the leading term in expansion in g/c_s is retained for simplicity. Note that the final derivative $\partial g/\partial f$ is just the extension per contour length of *untwisted* DNA, and therefore that during extended-cruciform coexistence,

$$\frac{X(f,\sigma)}{X(f,0)} \approx 1+\sigma = 1+\frac{\text{Lk}}{\text{Lk}_0}. \tag{5.7}$$

This leading term of the expansion in g/c_s is essentially Eq. (1) of Ref. [52]. Since the linking number can be counted precisely in terms of magnet turns, the extension change per turn during cruciform formation can therefore be used to *measure* Lk_0, the relaxed linking number of DNA or equivalently the helix repeat of DNA, in solution [52]. The approach presented here provides a thermodynamic derivation which provides corrections due to thermal fluctuations to the analysis in Ref. [52] presented by Vincent Croquette at the workshop.

One thing that has been omitted in this "intensive" thermodynamic treatment of cruciform formation is the finite free energy cost associated with creating the cruciform; this could be included in the model by adding a constant "core energy" for the cruciform defect to (5.2). Care should be taken with the free energy in this case since it must then be compared to the free energy for pure extended DNA without this core energy present. Extensive pure state free energies should be used since the core energy is a boundary term of $\mathcal{O}(k_BT)$, scaling differently from the total polymer free energy $\mathcal{O}(k_BTL/\xi)$ where ξ is a length comparable to the persistence length. Including this would allow the barrier necessary for cruciform formation to be included in the resulting theory.

6. Conclusion. This paper has reviewed the theory [11, 27] for the elasticity of single twisted DNA molecules, aimed at describing single-DNA micromanipulation experiments. The shape of the extension-linking number curves is characterized by two regimes, a parabolic small-$|\sigma|$ "peak", and for larger $|\sigma$, linear "wings". The model discussed in this paper supposes that the peak is "pure" extended DNA, while the wings correspond to "phase coexistence" of an extended state and interwound plectonemic supercoiling [10, 11].

While the language of first-order phase transitions is used in this theory, formally there can be no true phase transitions in real experiments on finite-length one-dimensional molecules. However, due to relatively short thermal correlation lengths and appreciable cooperativity, the transitions seen experimentally are remarkably sharp and well described by what amounts to a mean-field theory of first-order transitions.

As discussed in more detail elsewhere [27] this theory gives a good account of experiments where DNA is twisted and pulled [16, 6, 9], and is easily extended to include effects of stress-driven DNA "melting" (strand separation). This paper has reviewed the theory and has presented new results for effects of anharmonicity in the plectoneme free energy, and for the free energy balance during stress-induced formation of cruciform structures in palindromic DNA.

At this point there remain questions concerning the application of this model to experimental data. Careful fits of theory to experiment have not yet been done, and likely will require use of the anharmonic free energy for the plectonemic state. The payoff should be appreciable thanks to the availability of a microscopic model for the extended state of Moroz and Nelson [36, 37] which allows measurements of the force-dependent twist rigidity c_s to be linked to the microscopic twist elastic constant C.

The coexistence model discussed here allows one to then determine the absolute free energy of plectonemically supercoiled DNA over a very wide range of σ. The results of Sec. 4 show that even small changes in the shape of the plectonemic free energy $\mathcal{P}(\sigma)$ (Fig. 7) lead to changes in the of the linear portions of the extension *vs.* linking number curves (compare

Figs. 3 and 8) that should be readily observable experimentally. Given that salt concentration changes should change the plectonemic free energy more than the extended state free energy, it ought to be possible to use the theory of this paper for stretched twisted DNA to rather comprehensively study the free energy of the interwound plectonemic domains.

The third key output of this theory is prediction of the torque in stretched twisted DNA, and therefore inside plectonemic DNA. Constant torques generated during plectoneme-extended coexistence are starting to be used to drive rotary relaxation experiments on topoisomerases [21, 23], but those torques have proven much more difficult to calibrate than forces. The theory presented here allows one to more directly determine torques in experiments, from extension measurements as a function of force and linking number. Furthermore, molecules containing palindromic sequences suggests may provide a kind of "torque standard", as discussed in Sec. 5.

Many additions, improvements and generalizations of this type of theory are possible. At the workshop Vincent Croquette commented that the behavior of the crude DNA denaturation model [27] used in the model to predict the "phase diagram" and the properties of the denatured states is likely too simple to completely describe experimental data. This issue will require further study, especially of comparison of theory with experimental data, to understand completely. The effects of domain fluctuations and sequence-dependence of denaturation and "pinning" of positions of the plectonemic domains are poorly understood, both experimentally and theoretically. Finally, understanding how the presence of DNA-binding proteins, as found along chromosomes *in vivo*, modify extension-linking number curves is a subject of theoretical [53, 54] and experimental [55] interest.

Acknowlegements. I am indebted to Nick Cozzarelli, Eric Siggia, Vincent Croquette, David Bensimon, Jean-Francois Allemand, Terence Strick, Alfonso Mondragón and Phil Nelson for many collaborations and conversations concerning the topics discussed in this paper, over a period of many years. I am also grateful to Wilma Olson, De Witt L. Sumners, Craig Benham, and the Institute for Mathematics and its Applications for organizing and sponsoring this workshop in honor of Nick, who was an inspiring leader and colleague. This research was supported by the US National Science Foundation through Grants DMR-9734178, DMR-0203963, MCB-0240998, PHY-0445565, and DMR-0715099, by a subcontract to NIH Grant 1-R21-GM-71019-1, and by awards from Research Corporation, the Petroleum Research Foundation of the American Chemical Society, the Chicago Biomedical Consortium, the University of Illinois Foundation, and the Focused Giving program of the Johnson and Johnson Corporation.

REFERENCES

[1] S. Smith, L. Finzi, and C. Bustamante, *Science* **258**, 1122–1126 (1992).

[2] C. Bustamante, J.F. Marko, S. Smith, and E.D. Siggia, *Science* **265**, 1599–1600 (1994).

[3] A.V. Vologodskii, *Macromolecules* 27, 5623–5625 (1994).

[4] T. Odijk, *Macromolecules* **28**, 7016–7018 (1995).

[5] J.F. Marko and E.D. Siggia, *Macromolecules* **28**, 8759–8770 (1995).

[6] T.R. Strick, J.-F. Allemand, D. Bensimon, A. Bensimon, and V. Croquette, *Science* **271**, 1835–1837 (1996).

[7] T.R. Strick, J.-F. Allemand, D. Bensimon, and V. Croquette, *Biophys. J.* **74**, 2016–2028 (1998).

[8] T. Strick, J.F. Allemand, D. Bensimon, R. Lavery, and V. Croquette, *Physica A* **263**, 392–404 (1999).

[9] T.R. Strick, J.-F. Allemand, V. Croquette, and D. Bensimon, *Prog. Biophys. Mol. Biol.* **74**, 115–140 (2000).

[10] J.F. Marko and E.D. Siggia, *Science* **265**, 506–508 (1994).

[11] J.F. Marko and E.D. Siggia, *Phys. Rev. E* **52**, 2912–2938 (1995).

[12] J.F. Marko and A. Vologodskii, *Biophys. J.* **73** 123–132 (1997).

[13] B. Fain, J. Rudnick, and S. Ostlund, *Phys. Rev. E* **55**, 7364–7368 (1997).

[14] J.F. Marko, *Phys. Rev. E* **55**, 1758–1772 (1997).

[15] J.F. Marko, *Phys. Rev. E* **57**, 2134–2149 (1998).

[16] S. Neukirch, *Phys. Rev. Lett.* **93**, 198107 (2004).

[17] T.R. Strick, V. Croquette, and D. Bensimon, *Nature* **404**, 901–904 (2000).

[18] N.J. Crisona, T.R. Strick, D. Bensimon, V. Croquette, and N.R. Cozzarelli, *Genes. Dev.* **14**, 2881–2892 (2000).

[19] N.H. Dekker, V.V. Rybenkov, M. Duguet, N.J. Crisona, N.R. Cozzarelli, D. Bensimon, and V. Croquette, *Proc. Natl. Acad. Sci. USA* **99** 12126–12131 (2002).

[20] G. Charvin, T.R. Strick, D. Bensimon, and V. Croquette, *Biophys. J.* **89** 384–392 (2005).

[21] D.A. Koster, V. Croquette, C. Dekker, S. Shuman, and N.H. Dekker, *Nature* **434** 671–674 (2005).

[22] N.H. Dekker, T. Viard, C.B. de la Tour, M. Duguet, D. Bensimon, and V. Croquette, *J. Mol. Biol.* **329** 271–282 (2003).

[23] B. Taneja, B. Schnurr, A. Slesarev, J.F. Marko, and A. Mondragon, *Proc. Natl. Acad. Sci. USA* **104**, 14670–14675 (2007).

[24] Z. Bryant, M.D. Stone, J. Gore, S.B. Smith, N.R. Cozzarelli, and C. Bustamante, *Nature* **424**, 338–341 (2003).

[25] J. Gore, Z. Bryant, M.D. Stone, M.N. Nollmann, N.R. Cozzarelli, and C. Bustamante, *Nature* **439** 100–104 (2006).

[26] C. Deufel, S. Forth, C.R. Simmons, S. Dejgosha, and M.D. Wang, *Nature Methods* **4** 223–225 (2007).

[27] J.F. Marko, *Phys. Rev. E* **76**, 021926 (2007).

[28] J.F. Marko, in *Les Houches Session LXXXII, Multiple aspects of DNA and RNA from biophysics to bioinformatics*, ed. D. Chatenay *et al.*, pp. 248–250 (Elsevier, San Diego CA, 2005).

[29] P.J. Hagerman, *Ann. Rev. Biophys. Biophys. Chem.* **17** 265–86 (1988).

[30] P. Cluzel, A. Lebrun, C. Heller, R. Lavery, J.L. Viovy, D. Chatenay, and F. Caron, *Science* **271**, 792–794 (1996).

[31] S.B. Smith, Y. Cui, and C. Bustamante, *Science* **271**, 795–9 (1996).

[32] R.W. Ogden, G. Saccomandi, and I. Sgura, *Comp. Math. Appl.* **53**, 276–286 (2007).

[33] L.D. Landau and E.M. Lifshitz, *Theory of Elasticity*, Ch. II (Pergamon, New York NY, 1986).

[34] D.M. Crothers, J. Drak, J.D. Kahn, and S.D. Levene, *Meth. Enzym.* **212** 3–29 (1992).

[35] V. Rossetto, *Europhys. Lett.* **69** 142–148 (2005).

[36] J.D. Moroz and P. Nelson, *Proc. Natl. Acad. Sci. USA* **94**, 14418–14422 (1997)

[37] J.D. Moroz and P. Nelson, *Macromolecules* **31**, 6333–6347 (1998).

[38] J.F. Marko, *Europhys. Lett.* **38**, 183–188 (1997).

[39] R.D. Kamien, T.C. Lubensky, P. Nelson, and C.S. O'Hern, *Europhys. Lett.* **38**, 237–242 (1997).

[40] T. Lionnet, S. Joubaud, R. Lavery, D. Bensimon, and V. Croquette, *Phys. Rev. Lett.* **96**, 178102 (2006).

[41] J. Gore, Z. Bryant, M. Nollmann, M.U. Le, N.R. Cozzarelli, and C. Bustamante, *Nature* **442**, 836–839 (2006).

[42] A.V. Vologodskii, S.D. Levene, K.V. Klenin, M. Frank-Kamenetskii, and N.R. Cozzarelli, *J. Mol. Biol.* **227**, 1224–1243 (1992).

[43] W. Bauer and J. Vinograd, *J. Mol. Biol.* **47**, 419–35 (1970).

[44] K.V. Klenin, A.V. Vologodskii, V.V. Anshelevich, A.M. Dykhne, and M.D. Frank-Kamenetskii, *J. Mol. Biol.* **217**, 413–419 (1991).

[45] V.V. Rybenkov, A.V. Vologodkskii, and N.R. Cozzarelli, *Nucl. Acids Res.* **25** 1412–1418 (1997).

[46] A.E.H. Love, *A Treatise on the mathematical theory of elasticity*, pp. 417–419 (Dover, New York NY, 1944)

[47] A. Sarkar, J.F. Léger, D. Chatenay, and J.F. Marko, *Phys. Rev. E* **63** 051903 (2001).

[48] S. Cocco, J. Yan, J.F. Léger, D. Chatenay, and J.F. Marko, *Phys. Rev. E* **70** 011910 (2004).

[49] J.F. Allemand, D. Bensimon, and V. Croquette, *Proc. Natl. Acad. Sci. USA* **95**, 14152–14157 (1998).

[50] T.R. Strick, V. Croquette, and D. Bensimon, *Proc. Natl. Acad. Sci. USA* **95** 10579–10583 (1998).

[51] S. Kutter and E.M. Terentjev, *Eur. J. Phys. B* **21**, 455–462 (2001); S. Kutter, Ph.D. Thesis, (University of Cambridge, England UK, 2002).

[52] A. Dawid, F. Guillemot, C. Breème, V. Croquette, and F. Heslot, *Phys. Rev. Lett.* **96**, 188102 (2006).

[53] J. Yan and J.F. Marko, *Phys. Rev. E* **68**, 011905 (2003).

[54] S. Cocco, J.F. Marko, R. Monasson, A. Sarkar, and J. Yan, *Eur. Phys. J. E* **10**, 249–263 (2003).

[55] B. Schnurr, C. Vorgias, and J. Stavans, *Biophys. Rev. Lett.* **1**, 29–44 (2006).

FLEXIBILITY OF NUCLEOSOMES ON TOPOLOGICALLY CONSTRAINED DNA

ANDREI SIVOLOB*, CHRISTOPHE LAVELLE†, AND ARIEL PRUNELL‡

Abstract. The nucleosome plays an ever increasing role in our comprehension of the regulation of gene activity. Here we review our results on nucleosome conformational flexibility, its molecular mechanism and its functional relevance. Our initial approach combined both empirical measurement and theoretical simulation of the topological properties of single particles reconstituted on DNA minicircles. Two types of particles were studied in addition to the conventional nucleosome: a subnucleosome consisting of DNA wrapped around the $(H3\text{-}H4)_2$ histone tetramer, now known as a tetrasome, and the linker histone H5/H1-bearing nucleosome, or chromatosome. All particles were found to thermally fluctuate between two to three conformational states, which differed by their topological and mechanical characteristics. These findings were confirmed for the nucleosome and the tetrasome by the use of magnetic tweezers to apply torsions to single arrays of these particles reconstituted on linear DNA. These latter experiments further revealed a new structural form of the nucleosome, the reversome, in which DNA is wrapped in a right-handed superhelical path around a distorted octamer. This work suggests that the single most important role of chromatin may be to considerably increase overall DNA flexibility, which might indeed be a requirement of genome function.

Key words. Nucleosomes, DNA minicircles, DNA supercoiling, conformational flexibility, chiral transition, magnetic tweezers, single molecules, chromatin fibers, chromatin superstructure.

AMS(MOS) subject classifications. 92C05 Biophysics, 92C40 Biochemistry, molecular biology.

1. Introduction. DNA in the cell nucleus is bound to basic proteins, the histones, to form chromatin, whose repeat unit is the nucleosome. The core of the nucleosome (the core particle) contains 147 bp of DNA wrapped in ~1.7 turns of a left-handed superhelix around an octamer of two copies each of the four core histones H2A, H2B, H3 and H4. Its high-resolution crystallographic structure [1–4] (Fig. 1a) is characterized by a pseudo two-fold axis of symmetry that passes through the H3/H3 interface (the four-helix bundle) and the central base pair of the 147 bp DNA fragment where the major groove faces the octamer. That point is defined as superhelix location zero, SHL0, and for each successive turn of the double helix the SHL number increases positively or negatively up to ±7 (Fig. 1a). The histone octamer is tripartite, being made of a $(H3\text{-}H4)_2$ tetramer flanked by two H2A-H2B dimers. The $(H3\text{-}H4)_2$ tetramer organizes the central 3/4

*Department of General and Molecular Genetics, Taras Shevchenko National University, 64 Vladimirskaya street, 01033 Kiev, Ukraine (sivolob@univ.kiev.ua).

†Laboratoire Physico-Chimie Curie, UMR CNRS 168, Institut Curie, 11 rue P. et M. Curie, 75231 Paris Cedex 05, France (lavelle@ihes.fr).

‡Former affiliation: Institut Jacques Monod, Centre National de la Recherche Scientifique, Université Denis Diderot Paris 7 et Université P. et M. Curie Paris 6, 2 place Jussieu, 75251 Paris Cédex 05, France. (ariel.prunell@paris7.jussieu.fr).

C.J. Benham et al. (eds.), *Mathematics of DNA Structure, Function and Interactions*, The IMA Volumes in Mathematics and its Applications 150, DOI 10.1007/978-1-4419-0670-0_13, © Springer Science+Business Media, LLC 2009

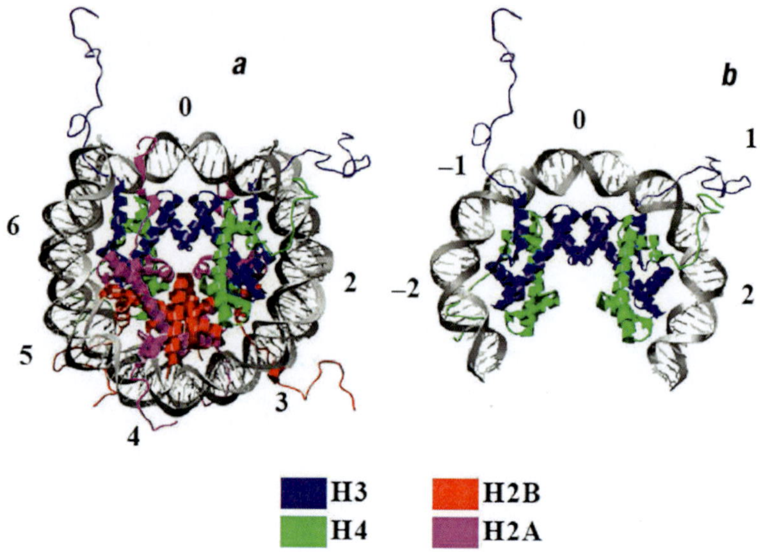

FIG. 1. *1.9 Å-resolution crystal structures. a) The 147 bp nucleosome core particle [PDB ID # 1KX5]. b) The tetrasome (central 55 bp on the (H3-H4)$_2$ tetramer) extracted from the core particle. Numbers indicate the SHLs (SuperHelix Locations). Images were created using UCSF Chimera (http://www.cgl.ucsf.edu/chimera).*

turn of the superhelix in between SHL±2.5 (Fig. 1b). This subnucleosome particle, called a tetrasome, or its precursor, the hexasome, may occur transiently through H2A-H2B dimer release during nucleosome remodeling [5] and/or transcription elongation [6–10]. H2A-H2B dimers complete the nucleosome by interacting with the two distal DNA regions from SHL+3.5 to +5.5 and SHL−3.5 to −5.5. Binding of the DNA ends at SHL±6.5 to the H3 α N extensions finally seals the DNA wrapping. The specific arrangement of α-helices in each histone, called the histone fold, not only insures the above described histone-DNA interactions, but also the histone-histone interactions within the octamer. The positively charged N-terminal tails of the histones protrude out from the particle, with H2B and H3 tails passing between the two gyres of the DNA superhelix through the channels formed by the aligned minor grooves [1]. The tails of H3, which are especially long, are appropriately located to interact with nucleosome entry/exit DNAs (Fig. 1a) and reduce their electrostatic repulsion. The tails, which are the substrate for various post-translational modifications [11], may also serve as platforms for the binding of specific activities (i.e. the so-called histone-code; [12, 13]). Among the various tail modifications, acetylation of lysine residues is associated with transcriptional activation (reviewed in [14, 15]). The pleiotropic roles of the tails also appear to include the modulation of nucleosome sliding and remodeling [13, 16, 17].

A single copy of the fifth histone, also known as the linker histone, H1 or H5 (H1 homologue in avian erythrocytes), interacts with the nucleosome. The H1/H5 molecule has an N-terminal tail, a globular domain and a long, highly positively charged, C-terminal tail (84 residues out of a total of 149 for H5) [18]. The globular domain seals the two superhelical turns at the DNA entry-exit, while the C-tail interacts further along these DNAs (see Section 2.3, below) [19–21]. The particle formed by the histone octamer, ~166 bp of DNA, and the H1/H5 histone is the chromatosome [19].

Nucleosomes in chromatin are connected by ~ 20–70 bp linker DNAs, resulting in an extended bead-on-a-string arrangement. This structure condenses at physiological ionic strength to resemble a zigzag by a process that is strictly dependent on the core histone tails [22–26]. At the next level of condensation, H1/H5 is required to stabilize a compact 30 nm chromatin fiber [27]. Microscopic techniques [28–30] and X-ray crystallography [31] have shown that the irregular 3D zigzag has nucleosomes with straight linkers projecting toward the fiber interior. Such a cross-linker model was also predicted by theoretical modeling [32–35], and is consistent with the internal location of H1/H5 [36–38] and the bridging together of nucleosome entry/exit DNAs into a stem through interactions with H1/H5 C-terminal tail (see Fig. 9, below) [21]. This stem could be recognized a posteriori in electron micrographs of native chromatin fragments [39] and it was subsequently considered as a unique structural motif directing chromatin higher order folding [40].

In the past decade, new concepts have emerged to illuminate the role of chromatin in regulating the access of transcriptional factors to their target sites. The central mechanism appears to be chromatin remodeling, both chemical, through covalent histone modifications (in particular acetylation; see above) and physical, whereby the energy of ATP hydrolysis is used to mobilize and structurally alter nucleosomes (reviewed in [41–45]). The latter mechanism may take advantage of *inherent* nucleosome dynamics, as shown by the spontaneous accessibility of nucleosomal DNA to binding proteins [46–48], and by the fluctuations of the fluorescence resonance energy transfer (FRET) between an acceptor and a donor fluorophores. These fluorophores, whose FRET efficiency is dependent on the distance, revealed dynamic modes when they were located i) either 75 bp apart in the same DNA fragment, so that DNA wrapping would bring them in register close to the dyad axis [49, 50]; ii) in the DNA and in the histones; or iii) both in the histones [48, 51, 52]. Other evidence for dynamic behavior of nucleosomes can be found in their ability to slide along the DNA at higher temperatures and salt [53], in the dependence of their overall structure on ionic strength, as again observed by FRET [50, 51], and in the extensive differences in DNA distortions observed between crystallized core particles on 146 and 147 bp of the same α-satellite sequence [1, 3].

This review is devoted to our studies of the topological manifestation of intrinsic nucleosome dynamics, which could be more relevant to their situation *in vivo*. It may be that nucleosomes with free DNA ends display artificially enhanced dynamics compared to nucleosomes that usually are, like ours, topologically constrained. Our results derive from two different substrates: single particles assembled on supercoiled DNA minicircles, and nucleosome arrays reconstituted on linear DNA with both ends attached. Minicircles were relaxed with topoisomerase I, and the products were analyzed and brought to simulations. Nucleosome arrays were subjected to rotational constraints using magnetic tweezers, and their length-vs.-torsion response was used to analyze nucleosome behavior in the context of the fiber. The following sections describe the methods, the results, and their potential physiological relevance.

2. A particle on a DNA minicircle. DNA topoisomers are identified by their linking number, Lk. Lk satisfies the well-known equation [54–56]:

$$Lk = Tw + Wr, \tag{2.1}$$

where $Tw = N/h$ is the twist of the double helix, with N being the number of base pairs and h the helical periodicity, and Wr the writhing of the closed curve formed by the double helix axis. Note that here and below the helical periodicity h is the so-called *intrinsic* or twist-related helical periodicity, i.e. the periodicity of the double helix in the laboratory frame. Generally, the linking number Lk does not coincide with the most probable twist $Tw_0 = Lk_0 = N/h_0$, where h_0 is the most probable helical periodicity for given conditions. This results in an elastic constraint in the circular DNA, which is measured by the linking number difference

$$\Delta Lk = Lk - Lk_0. \tag{2.2}$$

One also has

$$\Delta Lk = \Delta Tw + Wr, \tag{2.3}$$

where $\Delta Tw = Tw - Tw_0$.

The appearance of the constraint leads to an increase in the so-called supercoiling free energy. That energy, G_{sc}, depends quadratically on the linking number difference (with k_BT as the energy unit):

$$G_{sc} = (K_{sc}/N)(\Delta Lk)^2, \tag{2.4}$$

where K_{sc} is the supercoiling force constant [57].

A minicircle bearing a particle can be divided into two topologically distinct domains: the wrapped DNA, whose conformation is defined by histone interactions, and a free loop that is restricted only at its ends and adopts an equilibrium conformation elsewhere. In this case, the minicircle linking number difference becomes:

$$\Delta Lk = \Delta Lk_p + \Delta Lk_l. \tag{2.5}$$

This equation shows that ΔLk_p, the ΔLk associated with the particle, is the ΔLk of the topoisomer when the loop is relaxed ($\Delta Lk_l = 0$). It is easy to see that ΔLk is also equal to

$$\Delta Lk = \Delta Tw_p + \Delta Tw_l + Wr, \tag{2.6}$$

in which ΔTw_p and ΔTw_l are the twist changes on the histone surface and in the loop, respectively, and Wr the total writhe. Upon variations in ΔLk, ΔTw_p remains constant but the other two terms change. When the loop is relaxed ($\Delta Lk_l = \Delta Tw_l = 0$), $Wr = Wr_0$, and Eqs. (2.5) and (2.6) combine into

$$\Delta Lk_p = \Delta Tw_p + Wr_0. \tag{2.7}$$

The twist change in the particle is:

$$\Delta Tw_p = N_p(1/h_p - 1/h_0), \tag{2.8}$$

where N_p is the number of wrapped base pairs, and h_p their *intrinsic* helical periodicity. In general, this periodicity shall not coincide with the periodicity of the DNA contacts with the surface [58] (but see below), which will be referred to as the *local* periodicity h_{loc} (the periodicity in a local frame). If the vector normal to the double helix axis coincides with the normal to the surface (as is the case for the nucleosome), the relation between the two periodicities is [59]:

$$Tw_p = N_p/h_{loc} + \Theta_p/2\pi, \tag{2.9}$$

where Θ_p is the total geometrical torsion of the double helix axis in the particle. Because DNA wraps into a superhelix, the Frenet formulae of differential geometry can be used to give

$$\Theta_p = \frac{2\pi wp}{\sqrt{(2\pi r)^2 + p^2}} \tag{2.10}$$

where w is the number of turns of the superhelix, p its pitch ($p < 0$ for a left-handed superhelix), and r its radius.

Eqs (2.9) and (2.10) imply that the inequality $h_{loc} \neq h_p$ is a direct consequence of a superhelix with a non-zero pitch. However, it was recently recognized that this pitch is mostly defined by base pair longitudinal *slides* between successive, almost straight, DNA stretches [60]. Slide is here opposed to *shift*, which is the base pair lateral displacement. Strikingly, reconstruction of the superhelix with all base pairs parameters, except a zero shift, had little consequence on its geometry. In contrast, zeroing the slide resulted in a flattened superhelix (3 Å pitch, against 30 Å for the real superhelix) [60]. As a result, h_{loc} increases to nearly the level of h_p. $h_{loc} \sim h_p$ does not impinge on nucleosome and chromatosome calculations, which do not use h_{loc}, although there is a small effect on the geometry of the right-handed tetrasome (see Section 2.1, below).

The free energy of the particle-bearing minicircle is given by the same quadratic dependence as in Eq. (2.4), except that ΔLk in this equation is replaced by ΔLk_l, giving:

$$G_{sc} = (K_{sc}/N_l)(\Delta Lk - \Delta Lk_p)^2 + G_p, \qquad (2.11)$$

where N_l is the number of bp in the loop, and G_p describes the free energy of bending in the relaxed loop and additional contributions from the particle (various DNA distortions on the histone surface, histone-DNA and histone-histone interactions, etc.).

The experimental approach (Fig. 2, top), described in [61–63], involves first, the reconstitution of the particle on a negatively supercoiled topoisomer, and second, its relaxation with topoisomerase I. The result is an equilibrium mixture of particles on adjacent topoisomers (the starting topoisomer is not supposed to be a member of the equilibrium), and this mixture is electrophoresed in a polyacrylamide gel (Fig. 2, bottom right). The relaxed material is cut out from the gel (brackets), and eluted naked DNAs are electrophoresed in a second gel (Fig. 2, bottom left) to identify the topoisomers and quantify their relative amounts in the distributions (Fig. 2, profiles). The DNA length was changed by 1–2 bp increments at a time in order to get a rather continuous spectrum of Lk and ΔLk (see Eq. 2.2). This was accomplished for three unique DNA sequences derived from a fragment of plasmid pBR322, the 5S rDNA nucleosome positioning sequence [64], and a fragment of human α-satellite (centromeric) DNA. This resulted in three respective DNA minicircle series, the 351–366 bp pBR series [65], the 349–363 bp 5S series [66], and the 346–358 bp α-satellite series [67]. DNA most probable helical periodicities, h_0, were measured (together with K_{sc}) through relaxation of two naked minicircles of selected sizes within the series [66, 67].

Results are presented as a plot of the relative amount of each topoisomer in the equilibria as a function of ΔLk, for all DNA minicircle sizes of a series (see Figs 3b, 6a–c and 8, below). The usual multimodality of that plot reflects the possibility for the particle to exist in 2 or 3 discrete conformational states characterized by specific values of ΔLk_p, G_p and, in general, K_{sc}. According to Boltzmann law, the probability of a particle in state i on topoisomer ΔLk is proportional to

$$f(i, \Delta Lk) = exp(-G_{sc}(i, \Delta Lk)) \qquad (2.12)$$

where $G_{sc}(i, \Delta Lk)$ depends on $\Delta Lk_p(i)$, $G_p(i)$ and $K_{sc}(i)$ through Eq. (2.11). Neglecting the $\pm 2\%$ variation in N between 346 and 366 bp, N can be replaced by its mean, which gives:

$$F(\Delta Lk) = \frac{\sum_i f(i, \Delta Lk)}{\sum_{j=-3}^{3} \sum_i f(i, \Delta Lk + j)} \tag{2.13}$$

where $F(\Delta Lk)$ is the ordinate in the experimental topoisomer-relative amounts-*versus*-ΔLk plot. Eq. (2.13) was fitted to that plot to find the values of ΔLk_p and ΔG_p (ΔG_p is measured by reference to one of the states).

With two states, K_{sc}/N_l values in Eq. (2.11) can also be obtained from the fitting. With three states, however, the accuracy would decrease due to the larger number of parameters, and K_{sc} values were instead calculated using the explicit solutions to the equations of the equilibrium in the theory of the elastic rod model for DNA (referred to below as the "exact solutions theory" [68–70]). In this theory, the loop domain is treated as a segment with specified conditions at its end points where it contacts the protein surface. Because particles have a two-fold symmetry, tangent vectors to the end points are symmetrical to each other with respect to the dyad axis. The end conditions are then defined solely by the distance between these two end points and the relative orientation of these two vectors, both of which depend upon the geometry of the histone-bound DNA. This geometry can be approximated by an ideal superhelix of pitch p and radius r.

The DNA segment is treated in the theory as an inextensible, homogeneous body whose behavior can be described by the rod theory of Kirchhoff. The solutions to the equations of the equilibrium lead to the most probable conformation of the loop, with or without self-contacts, which minimizes the elastic free energy for specified end-conditions (in particular the pair of superhelix parameters p and r), and the loop torsional constraint, ΔTw_l (see Eq. 2.6). Once such a conformation is found, the elastic energy of the loop and the writhing of the whole minicircle can be calculated knowing the geometry of DNA in the particle (see [68] for details). The topoisomer ΔLk can subsequently be calculated when the DNA twist in the particle, ΔTw_p, is specified (see Eq. 2.6). The elastic energy was found to vary with ΔLk approximately according to a second-degree polynomial, which gives K_{sc} after identification with G_{sc} in Eq. (2.11) (neglecting thermal fluctuations, which is a reasonable approximation for a small loop).

Applications of these experimental and theoretical tools to the different particles are presented in the following subsections.

2.1. The tetrasome chiral transition. Examples of tetrasomes reconstituted on $\Delta Lk = \pm 1$ topoisomers of a 359 bp minicircle are shown in electron micrographs of Fig. 3a. In contrast to nucleosomes (Fig. 5, below), there is no hidden DNA turn wrapped around the histones, and the contour length of the particles is identical to that of the naked DNA (Fig. 3a). This is consistent with a horseshoe-shaped tetramer with ~ 55 bp of DNA wrapped in $\sim 3/4$ turn of a superhelix, as derived from the nucleosome crystal structure in Fig. 1b. Tetrasomes reconstituted on a short

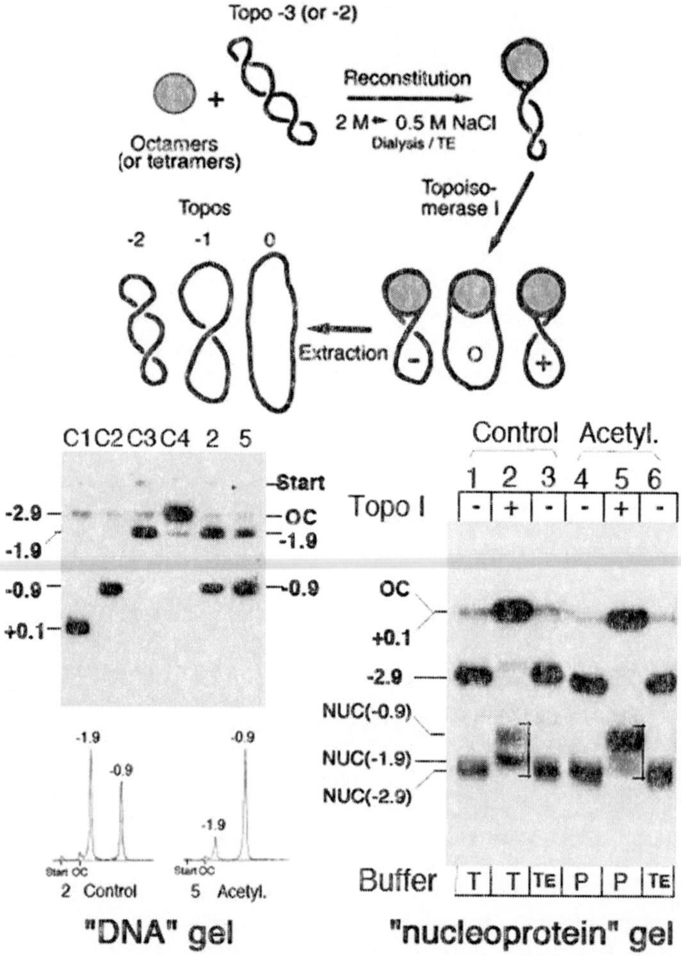

FIG. 2. *The minicircle approach and its illustration for nucleosomes on pBR 356 bp DNA minicircle. Top scheme: Mainsteps involve: 1) reconstitution; 2) relaxation with a topoisomerase; 3) gel electrophoreis of chromatin products; and 4) extraction of DNA products and their gel electrophoresis (from Fig. 4 in [67]). Bottom: Reconstitutions were performed on a* ^{32}P*-labeled topoisomer of* $\Delta Lk = -2.9$ *(see Eq. (2.2)) with control or acetylated (Acetyl.) core histones. Samples were incubated at 37 °C in relaxation buffer, either Tris (T: 50 mM Tris-HCL) (pH 7.5), 0.1 mM EDTA, 50 mM KCL, 5 mM* $MgCL_2$ *and 0.5 mM dithiohreitol) or phosphate (P: same as Tris buffer with 50 mM potassium phosphate (pH 7.5) instead of 50 mM Tris-HCL), with (Topo I+) or without topoisomerase I (Topo I−). Electrophoreses were in polyacrylamide gels at room temperature. OC: open (nicked) circular DNA. TE: starting chromatin in TE buffer (10 mM Tris-HCL (pH 7.5), 1 mM EDTA]. Note the two bands in nucleosome relaxation products (NUC), and the shift in their stoichiometry from the first conditions to the second. Eluted DNA products (brackets) were electrophoresed in a chloroquine-containing polyacrylamide gel, together with control topoisomers (C1–C4). Radioactivity profiles allow quantification of the topoisomers. The gel autoradiographs are shown (from Fig. 4 in [129]).*

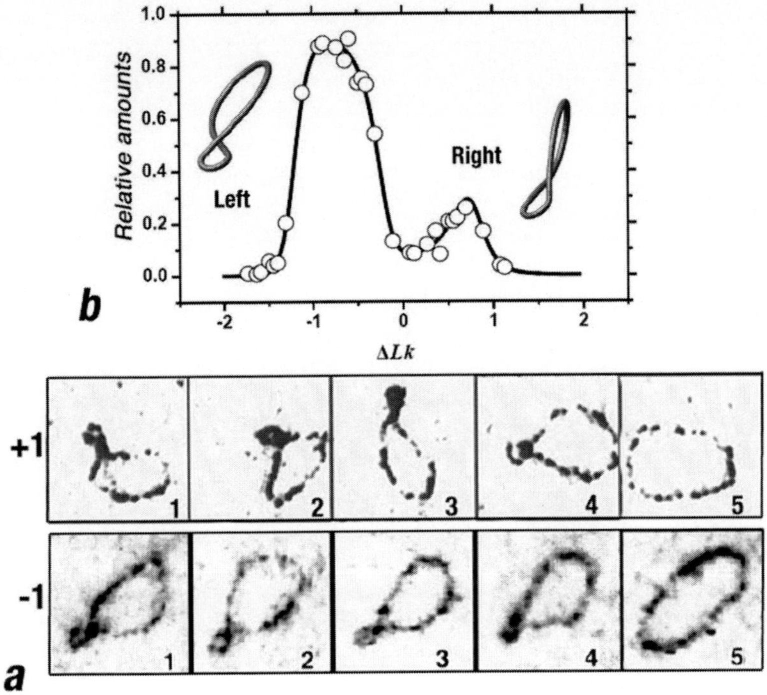

FIG. 3. *Tetrasomes visualized on* $\Delta Lk = \pm 1$ *topoisomers, and their relaxation data on pBR DNA minicircle series. a) Panels 1–4: electron micrographs of tetrasomes on 5S 359 bp DNA minicircle. Panels 5: naked topoisomers (from Figs. 10a and 10b in [71]). b) Relaxation data acquired as shown in the scheme of Fig. 2 are shown as topoisomer relative amounts versus* ΔLk *(*ΔLk *uses Eq. (2.2) with* $h_0 = 10.49_4(\pm 0.003)$ *bp/turn in Tris buffer; see legend to Fig. 2). The smooth curve was obtained from the fitted two-state model [Eqs. (2.11)–(2.13)] (from Fig. 5a in [73]). Relaxed left- and right-handed tetrasome DNA conformations were calculated using the exact solutions theory (from Fig. 3 in [80]).*

DNA fragment [71] or tandem repeats of 5S DNA [72] had a similar, although often uncrossed, hair-pin-like appearance. Results from relaxations of tetrasomes on the pBR minicircle series are shown in Fig. 3b. According to Eqs (2.11) and (2.13), a maximum in topoisomer probability should be observed when the minicircle ΔLk coincides with ΔLk_p, i.e. when the loop is relaxed. It follows from this that the bimodal profile in Fig. 3b should reflect tetrasome access to two alternative DNA conformations, around $\Delta Lk = \Delta Lk_p = -0.7$ and $+0.6$, respectively.

Fitting the plot in Fig. 3b to a two-state model produces the linking number difference, ΔLk_p, of each state, their free energy difference, ΔG_p, and their associated supercoiling force constant, K_{sc} [73]. ΔLk_p values, -0.74 and $+0.51$ for left- and right-handed states, approximately correspond to the center of the peaks in Fig. 3b, as expected. The right-handed state is energetically unfavorable by 1.9 k_BT relative to the left-handed state. K_{sc} values, 2400 and 1300 for the left- and right-handed states respectively, are quite different and both much lower than the naked minicircle value (4000). The naked DNA value was obtained around the relaxation point, i.e. when a change in the minicircle topological constraint should be stored almost entirely as torsion. It has been shown both theoretically [70, 74–76] and experimentally [77–79], that a threshold constraint is required before the onset of writhing in a minicircle, on the way to a figure-eight conformation. Here the loop is beyond the onset of writhing, and the low K_{sc} value simply reflects the fact that changing the writhe is easier in terms of energy than changing the twist by the same amount. However, an initial writhing of high energetic cost is required, and this energy is provided by histone-DNA interactions upon DNA wrapping. Interestingly, therefore, packing of DNA into a particle leads to a large increase in DNA conformational flexibility by overcoming this initial energetic barrier. Considering the existence of two states, the overall DNA flexibility is even larger.

The exact solutions theory explains why the loop can be more flexible in the right-handed state than in the left-handed state (see their K_{sc} values above), or, more precisely, why a given topological constraint should change the writhe of the loop more, and its twist less, in the right- *versus* the left-handed state. The reason is that the loop end-conditions change from one state to the other. Our reconstructions in Fig. 3b have a DNA superhelix radius of 5.1 nm in the right-handed state *versus* 4.7 nm in the left-handed state (against 4.3 nm in the nucleosome crystal structure). Such a lateral opening of right-handed particles was supported by electron microscopic visualization of a large number of tetrasomes on both linear and circular DNAs [71].

With $h_p = h_{loc}$ (see above), the DNA helical periodicity on the tetrasome changes slightly, as well as the radius estimate in the right-handed state. One obtains $h_p = h_{loc} = 10.3\pm\ 0.1$ bp/turn and $Wr_0 = 0.43 \pm 0.05$, from $h_{loc} = 10.2\pm\ 0.1$ bp/turn and $Wr_0 = 0.31\pm\ 0.05$ in [73, 80]. Such an h value, compared to 10.49 bp/turn for naked pBR DNA [63, 66], points to a significant DNA overtwisting in pBR tetrasomes. DNA is even more overtwisted on 5S tetrasomes, as indicated by $h_p = h_{loc} = 10.2\pm\ 0.1$ bp/turn, against 10.54 bp/turn for naked 5S DNA [80]. The 5S topoisomer amounts-*versus*-ΔLk profile (not shown) is similar to that in Fig. 3b. A shift along the ΔLk axis is observed, however, as a consequence of the larger overtwisting, resulting in $\Delta Lk_p = -0.68$ and $+0.60$ for left- and right-handed states. Moreover, the relative area of the "positive" peak is reduced compared to the pBR profile, reflecting a $\sim$ 50% higher transition free energy, ΔG_p.

Trypsinized tetramers, with H3, H4, or both H3 and H4 tails removed, where also studied [73]. Tail removal (especially H3's) decreases the proportion of negatively supercoiled topoisomers in the relaxation equilibria, indicating a facilitation of the tetrasome chiral transition. A similar trend was observed with tetrasomes reconstituted with moderately acetylated tetramers [73], but hyperacetylation turned out to be just as efficient in facilitating the transition as tail removal [81]. Trypsinized tetrasomes showed considerable changes in all parameters of the two conformational states. The transition free energy decreased by two-thirds, and a 10% lateral opening occurred in the left-handed conformation. These results reflect a regulatory role for the tails in the chiral transition.

A hint at the mechanism of this regulation can be found in the nucleosome crystal structure, which shows the histone fold-proximal domain of the H3 tails passing through channels provided by the aligned minor grooves of the two gyres at superhelix locations SHL+7 and −1 and SHL−7 and +1 [1]. In the absence of the second gyre, these interactions may still occur at SHL±1. At such locations, H3 tail proximal domains may act as wedges against the narrowing of the minor groove, i.e. the local straightening of the DNA, resulting from the transition-associated opening. Then only upon their release could the tetrasome open and the transition to the right-handed conformation occur [73]. The spontaneous occurrence of the transition under physiological conditions, i.e. the lateral opening, suggests that the tails are transiently released (or destabilized) due to thermal motions. Such a release can only become more frequent upon a decrease in the tail/DNA interactions resulting from acetylation.

The occurrence of a transition was initially proposed on the basis of tetrasome ability to assemble with similar efficiencies on both negatively and positively supercoiled DNA minicircles [82]. Negative and positive tetrasomes also had a similar appearance under electron microscopy, with a less-than-a-turn wrapping and crossed entry-exit DNAs (Fig. 3a). From this, the transition was thought to involve a change in chirality of the wrapped DNA, accompanied by a 360° rotation of the loop around the particle dyad axis and by a reversion of the crossing polarity from negative to positive. A reorientation of the two H3-H4 dimers in the H3/H3 four-helix bundle interface (Fig. 1b) was further suggested to mediate the change in the wrapping chirality. The involvement of the protein was directly demonstrated by the observation that a steric hindrance at the H3/H3 interface interferes with the transition. Bulky adducts introduced through thiol oxidation of H3 cysteines 110 (located on the interface) indeed oppose the transition or, on the contrary, block the tetramer right-handed [71, 82, 83]. This is the case of 5,5'-dithio-bis(2-nitrobenzoic acid) (DTNB), which was recently found to break the tetramer into its H3-H4 dimers [84]. This indicates that the stable positive supercoiling provided by DTNB-modified histones is acquired through a destabilized H3/H3 interface which reestablishes upon binding to DNA [84]. Similar results were obtained with the archeal histone-like HMf through mutagenesis at the HMf/HMf interface [85]. However, all our results with unmodified tetramers amply demonstrate that their chiral transition is smooth and does not require breaking them into dimers.

The proposed tetrasome chiral transition later received further experimental and theoretical support: i) ethidium bromide was found to hamper the transition, suggesting that the local base pair undertwisting resulting from its intercalation opposes DNA overtwisting in the dyad region that normally accompanies H3-H4 dimer reorientation [79]; ii) the neutron scattering pattern of tailless octamers exactly matches that predicted from the crystal structure, but the pattern of tailless tetramers does not [86], possibly as a reflection of the tetramer in solution being a mixture of left- and right-handed conformations [87]; iii) torsion of single tetrasome fibers in low salt revealed a centre of rotation similar to that of naked DNA (Fig. 13b, below), indicating that tetrasomes equilibrate equally between their two chiral forms; and iv) a molecular dynamics study (Normal Mode analysis) of the tetrasome revealed three lowest-frequency, i.e. most cooperative, vibrational modes, corresponding to movements of the whole H3-H4 dimers about each other (Fig. 4) [87]. The second of these modes involves dimer reorientation around an axis going through the two cysteines 110, while the third mode describes a lateral opening around an axis orthogonal to the former axis and intersecting it. These results explain our initial observation that the transition can occur unabated after cross-linking of these two cysteines through disulfide bridge formation [82].

2.2. Nucleosome conformational flexibility. Monte-Carlo calculations [59], and later the exact solutions theory [68], showed that a canonical $\sim$ 1.7-turn nucleosome on a DNA minicircle with a relaxed loop has a writhe $Wr_0 \sim -1.7$, while a $\sim$ 1.4-turn uncrossed nucleosome has $Wr_0 \sim -1.0$. Such nucleosomes were visualized by electron microscopy on $\Delta Lk = -1$ and -2 topoisomers of a pBR 359 bp fragment [88]. Interestingly, nucleosomes on the latter topoisomer fluctuate about equally between *closed negative* and *open* conformations in low salt (TE: 10 mM Tris-HCl and 1 mM EDTA, pH 7.5) (Fig. 5a), with the closed negative conformation being stabilized upon addition of 100 mM NaCl [88]. In contrast, nucleosomes on the $\Delta Lk = -1$ topoisomer were frozen in the open conformation regardless of the salt concentration [88]. Moreover, most of nucleosomes on the $\Delta Lk = 0$ topoisomer also had a crossed appearance [88], although their crossing must have been *positive* in order to compensate for the negative crossing inside the particle and minimize overtwisting of loop DNA.

Nucleosome relaxation and subsequent gel electrophoretic fractionation of nucleoprotein and DNA products is illustrated in Fig. 2, bottom, for the particular example of 356 bp pBR minicircle. The resulting topoisomer relative amounts-*versus*-ΔLk plot of these nucleosomes (Fig. 6a) shows shoulders or peaks centered at ΔLk values around -1.7, -1 and -0.5, which reflect nucleosome access to three distinct DNA conformations [63, 66]. As for tetrasomes, these peaks or shoulders must result from the relative energy benefit of relaxing into these particular topoisomers of ΔLk

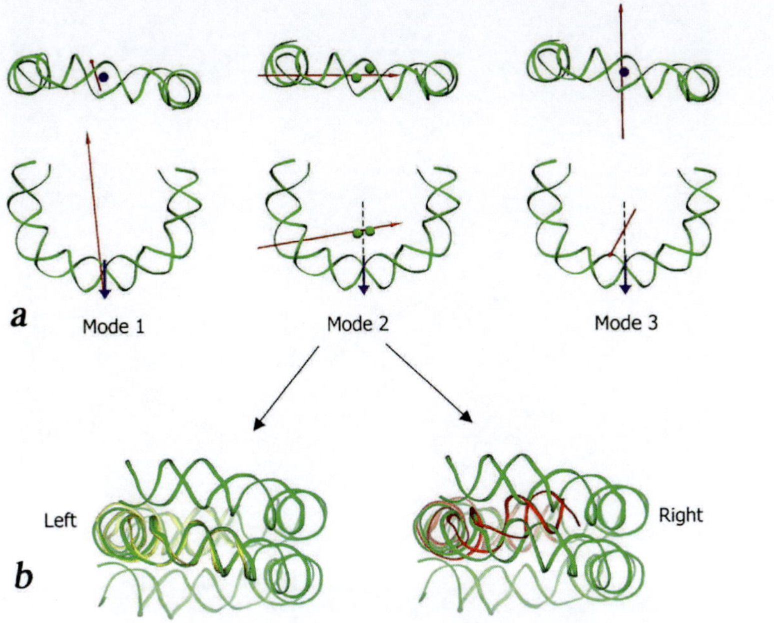

FIG. 4. *Normal mode analysis of the tetrasome. a) The axes (thin red arrows) of the rotation components of the three main vibrational modes are shown for each mode, superimposed on the tetrasome DNA superhelix viewed along the dyad (blue dots and thick blue arrows) or the superhelical axis. The axis of mode 1 runs close to the dyad, and the axis of mode 3 is approximately parallel to the superhelical axis. Mode 2 axis is approximately perpendicular to both dyad and superhelical axes. All three axes traverse cysteines 110 (green balls). b) The tetrasome was perturbed along the direction of mode 2 toward a positive superhelical pitch (right) and allowed to relax without constraint until its energy reached a local minimum. The resulting tetrasome DNA superhelix (red; note its right-handedness) is shown superimposed onto the background side of the nucleosomal superhelix (green) viewed perpendicular to both dyad and superhelical axes. In contrast, a perturbation along the same mode toward a more negative pitch (left) does not lead to a local energy minimum, and the tetrasome returns to its initial conformation (yellow) (the figure is Fig. 4 in [87]).*

$= \Delta Lk_p$, because only these topoisomers can provide a relaxed loop to the nucleosomes in these particular conformations. The $\Delta Lk \sim -0.5$ figure readily suggests that the crossing in the *closed positive* conformation is not complete, and stops about half-way (see below).

Application of the exact solutions theory to 1.45- and 1.7-turn nucleosomes led to K_{sc}/N_l estimates of 12 (± 1) ($K_{sc} \sim 2500$), only slightly different between the states [66]. Eqs (2.11–2.13) were fitted to the topoisomer relative amounts-*versus*-ΔLk plot in Fig. 6a, resulting in $\Delta Lk_p(i)$ and $\Delta G_p(i)$ values listed in Table 1 ($+Mg^{2+}$). The closed negative state is the most favorable, and the closed positive state the least, as expected, while the open state, taken as a reference of energy, is intermediate.

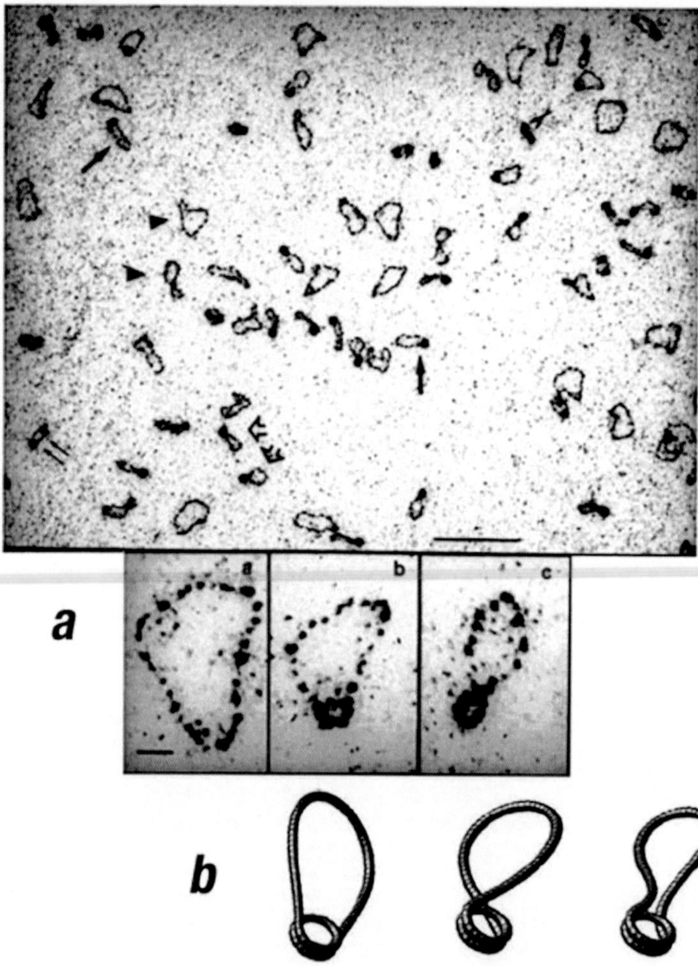

FIG. 5. *Nucleosomes visualized on* $\Delta Lk = -2$ *topoisomer, and model.* **a**) *Electron micrograph of chromatin on pBR 359 bp DNA minicircle in TE buffer. Gallery: a, naked topoisomer; b, open conformation; c, closed negative conformation (from Fig. 5 in [88]).* **b**) *Relaxed open, closed negative and positive 359 bp DNA conformations calculated using the exact solutions theory, with wrappings of 1.7 and 1.45 turns in closed and open states. Corresponding* W_r *values are* $-1.6_5, -1.0$, *and* -0.3 *(from Fig. 4 in [80]).*

$\Delta G_p(i)$ can be used to calculate the relative steady-state occupancy of state i, f_i, by a nucleosome with a nicked loop, i.e. free from torsional constraint. Using the equation

$$f_i = \frac{\exp(-\Delta G_p^i)}{\sum_i \exp(-\Delta G_p^i)} \tag{2.14}$$

TABLE 1

Nucleosome conformational state parameters on the three DNA series. $\pm Mg^{2+}$ *refers to the presence or absence of* $MgCl_2$ *in the relaxation buffer.* $\boldsymbol{h_p}$ *was calculated in the open state.*

DNA (histones)	Mg^{2+}	State	ΔLk_p ± 0.02	$\Delta G_p(k_BT)$ ± 0.1	$\boldsymbol{h_p}$ (± 0.03)/ $h_0(\pm 0.005)$ (bp/turn)
pBR (control)	+	negative	−1.69	−0.8	**10.49**/10.49
		open	−1.04	0	
		positive	−0.56	1.2	
	−	negative	−1.69	0.4	
		open	−1.04	0	
		positive	−0.56	1.7	
pBR (acetylated, phosphate)	+	negative	−1.73	0.8	
		open	−1.02	0	
		positive	−0.61	3.6	
5S (control)	+	negative	−1.40	−1.7	**10.30**/10.54
		open	−0.72	0	
		positive	−0.41	≥ 2.2	
	−	negative	−1.40	−0.6	
		open	−0.72	0	
		positive	−0.41	∞	
α-satellite (control)	+	negative	−1.55	−1.5	**10.30**/10.49
		open	−0.79	0	
		positive	−0.47	0.8	
α-satellite (CENP-A)		negative	−1.55	−0.1	
		open	−0.79	0	
		positive	−0.47	2.7	

one obtains 63%, 28% and 9% of pBR nucleosomes in the closed negative, open and closed positive states, respectively. This provides a concrete picture of the energy dependence of the equilibrium.

Interestingly, the above calculated Wr_0 is virtually identical to the fitted ΔLk_p for both closed negative and open states (Table 1). This coincidence reflects the absence of mean DNA overtwisting upon wrapping in pBR nucleosomes ($\Delta Tw_p = 0$ in Eq. 2.7), which results in h_p (the mean DNA helical periodicity on the histone surface; see Eq. 2.8) $= h_{loc} = h_0 =$ 10.49 bp/turn (Table 1). This result, together with the above reported overtwisting on the tetramer surface, would suggest that the DNA wrapped on H2A-H2B dimers is undertwisted in the pBR nucleosome (see below). In contrast, 5S nucleosomes (Fig. 6c) show a ~ 0.3 increase in ΔLk_p of both closed negative and open states, relative to Wr_0 values (Table 1; $+Mg^{2+}$). This reflected a $\Delta Tw_p \sim 0.3$ overtwisting relative to the naked DNA ($h_p =$ 10.30 bp/turn; Table 1), and a ~ 0.2 overtwisting, i.e. ~ 2 bp, relative to

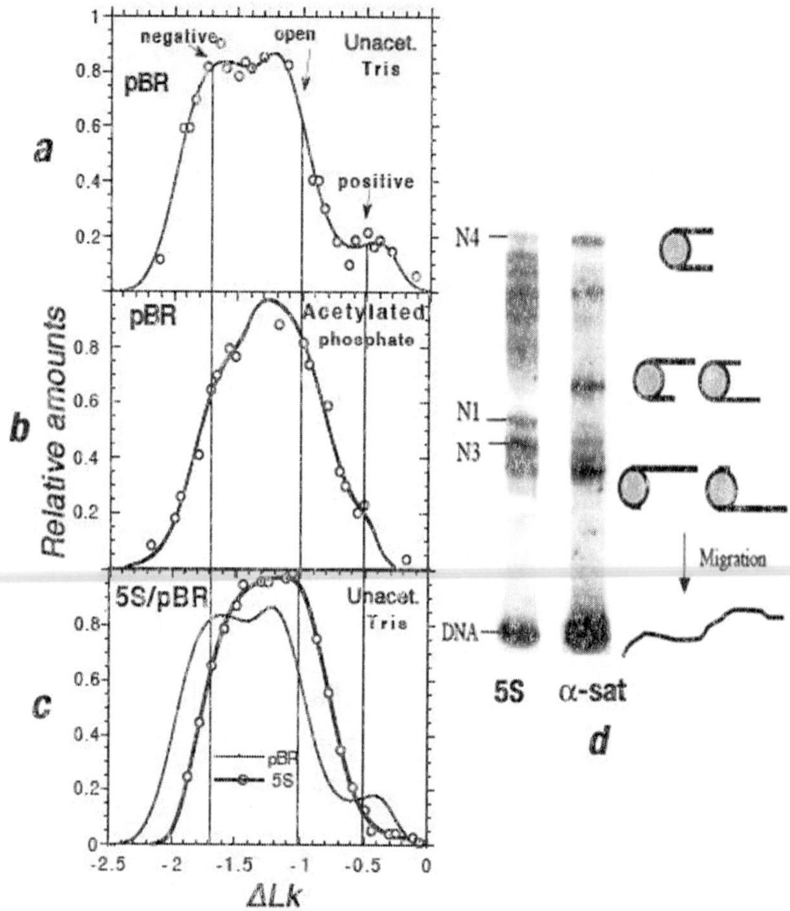

FIG. 6. *Relaxation data of nucleosomes on pBR and 5S DNA minicircle series, and their alternative DNA positions. a)–c) Data were acquired as shown in Fig. 2, bottom, using* $h_0 = 10.47_5(\pm 0.003)$ *bp/turn for pBR DNA in phosphate buffer (see legend to Fig. 3b for* h_0 *in Tris buffer), and* $h_0 = 10.53_8(\pm 0.006)$ *bp/turn for 5S DNA in Tris buffer. Smooth curves were obtained from the fitted three-state model (the figure is Fig. 5 in [129]). d) Electrophoretic fractionation in polyacrylamide gels of mononucleosomes on 357 bp 5S and 350 bp α-satellite DNA fragments. A subset of 5S nucleosome positions is marked (see their complete map in [21]). The diagram schematizes nucleosome position-dependent migration [53, 91] (from Fig. 2 in [67]).*

pBR nucleosomes (taking into account the h_0 difference, in the opposite direction, between the naked DNAs; Table 1). Consistent with this discrepancy, a comparison of DNase I footprints of the two nucleosomes trimmed to core particles revealed the same local periodicity everywhere except for a ~ 1 bp untwisting of pBR DNA relative to 5S DNA at each of the two dyad-distal sites (SHL±5) where H2B N-terminal tails pass between the two gyres (Fig. 1a) [66]. α-satellite nucleosomes also show an overtwisting ($\Delta Tw_p \sim 0.2$) relative to naked DNA (Table 1) [67].

5S nucleosomes access the negative state more frequently than do pBR nucleosomes (83% against 63% in the steady state equilibrium, respectively, calculated from Eq. (2.14) with corresponding ΔG_p values in Table 1), but about the same as do α-satellite nucleosomes (76%). Their unique feature, however, is to hardly access the positive state ($\leq 2\%$), in contrast to the other two (9% and 8%, respectively). Interestingly, this behavior is predicted by the loop elastic energy, G_{sc}, plotted as a function of ΔLk in Fig. 7a (straight). The theoretical $\Delta G_{sc} \sim 6k_BT$ between positive and negative states is indeed similar to the 5S $\Delta(\Delta G_p) \geq 4\ k_BT$ (Table 1). A closer look at the curve in Fig. 7a shows that the energy minimum of the negative state is located at the expected $\Delta Lk = -1.7$ (in the absence of overtwisting), whereas the positive state minimum, at $\Delta Lk \sim -0.3$ (against $\Delta Lk_p \sim -0.6$ for pBR nucleosome in Table 1), is not. This discrepancy may originate from the unfavorable position of the DNA self-contact in the loop (circles in Fig. 7a; straight), which prevents the true positive minimum to be reached, whereas the self-contact is too far on the left side of the curve to interfere with the negative state. Theoretical conformations for the three states are displayed in Fig. 5b [66].

With $\Delta Tw_l = 0$ at or around the G_{sc} minima in Fig. 7a, the twist contribution is cancelled and the entire loop elastic energy is in bending. It should be noted that there are other contributions to G_p in Eq. (2.11). Two of them originate from the DNA and favor the open state: an electrostatic repulsion between entry/exit DNAs, which is lower in the open state; and the straightening of the unwrapped DNA at the edges upon breaking of the contacts at SHL±6.5 (Fig. 1). Another contribution originates from the protein through these contacts, which stabilize both closed states (see below). The bending energy (ΔG_{sc}) and the electrostatic repulsion can then be considered as the sole contributors to G_p. Due to the early DNA self-contact described above, electrostatic repulsion should contribute more to the energy of the positive state, and $\Delta G_{sc} \sim 6\ k_BT$ in Fig. 7a should be considered as a lower bound for the free energy difference between the two states.

So why are the corresponding $\Delta(\Delta G_p)$ differences of pBR and α-satellite nucleosomes ($\sim 2\ k_BT$; Table 1) much smaller than the predicted value, allowing their easy access to the positive state? A simple answer to this question is to suppose that the relative orientation of entry/exit DNAs can vary. If they are slightly less divergent than expected from the standard superhelix, the positive crossing would indeed become easier and the negative crossing more difficult, as observed. To quantify the effect, we curved the superhelix axis in order to bring the two DNA gyres in contact at the entry-exit points (Fig. 7b). As shown in the profile (Fig. 7a; curved), the difference in the state energies, 2 k_BT, is now close to that of pBR and α-satellite nucleosomes. This curvature, called gaping, has subsequently

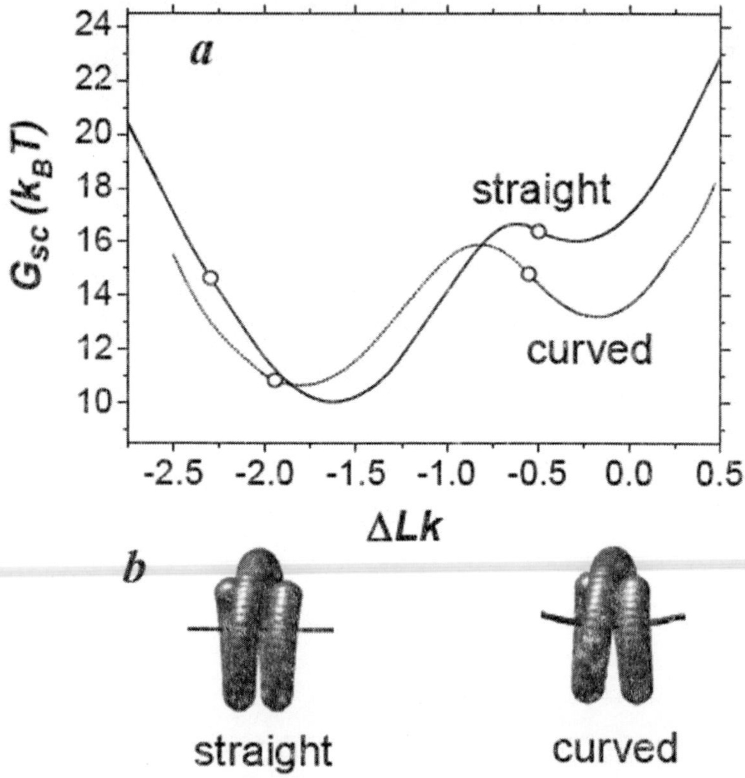

FIG. 7. *Loop elastic energy for two models of the nucleosome. a) The loop elastic energy, G_{sc}, was calculated as a function of the topoisomer ΔLk using the exact solutions theory for 1.7 turn DNA superhelices ($\Delta Tw_p = 0$) with straight or curved axes, as indicated in b). G_{sc} minima at $\Delta Lk \sim -1.7$ and -0.3 correspond to closed negative and positive states. Starting from the midregion of the energy profiles, the points at which a DNA self-contact first occurs in the loop are indicated by empty circles. A similar energy profile (straight) had previoulsy been reported in [69] (from Fig. 6 in [66]).*

been explored as a possibility to improve nucleosome-stacking properties of the 30 nm chromatin fiber [89] and condensation of mitotic chromosomes [90]. The process requires a rotation of the two H3-H4 dimers around their H3/H3 interface in a clockwise direction that increases the pitch of the negative superhelix (Fig. 4b). This not only incurs at high energetic cost ($\sim 20\ k_BT$; [89]), but is not supported by Normal Mode analysis of tetrasome structural dynamics (Fig. 4b, left).

For these reasons, reorientation of entry/exit DNAs in pBR nucleosomes has probably little to do with gaping, but is more likely a consequence of the 1 bp undertwistings at SHL±5 where H2B tails pass in between the two gyres (see above and Fig. 1) [66]. Other reorientation mechanisms may exist, however, as suggested by the similar ability of α-satellite nucleosomes to cross positively in the likely absence of undertwistings at SHL±5 (α-satellite nucleosomes resemble 5S nucleosomes with respect to mean twist; Table 1). At this point, it is important to remember that single pBR, 5S and α-satellite nucleosomes occupy multiple alternative positions (~15 for 5S and pBR and ~6 for α-satellite DNAs), as illustrated in Fig. 6d on linearized 5S and α-satellite minicircles. [Note that the fractionation in the gel is due to DNA curvature by the histones, which affects the molecule overall dimensions differentially, depending on the nucleosome position relative to the fragment ends [53, 91], exactly as was first observed with curved DNA [92].] These alternative nucleosomes are different from each other in a number of criteria, including their h_{loc} [91], and the features investigated here are, therefore, averaged over those populations. In particular, if a relation exists between entry-exit DNA reorientation and undertwistings in pBR nucleosomes, it is, therefore, on a statistical, but not a one-to-one basis.

Nucleosome conformational dynamics depends, therefore, on the DNA sequence (see a recent confirmation of this sequence-dependent nucleosome polymorphism in [50]), but also on the histone modification state. Relaxation of pBR nucleosomes reconstituted with *acetylated* histones in the presence of phosphate (Buffer P in Fig. 2, bottom) substantially modifies the relative amounts-*versus*-ΔLk profile (Fig. 6b) and led to large increases in ΔG_p of both closed states, making the open state energetically more favorable (Table 1). The role of acetylation in favoring nucleosome opening is in keeping with H3 N-terminal tails interacting with entry/exit DNAs, as shown by UV laser-induced cross-linking of long mononucleosomes [93]. The tails contain most of the acetylatable lysine residues, and their acetylation decreases the tail's overall positive charge. This in turn weakens the tails' interactions with entry/exit DNAs, especially in the presence of phosphate [94], and the DNA mutual repulsion increases. Interestingly, a similar effect was obtained upon removal of $MgCl_2$ from the relaxation buffer (Table 1; $-Mg^{2+}$) (and addition of monovalent cations (K^+) to keep h_o constant; [66]). Mg^{2+} may stabilize tail interactions with entry-exit DNAs, or directly favor the closed states by cross-linking the DNAs at their points of contacts. The effects of acetylation and of mono- and divalent salts were recently analyzed in details using FRET to measure the distance of DNA ends of mononucleosomes reconstituted on short fragments [95]. Steady-state occupancies of closed negative and open states by acetylated nucleosomes in phosphate become 32% and 65% (as compared to the reverse figures, 63% and 28%, for control nucleosomes; see above) and only 3% (against 9%) for the closed positive state.

Some histone variants favor nucleosome opening, such as H2A.Bbd, an H2A alternative enriched in transcriptionally active chromatin [96]. This was initially observed through micrococcal nuclease cleavage and FRET [97], and more recently by cryoelectron and atomic force microscopies [98]. This is also the case of CENP-A, an H3 variant of centromeric nucleosomes [99, 100], although its effect is somewhat subtler. The main changes in the CENP-A histone fold domain are a 2-residue expansion in loop L1 (between helices $\alpha 1$ and $\alpha 2$; [1]) and a replacement of arginine residues at H3 equivalent positions 49 and 83 by a lysine and an asparagine, respectively. While the effect of the 2-residue expansion is not clear, the consequence of the replacements is straightforward. H3 arginines 49 (in the αN extension) and 83 (in L1) stabilize the DNA superhelix at entry/exit positions of the nucleosome and the tetrasome, respectively, through intercalation of their lateral chain into the small groove at SHL6.5 and 2.5 [1], which lysine and asparagine will not do. A destabilization at the entry-exit was indeed observed in CENP-A nucleosomes (where H3 was substituted for CENP-A), as the energy of both negative and positive states was increased by 1.5–2 k_BT (Table 1). This further indicates that αN-DNA binding sites at SHL$\pm$6.5 are similarly effective in both conformations. The state occupancy can again be calculated using Eq. (2.14), and in turn the mean dynamic wrapping from wrappings in closed and open states (147 and 126 bp, respectively). When compared to H3 nucleosomes, CENP-A nucleosomes showed a 7($\pm$2) bp steady state unwrapping, which is sufficient to compromise the binding of a linker histone and to promote dissociation of H2A-H2B dimers by nucleosome assembly protein 1 (NAP-1) [67]. NAP-1 is ineffective to remove tetramers, and it was replaced by heparin, a strong acidic polyelectrolyte. The (CENP-A-H4)$_2$ tetramer was found much easier to release than the (H3-H4)$_2$ tetramer, consistent with replacement at position 83. Such a preferential two-stage disassembly of CENP-A nucleosomes relative to conventional nucleosomes was proposed to promote their observed progressive clearance from the chromosome arms by proteolysis following CENP-A transient over-expression [101, 102]. If applicable to CENP-A normal expression, this mechanism may be relevant to the problem of CENP-A exclusive centromeric localization (reviewed in [103]).

2.3. Chromatosome enhanced conformational flexibility. Relaxations of H5-containing pBR and 5S nucleosomes in the absence of Mg^{2+} (Mg^{2+} caused their precipitation) resulted in bi-modal plots with two well-separated peaks for negative and positive states, and no peak for the open state (Fig. 8, bottom). Fitting of the plots with the two-state model led to the values listed in Table 2. The two peaks are still observed with H5 globular domain (GH5 lacks both N- and C-terminal tails), although the positive peak is now substantially reduced in the 5S plot compared to the pBR plot (Fig. 8, top). Moreover, the peaks now partially overlap due to the smaller difference between their ΔLk_p values ($\sim$ 1, against $\sim$ 1.5 with H5; Table 2). Such a rescue by GH5 of the positive crossing in 5S nucleosomes presumably results from a normalization, albeit incomplete, of the relative orientation of entry-exit DNAs following GH5-induced increase in wrapping (compare linear −H5 and +GH5 nucleosomes in the gallery of Fig. 9, top). Surprisingly, GH5 generally decreases the amplitude of the crossings relative to control nucleosomes, as reflected by a mean shift of

$\sim +0.2$ in ΔLk_p (except for the positive crossing of 5S nucleosome, the ΔLk_p of which is instead shifted by ~ -0.1; Tables 1 and 2). The opposite is observed upon addition of the tails, i.e. the whole H5 amplifies the crossings relative to the controls (mean ΔLk_p shift of -0.2_5). H5 also increases the loop flexibility in both states, as indicated by the low K_{sc}/N_l values, 4–6 (Table 2), against 12 in control nucleosomes (see above). Relaxation experiments conducted with engineered H5 tail-deletion mutants [104] made it clear that the N-terminal tail plays a negligible role in the observed features, and that they are entirely due to the long, highly positively charged, C-terminal tail.

H5 C-terminal tail appears to act through the *stem* formed upon joining entry/exit DNAs together [21]. Mean stem lengths, measured on the molecules shown in Fig. 9 and others, were $\sim$ 10 bp in circular nucleosomes, and $\sim$ 30 bp in linear nucleosomes. With 10 bp, the contour length of the loop is $360 - 160 - 2 \times 10 = 180$ bp (360 bp is the minicircle size and 160 bp the length of wrapped DNA). The question then is how such a short loop can be that flexible. The exact solutions theory again gives the answer. Calculations showed that a 180 bp loop with its ends in contact reaches the observed mean value of $K_{sc}/N_l = 4.5$ only when the ends were parallel. In contrast, the rigidity increased rapidly upon introduction of an angle, or if the ends are moved apart from each other. The calculation further showed that the loop could not be significantly smaller than 180 bp, that is, the stem could not be significantly longer than 10 bp, if the large flexibility were to be preserved [104]. The occurrence of the stem also explains the extensive crossings observed. Indeed, the entry/exit duplexes are expected to be at an angle when they first come into close contact, so that they will tend to wind around each other along the stem to minimize bending. The winding will increase the loop net rotation angle around the dyad axis, shifting ΔLk_p of both states accordingly.

Building on this structural information, a model of the H5-containing nucleosome was constructed, which provided a physical and mathematical continuity to the DNA from the histone surface to the loop. In the junction domain, nucleosome entry/exit DNAs come into contact under a chosen angle, and cross negatively or positively. A right-handed or left-handed, respectively, double helix then insures the additional rotation of the loop around the dyad axis, and eventually brings the two duplexes into parallelism [104]. Fig. 9, bottom right, shows chromatosomes in the two states (with a relaxed loop). With small K_{sc} values, the loop rotates easily around the stem axis when submitted to a constraint (depending on the topoisomer ΔLk), keeping the supercoiling energy low.

3. The chromatin fiber. Nucleosome arrays were reconstituted on 2×18 tandem repeats of a 190 bp or 208 bp 5S nucleosome positioning sequence. They were subsequently ligated to one DNA spacer plus one DNA sticker at each end (Fig. 10a), and attached to the coated bottom of

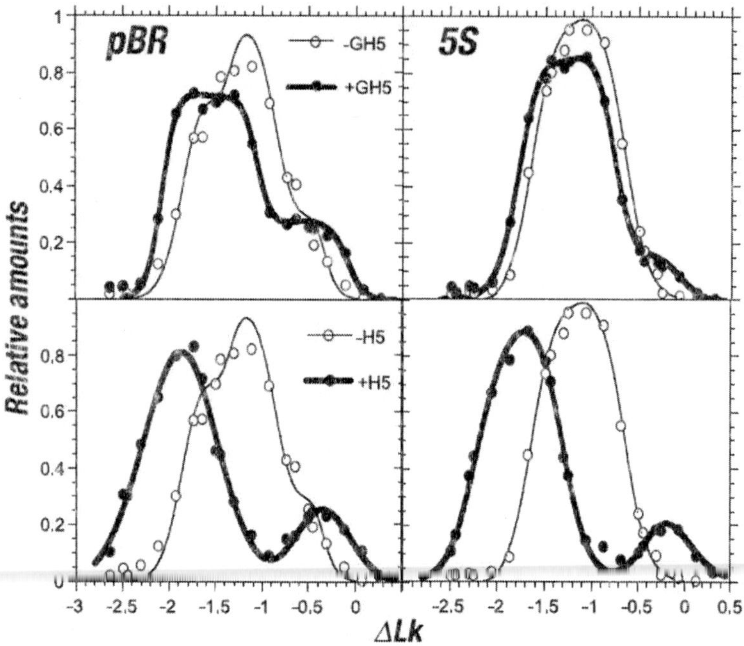

FIG. 8. *Relaxation data of GH5-and H5-containing nucleosomes on pBR and 5S DNA minicircle series. Relaxations were in $MgCl_2$-deprived Tris buffer (see legend to Fig. 2), with increased KCl concentration to keep DNA helical periodicities unchanged (see text). Smooth curves were calculated using the fitted two-state model (from Figs. 2 and 3 in [104]).*

the flow cell of a "magnetic tweezers" set-up at one end and to a paramagnetic bead at the other end (Fig. 10b). The rotation of the magnets, and hence of the bead, exerts torsion on a chosen fiber. The fiber extension and the force exerted on it are measured from the recorded three-dimensional position of the bead [87, 105].

3.1. Structural plasticity. Torsional behaviors are entirely described by the length-*versus*-rotation plots (Fig. 11) [106]. The response of the naked DNA (red in Fig. 11a) was obtained following chemical dissociation of the histones *in situ*. Its upper part corresponds to the elastic regime, and the quasi-linear compactions on both sides to the plectoneme regimes. The slope in these regimes is related to the radius and pitch of the plectoneme superhelical structures [106, 107]. The lower compaction on the negative side is due to force-dependent strand melting at high negative torsions, which relaxes the molecule. Compared to DNA, chromatin (blue in Fig. 11a) is shorter and its centre of rotation is shifted to negative values. These are the consequences of wrapping $\sim$ 50 nm of DNA, i.e. 150 bp, per nucleosome in a left-handed superhelix of $\Delta Lk_p \sim -0.8 \pm 0.1$

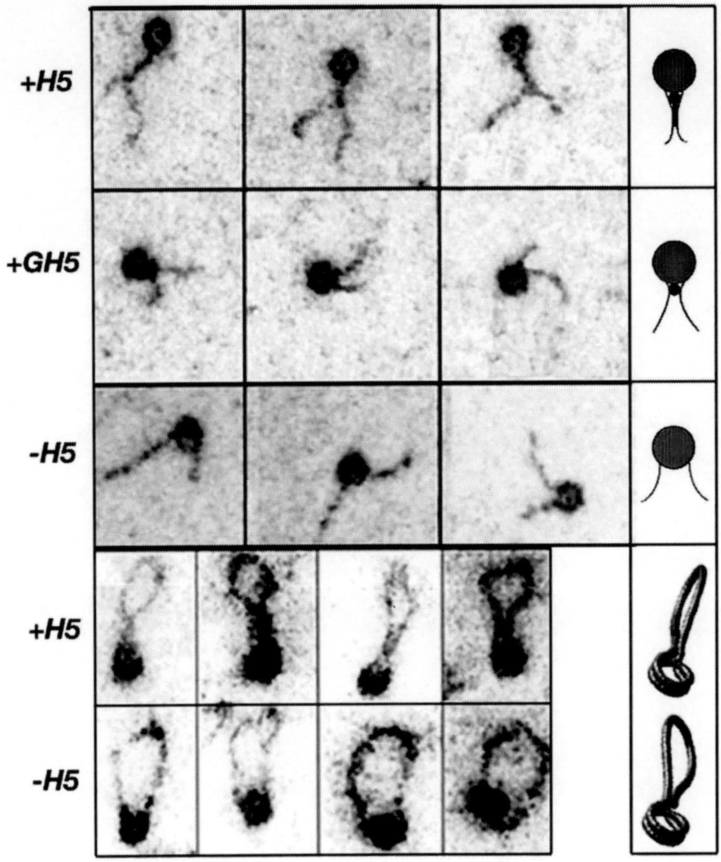

FIG. 9. *GH5/H5-containing nucleosomes visualized on linear and circular DNAs, and model. Electron micrographs of nucleosomes on 5S 256 bp DNA* ($-H5$ *and* $+GH5$), *5S 357 bp DNA (+H5) (top), or* $\Delta Lk = -1$ *topoisomer of pBR 359 bp minicircle* ($\pm H5$). *Linear nucleosomes were in TE buffer plus 50mM NaCl and 5mM* $MgCl_2$, *and circular nucleosomes in TE or TE plus 50–100 mM NaCl (with the same results; from Fig. 3 in [129] for linear nucleosomes and Fig. 5 in [62] for circular nucleosomes). Linear nucleosomes are schematized. H5-containing nucleosomes (with relaxed loops) were modeled using the exact solutions theory* ($Wr = -1.89$ *and* -0.39 *in negative (lower) and positive (upper) states; from Fig. 8 in [104]).*

(see below). Further comparison of DNA and fiber profiles with respect to their breadth requires the two have the same maximal extension under the same force. Taking advantage of the invariance in length of the DNA rotational response [106], the DNA profile was renormalized by dividing all lengths and rotations by the ratio of the maximal lengths, and shifted in order for its center of rotation to coincide with that of the fiber (red crosses in Fig. 11b). Compared to DNA of the same length, therefore, the fiber

TABLE 2
Conformational state parameters of H5- and GH5-containing nucleosomes.

DNA series	Linker histone	State	ΔLk_p ± 0.02	K_{sc}/N_l ± 1	$\Delta G_p(k_BT)$ ± 0.1
pBR	H5	negative	−1.89	3	0
		positive	−0.34	6	1.1
	GH5	negative	−1.57	6	0
		positive	−0.65	7	0.9
5S	H5	negative	−1.76	4	0
		positive	−0.16	6	1.6
	GH5	negative	−1.26	4	0
		positive	−0.29	12	1.9

appears extremely flexible in torsion, i.e. it can absorb large amounts of torsion without much shortening. Consistently, the worm-like rope elasticity model [108, 109] gives a rotational persistence length of 5 nm, much smaller than the 80 nm of DNA (smooth black curves in Fig. 11b). Moreover, the fiber is also more flexible in bending, with a persistence length of 28 nm, against 53 nm for DNA. Except for the fiber rotational persistence length, obtained for the first time, all values are similar to those obtained by others [106, 110–112]. Interestingly, the fiber plectoneme regimes are less steep than those of DNA, with a slope of 25 nm/turn, against 90 nm/turn for DNA. A smaller pitch and radius of the fiber plectonemes would be expected from its smaller bending stiffness. Partial neutralization of DNA phosphates by the highly positively charged histone tails could also result in a closer DNA/DNA approach of the linkers, or of nucleosome-free gaps.

This large torsional resilience of the fiber was interpreted as a reflection of nucleosome dynamic equilibrium between the three conformational states previously identified. A molecular model of the fiber architecture in the elastic regime was designed (Fig. 12), which quantitatively accounted for the upper part of the profile. The topological parameters derived from the model were actually close to those found above for 5S DNA (Table 1). The energy parameters showed an open state favored by ~ 1 and ~ 2 k_BT over the negative and positive states, respectively [105], quite similar to the situation encountered with acetylated histones in phosphate (Table 1). The reason is the low salt buffer (TE is used to minimize artifacts of nucleosome attractive interactions [113]), which also enhances entry/exit DNA repulsion. Nucleosomes in the open state must then predominate in the relaxed fiber at, or close to, the center of rotation, while the equilibrium is displaced toward negatively or positively crossed nucleosomes upon application of negative or positive torsions. The plectonemic regime is entered after all (negative) or most (positive) nucleosomes are in the crossed conformations.

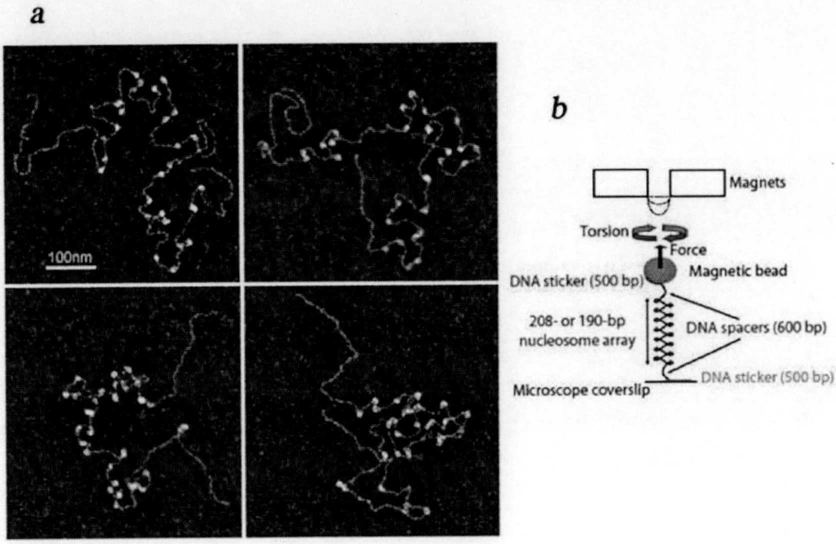

FIG. 10. *Chromatin fibers and their micromanipulation with magnetic tweezers. a) Electron micrographs of typical fibers reconstituted on* 2×18 *tandem repeats of a 190 bp 5S DNA fragment before their attachment. Red arrowheads indicate the occasional presence of clusters of two or three close-packed nucleosomes devoid of linker DNA. Nucleosome-free DNA spacers and stickers (*$\sim$ *1100 pb total, ligated onto the fibers after reconstitution) flanking the arrays are well visible. b) Scheme of the fiber and the magnetic tweezers setup (the figure is Fig. 1 in [87]).*

3.2. The nucleosome chiral transition. Provided that the torsion is not increased much beyond the zero-length limit on the positive side, forward and backward curves obtained upon increase or decrease of the torsion, respectively, more or less coincide (not shown). Beyond this limit, i.e. upon the application of typically +70 turns, the backward curve (green in Fig. 13a) departs from the forward curve (blue) on the positive side, revealing a hysteresis.

The hysteresis was argued to reflect the trapping of positive turns in individual nucleosomes, through their transition to an altered form called *reversome* (for chirally-*reverse* nucleo*some*), rather than collective effects (e.g. chromatin loops stabilized by nucleosome/nucleosome attractive interactions) [87]. Shifts on the positive side were reproducible for any given fiber over many cycles of torsions/detorsions, and were directly proportional to the number of regularly-spaced nucleosomes it contained, with a rate of 1.3 ± 0.1 turns per such nucleosome [87]. [Close-packed nucleosomes in Fig. 10a appear rigid and do not participate in conformational [105] nor in chiral [87] dynamics.] With $\Delta Lk_p \sim -0.4$ for positively crossed nucleosomes in the plectonemic regime [105], it comes for the reversome: $\Delta Lk_p \sim -0.4 + 1.3 \sim +0.9$.

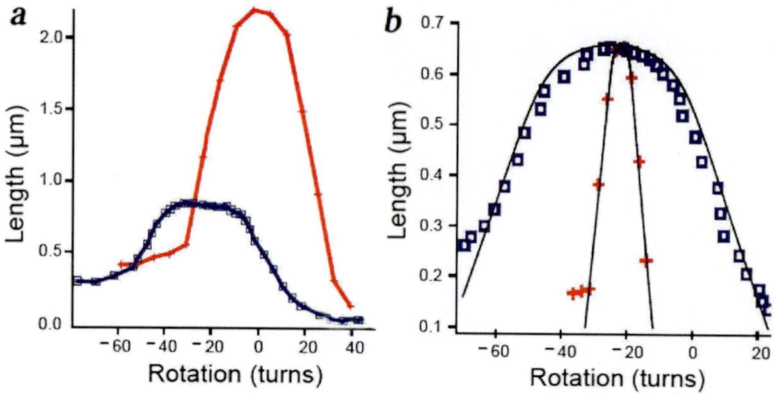

FIG. 11. *Fiber and DNA torsional responses. a) Extension-versus-rotation curve under a force of 0.35 pN in TE buffer of a chromatin fiber reconstituted on* 2×18 *tandem repeats of a 5S 208 bp DNA fragment (blue) and its corresponding naked DNA after complete nucleosome dissociation in the presence of 100 μg/mL heparin (red). b) Extension-versus-rotation curve of another chromatin fiber and of its corresponding DNA after renormalization. Smooth curves were obtained using the worm-like rope model (see text), assuming an elastic response in bending, stretching and twisting (from Fig. 2 in [105]).*

The hysteresis may then reflect the reversome metastability, due to a barrier in the energetic landscape between the two forms of the nucleosome. Consistently, when a fiber in the backward curve was allowed to relax in real-time, at constant force and rotation, a time-dependent shortening was observed which reflected reversome return to the canonical state. The proportions of each state were calculated as a function of time and used to estimate the energy parameters of the transition. We obtained an equilibrium energy difference of $\sim$10 k_BT relative to the ground state of the nucleosome (the open state) and an energy barrier of $\sim 30k_BT$ [87].

The hysteresis depends on the presence of H2A-H2B dimers. After their depletion upon successive treatments with heparin and core particles (NCPs), the resulting tetrasome fiber showed (Fig. 13b, purple): (i) an extended structure of maximal length intermediate between those of the initial nucleosome fiber and naked DNA; (ii) no hysteresis upon return from high positive torsions; and (iii) a center of rotation approximating that of the naked DNA. The first feature is consistent with the smaller wrapping in tetrasomes relative to nucleosomes, the second with the strict dimer requirement of the hysteresis, and the third with tetrasomes ability to fluctuate between left- and right-handed conformations of nearly equal and opposite ΔLk_p (see Section 2.1).

The requirement to break docking of dimers on the tetramer is expected to be a major contributor to the energy barrier. This view is in

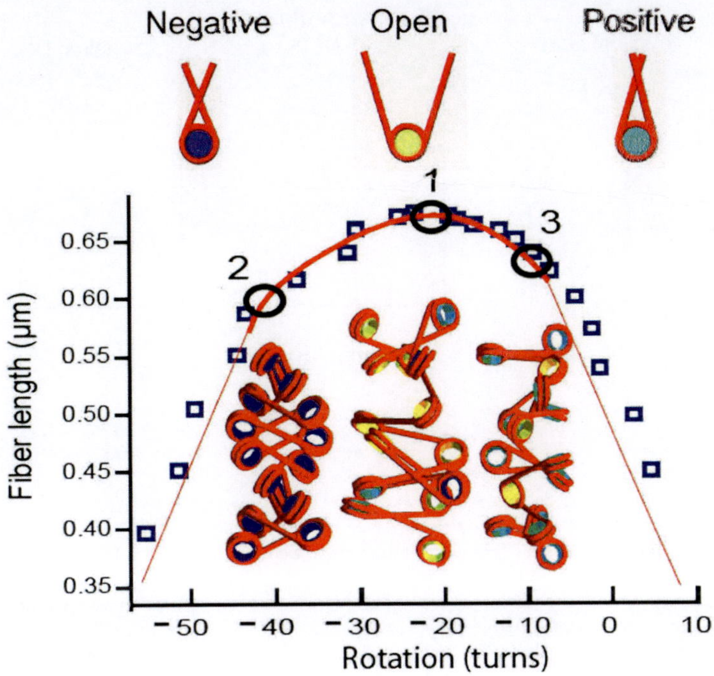

FIG. 12. *The fiber three-state molecular model. Top: diagrams of nucleosomes in negative, open and positive states. Bottom: the model, fitted to data of Fig. 11b, predicts the response over 30 turns around the apex (bold red curve). Beyond these torsions, the thin red straight line represent the best fit of a plectoneme model (not described). Under the curve are shown typical structures of the fiber at torsions marked by black circles (structures 1 at the apex, 2 and 3 at the thresholds on negative and positive plectonemic regimes). In structure 1, steady-state proportions of nucleosomes in open, positive and negative conformations are 65%, 20%, and 15%, respectively, in structures 2 and 3, 100% and 80% are negative and positive, respectively, the remaining 20% are in the open state (from Fig. 5 in [105]).*

keeping with an estimate of $\sim 17\ k_BT$ for the binding energy of each dimer onto the tetramer [114]. A mechanical (or elastic) barrier is also likely to exist beyond the point of dimers undocking: twist may accumulate at the expense of writhe and be suddenly released, generating an instability similar to that previously predicted for twisted rods [115]. The histone-imposed DNA curvature is expected to enhance the writhing instability, in conjunction with the extra lateral opening of the structure required at mid-transition to relieve the clash between entry/exit DNA arms [73].

The reversome ΔLk_p is close to that of the right-handed tetrasome (+0.9 against +0.6 for 5S tetrasomes; see Section 2.1). Based on the similarity between the torsional response of the tetrasome fiber (purple in Fig. 13b) and the backward curve of the nucleosome fiber (green) with respect to their breadth and center of rotation, we have proposed: 1) the

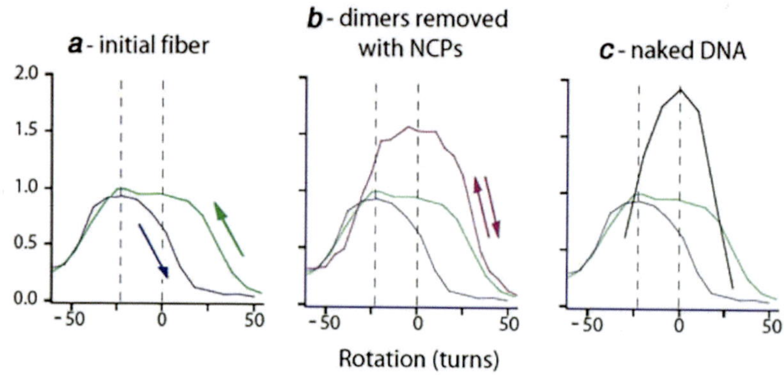

FIG. 13. *The fiber hysteretic response. a) Forward (blue) and backward (green) extension-vs-rotation curves in TE buffer of a 2 × 18 5S 190 bp fiber under a force of 0.35 pN after excursion at high positive torsions. b) Torsional response of the same fiber in TE (purple) after successive treatments with 1 μg/mL heparin in TE buffer, and 1 μg/mL nucleosome core particles (NCPs) in TE buffer plus 50 mM NaCl, under a steady high positive torsion. Similar results were obtained when H2A-H2B dimers were removed instead with 700 mM NaCl or upon transient application of a force of 3.5 pN (not shown). Moreover, the evidence that no (H3-H4)$_2$ tetramers were removed by the treatment was provided by the rescue of the initial fiber length and torsional behavior upon incubation with H2A-H2B dimers [87]. c) Corresponding naked DNA response after heparin-depletion of all histones and return to TE buffer (black) (from Fig. S1 in Supplemental Data to [87]).*

reverse, right- to left-handed, transition process to be common to both particles; and 2) the reversome core to be a right-handed tetrasome. The hysteresis observed for the nucleosome fiber, but not for the tetrasome fiber, may then solely reflect the H2A-H2B-linked energetic barrier in nucleosomes.

In the first step of the transition, dimers are expected to break their docking on the tetramer (Fig. 14). In the second step, the tetramer may undergo the chiral transition. We know that the right-handed 5S tetrasome partitions its $\Delta Lk_p = +0.6$ into $Wr = +0.4$ and $\Delta Tw = +0.2$ (see Section 2.1) [80]. Assuming a similar ΔTw on the reversome (if H2A-H2B dimers do not contribute), one gets $Wr = +0.7$ $(+0.9 - 0.2)$. This writhe is intermediate between that of the above tetrasome, +0.4, and that of a virtual right-handed nucleosome mirror image of the open-state nucleosome, +1. The reversome may then be substantially more open than the open nucleosome, although both particles fold a similar length of DNA (the similar maximal fiber extensions in forward and backward curves necessarily reflect similar length components along the direction of the force). As a consequence, dimers may not be strongly docked on the reverse tetramer, as expected from their less favorable new interface in reversomes (see arrows on H2As; Fig. 14). Moreover, H3 α_N-extensions (and N-terminal tails) are no longer appropriately located to interact with, and stabilize,

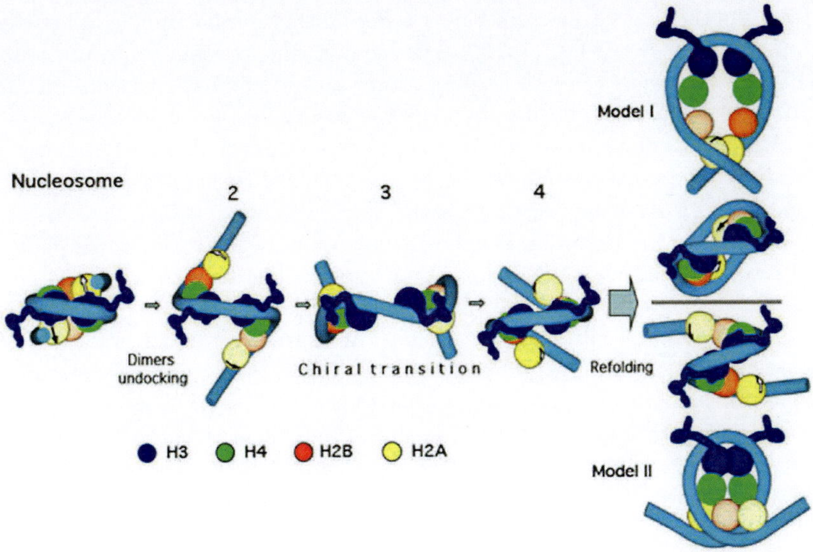

FIG. 14. *Scenario for the nucleosome-reversome transition. Individual H2As and H2Bs in nucleosome upper and lower faces are differentiated by light and dark colors and for H2A also by arrows. The two distal 10 bp DNAs are straight in step 2 as a result of the breaking of the H3 αN-entry-exit/DNA binding sites. Two alternative routes for refolding into the reversome are shown beyond step 4. In model I, entry-exit DNAs with bound H2A-H2B dimers tend to wind around each other. In model II, the DNAs plus the dimers tend to continue the tetrasome right-handed superhelix. The DNA diameter is not to scale to better show the histones (the figure is Fig. 7 in [87]).*

reversome entry-exit DNAs (Fig. 14). Two possible paths for those DNAs, which incorporate these features, are illustrated in Fig. 14. In model I, the dimer-bound DNA duplexes tend to wind around each other along the dyad axis. In model II, they instead try to continue the right-handed superhelix of the tetrasome, helped by the dimers that would somehow extend the tetramer's positive superhelical spool.

4. New solutions to old problems. The intricacies of DNA topology in chromatin. Reconstitutions of minichromosomes on DNA plasmids showed that the number of nucleosomes assembled did not depend significantly on the plasmid supercoiling [116]. With nucleosomes believed at that time to have a unique closed negative conformation, it was instead expected that the positive torsional stress resulting from their formation would hinder further reconstitution when the plasmids were relaxed or slightly positively supercoiled. At the same time, a number of physicochemical criteria indicated that the positively constrained nucleosomes were structurally identical to regular nucleosomes, raising the question of how so much stress could be dissipated. Moreover, whichever hidden alteration had occurred to the particles, it was entirely reversible upon release of the constraint, as shown by topoisomerase I relaxing them into canonical particles of mean $\langle\Delta Lk_n\rangle \sim -1$ [117–120] (see below). Nucleosome conformational dynamics provides a simple explanation to this enigma: the equilibrium shifts progressively to positively crossed nucleosomes upon reconstitution. Such almost topologically neutral nucleosomes (internal negative and external positive crossings compensate) lost much of their otherwise adverse influence on further nucleosome assembly.

In another experiment, negative supercoiling was introduced in naked and reconstituted plasmids using DNA gyrase. The maximal DNA supercoiling density reached ($\sigma = \Delta Lk/Lk_o$ (see Eq. 2.2) ~ -0.1) was nearly identical before and after reconstitution (measured in this latter case after deproteinization) [121]. Again, subsequent treatment with topoisomerase I resulted in canonical $\langle\Delta Lk_n\rangle \sim -1$ particles. Such a transparency of nucleosomes to DNA gyrase did not require DNA untwisting on the histone surface, as then hypothesized, but only a displacement of the equilibrium, now toward the negatively-crossed conformation, as quantitatively shown in [61].

The unit $\langle\Delta Lk_n\rangle$ value itself reflects an old problem: the so-called linking number paradox, which emerged from the necessity to reconcile topological and structural data of nucleosomes and chromatin [122–124]. With DNA assumed to continue the 1.75-turn left-handed superhelix revealed by the first crystal structure of the core particle [125], nucleosomes were viewed as two-turn particles, and as such should have reduce Lk by two turns (one-turn per negative crossing) instead of one. The early-proposed solution to the paradox was contained in Eq. (2.3): a positive ΔTw, i.e. a DNA overtwisting on the histone surface, if sufficiently large, can satisfy $\Delta Lk_n = -1$ [122, 126, 127]. Later on, this solution lost some of its luster when it was shown that the overtwisting observed was definitely too small [128]. Again, nucleosome conformational dynamics provides the explanation (reviewed in [129]): $\langle\Delta \mathrm{Lk}_n\rangle = -1$ simply reflects the steady-state proportions of nucleosomes with negative and positive crossings.

$\langle\Delta Lk_n\rangle$ in the minicircle system can be calculated from the state-averaged ΔLk_p pondered by the state occupancy (f_i; Eq. 2.14). It writes:

$$\langle\Delta Lk_p\rangle = \sum_i f_i \Delta Lk_p^i \tag{4.1}$$

where ΔLk_p^i are taken in Tables 1 and 2. Table 3 shows that $\langle\Delta Lk_p\rangle$ varies substantially from control to acetylated histones in phosphate, and from GH5 to H5. In contrast, it varies little between pBR and 5S nucleosomes (mean $\Delta\langle\Delta Lk_p\rangle = +0.07$), despite a more than 3-fold larger difference in ΔLk_p of the individual states (mean $\Delta(\Delta Lk_p) = +0.25$). $\langle\Delta Lk_p\rangle$ is found equal to -1.1_5 for 5S nucleosomes in the absence of Mg^{2+} (Table 3), not much different from $\langle\Delta Lk_n\rangle = -1.0$ for 5S minichromosomes relaxed under similar conditions [118, 119]. Moreover, the shift of $\langle\Delta Lk_p\rangle$ between control and acetylated histones in phosphate ($+0.2_5$), as well as upon Mg^{2+} depletion (mean = +0.2 over 5S and pBR nucleosomes; Table 3), is identical to that observed with minichromosomes from control to hyperacetylated histones ($-1.0_4 \pm 0.08$ to $-0.8_2 \pm 0.05$; [119]). The center of rotation of fibers micro-manipulated in magnetic tweezers also shifts by the same amount ($0.2_5 \pm 0.05$ turn per nucleosome) upon addition of 2 mM $MgCl_2$ and 40 mM NaCl [105]. It can be concluded that increasing the repulsion of nucleosome entry/exit DNAs, whether in a minichromosome, a fiber or a minicircle, either through a decrease in ionic strength or upon histone acetylation, similarly displaces the equilibrium toward the open state.

TABLE 3
$\langle \Delta Lk_p \rangle$ (± 0.05) *calculated from Eq. (4.1).*

	control $+Mg^{2+}$	acetylated/ phosphate	control $-Mg^{2+}$	+GH5	+H5
pBR	−1.4	-1.2_5	-1.2_5	−1.3	−1.5
5S	−1.3	-1.2_5	-1.1_5	-1.1_5	−1.5

We now believe, in view of the apparent absence of a relation between nucleosome DNA overtwisting and ability (or inability) for positive crossing (cf. contrasting data of 5S and α-satellite nucleosomes; Table 1), and contrary to a previous statement in [129], that $\langle \Delta Lk_n \rangle$ need not be an invariant, at least *in vitro*. $\langle \Delta Lk_n \rangle = -1.0$ was indeed obtained with 5S minichromosomes made of overtwisted nucleosomes of $\Delta Lk_p = -0.7$ in the open state (Table 1). In the absence of overtwisting and with ΔLk_p (open) $= -1$ (pBR in Table 1), the dominance of the negative state over the positive state should draw $\langle \Delta Lk_n \rangle$ below −1.0. A deviation of $\langle \Delta Lk_n \rangle$ of 10–20% from the unit value would hardly have been detected in reported experiments with SV40 or other non-5S minichromosomes [117, 130–132] in particular because the number of nucleosomes was not measured with sufficient precision. The influence of the linker histone on DNA topology in minichromosomes is also unclear. A series of measurements showed little effect of H1/H5 on $\langle \Delta Lk_n \rangle$ [117, 120, 130, 133, 134], but other data [131, 135] rather pointed to a large effect. It is also interesting that $\langle \Delta Lk_n \rangle$ was not shifted when hyperacetylated SV40 minichromosomes were assembled *in vivo* (and relaxed *in vitro*) [130]. These discrepancies suggest the existence of nucleosome interactions that interfere with the measurements by hindering their mutual rotation around the dyad axis, preventing the thermodynamic equilibrium to be reached. Similarly, minichromosomes show an abnormally low ability of their internucleosomal linker DNAs to untwist upon an elevation of their relaxation temperature (the so-called thermal flexibility) [133, 136], with the notable exceptions of yeast chromatin [137] and our single nucleosomes on DNA minicircles [65]. Nucleosome interactions would be expected to be negligible at low nucleosome density, and maximal at the saturated density achieved *in vivo*, explaining the SV40 data above [130]. The effect was directly observed in an experiment involving the binding of H5 to minichromosomes containing a variable number of nucleosomes. H5 again had little influence on $\langle \Delta Lk_n \rangle$ at high densities, but the shift at low densities was comparable to that observed with single nucleosomes in Table 3 [131].

5. Physiological relevance and prospects. The unique features of nucleosome conformational dynamics and chiral transition in chromatin fibers and DNA minicircles strongly appeal to their physiological relevance. Chromatin torsional resilience, mediated by the nucleosome conformational dynamics, may serve to cushion the supercoiling waves generated by polymerases upon replication or transcription (positive downstream and negative upstream; [138, 139]), and may actually be for these mechanisms the oil drop within the gear [140]. That resilience should even increase in the presence of the linker histone, as suggested by the enhanced loop flexibility resulting from stem formation between entry-exit DNAs (Section 2.3 and Table 2). This holds even if H1 binding is dynamic rather than static, as shown by its high exchangeability *in vitro* and *in vivo* [141–145]. With a deficit of H1 in active chromatin, nucleosomes should tend to adopt the open conformation (see [146, 147] for recent reviews of H1 role in regulating chromatin function). Consistently, transcriptional activity is tightly associated with histone acetylation [148, 149], which also favors the open state (Section 2.2 and Table 1). The open state facilitates the release of H2A-H2B dimers, as recently shown *in vitro* using NAP-1 (a histone chaperone) as a histone acceptor [67]. This further leads to additional unwrapping and to formation of single-turn tetrasomes [67], which expose more sites of potential binding to protein effectors.

Reversomes may be the last recourse when positive supercoiling waves can no longer be absorbed by the fiber. The formal condition for this is met since RNA polymerases exert a torque $> 1.25\ k_BT/\text{rad}$, equivalent to an energy $> 8\ k_BT$ over one turn [150], as compared to a transition free energy of $\sim 10\ k_BT/\text{turn}$ in TE (Section 3.2) and $\sim 6\ k_BT/\text{turn}$ in 50 mM salt [87]. The chiral transition may not, however, be a safeguard only, but may also be mechanically linked to transcription *in vivo*. We have proposed that the chiral-switching ability of the tetramer is used by the main polymerase to break docking of H2A-H2B dimers [87]. This idea is supported by the observation that a single nucleosome on a short DNA fragment, in which torsional constraints cannot develop due to free rotation of the ends, presents an almost absolute block to *in vitro* transcription by RNA polymerase II at physiological ionic strength [7]. The block is relieved in higher salt ($>$ 300 mM KCl), i.e. under conditions favoring dimer loss, and enzymes such as ACF or elongation factors such as FACT, which promote removal of a dimer, facilitate transcription elongation [151, 152]. Thus, dimers are likely to introduce a strong barrier to transcription also *in vivo*, and the tetramer chiral flexibility may, via the dynamic supercoiling, concur with local endogenous activities to destabilize them. Once reversomes are formed at a distance, they should be easily transcribed owing to their open structure and destabilized dimers (Section 3.2). Such reversomes may be viewed as transiently activated nucleosomes poised for polymerase passage.

Endogenous relaxing activities are not expected to interfere significantly with the above processes. Topoisomerase II (topo II) is notable since it was shown in yeast to relax chromatin five times as fast as topo I (topo I relaxes naked DNA twice as fast as topo II under the same conditions) [153]. The transcription-generated supercoiling was recently measured in B-cells using an activatable site-specific recombinase to excise a chromatin fragment positioned between two divergent promoters of a reporter gene (c-myc), which trapped the transient unrestrained negative supercoiling as chromatin circles. Before slowly decaying (in $\sim$ 30 min), that supercoiling was able to trigger non-B-DNA structure in a specific supercoiling-sensing sequence located within a linker six nucleosomes upstream of the promoters. This non-B-DNA structure in turn recruited two transcriptional factors essential for the expression of the gene [154]. Therefore, in addition to provide a cushion to transcription-induced supercoiling waves, and to be precisely tuned to polymerase passage, chromatin may also be the drive shaft in the modulated transmission of those waves for the dynamic control of gene expression [155].

Acknowledgements. This work, which spans twenty years or so, could not have been done without the enthusiastic help of many collaborators and co-authors of about the same number of papers referred to in the text. AP would like to express his gratitude to all of them, and especially to (by order of appearance) M. Le Bret, B. Révet, P. Furrer, V. Ramakrishnan, F. De Lucia, M. Alilat, N. Conde e Silva and A. Bancaud.

REFERENCES

[1] K. Luger, A.W. Mader, R.K. Richmond, D.F. Sargent, and T.J. Richmond, Crystal structure of the nucleosome core particle at 2.8 A resolution, *Nature*, **389** (1997), 251–260.

[2] J.M. Harp, B.L. Hanson, D.E. Timm, and G.J. Bunick, Asymmetries in the nucleosome core particle at 2.5 A resolution, *Acta. Crystallogr. D. Biol. Crystallogr.*, **56** (2000), 1513–1534.

[3] C.A. Davey, D.F. Sargent, K. Luger, A.W. Maeder, and T.J. Richmond, Solvent mediated interactions in the structure of the nucleosome core particle at 1.9 a resolution, *Journal of molecular biology*, **319** (2002), 1097–1113.

[4] R.K. Suto, R.S. Edayathumangalam, C.L. White, C. Melander, J.M. Gottesfeld, P.B. Dervan, and K. Luger, Crystal structures of nucleosome core particles in complex with minor groove DNA-binding ligands, *Journal of molecular biology*, **326** (2003), 371–380.

[5] L.A. Boyer, X. Shao, R.H. Ebright, and C.L. Peterson, Roles of the histone H2A-H2B dimers and the (H3-H4)(2) tetramer in nucleosome remodeling by the SWI-SNF complex, *J. Biol. Chem.*, **275** (2000), 11545–11552.
[6] B.W. Baer and D. Rhodes, Eukaryotic RNA polymerase II binds to nucleosome cores from transcribed genes, *Nature*, **301** (1983), 482–488.
[7] M.L. Kireeva, W. Walter, V. Tchernajenko, V. Bondarenko, M. Kashlev, and V.M. Studitsky, Nucleosome remodeling induced by RNA polymerase II: Loss of the H2A/H2B dimer during transcription, *Mol. Cell.*, **9** (2002), 541–552.
[8] B. Li, M. Carey, and J.L. Workman, The role of chromatin during transcription, *Cell*, **128** (2007), 707–719.
[9] O.I. Kulaeva, D.A. Gaykalova, and V.M. Studitsky, Transcription through chromatin by RNA polymerase II: histone displacement and exchange, *Mutat. Res.*, **618** (2007), 116–129.
[10] V. Jackson, In vivo studies on the dynamics of histone-DNA interaction: evidence for nucleosome dissolution during replication and transcription and a low level of dissolution independent of both, *Biochemistry*, **29** (1990), 719–731.
[11] R. Marmorstein, Protein modules that manipulate histone tails for chromatin regulation, *Nat. Rev. Mol. Cell. Biol.*, **2** (2001), 422–432.
[12] B.M. Turner, Cellular memory and the histone code, *Cell*, **111** (2002), 285–291.
[13] B.D. Strahl and C.D. Allis, The language of covalent histone modifications, *Nature*, **403** (2000), 41–45.
[14] M. Grunstein, Histone acetylation in chromatin structure and transcription, *Nature*, **389** (1997), 349–352.
[15] C.A. Mizzen and C.D. Allis, Linking histone acetylation to transcriptional regulation, *Cell. Mol. Life Sci.*, **54** (1998), 6–20.
[16] A. Hamiche, J.G. Kang, C. Dennis, H. Xiao, and C. Wu, Histone tails modulate nucleosome mobility and regulate ATP-dependent nucleosome sliding by NURF, *Proc. Natl. Acad. Sci. U.S.A.*, **98** (2001), 14316–14321.
[17] G.J. Narlikar, H.Y. Fan, and R.E. Kingston, Cooperation between complexes that regulate chromatin structure and transcription, *Cell*, **108** (2002), 475–487.
[18] V. Ramakrishnan, Histone H1 and chromatin higher-order structure, *Crit. Rev. Eukaryot. Gene. Expr.*, **7** (1997), 215–230.
[19] R.T. Simpson, Structure of the chromatosome, a chromatin particle containing 160 base pairs of DNA and all the histones, *Biochemistry*, **17** (1978), 5524–5531.
[20] J. Allan, P.G. Hartman, C. Crane-Robinson, and F.X. Aviles, The structure of histone H1 and its location in chromatin, *Nature*, **288** (1980), 675–679.
[21] A. Hamiche, P. Schultz, V. Ramakrishnan, P. oudet, and A. Prunell, Linker histone-dependent DNA structure in linear mononucleosomes, *Journal of molecular biology*, **257** (1996), 30–42.
[22] F. Thoma, T. Koller, and A. Klug, Involvement of histone H1 in the organization of the nucleosome and of the salt-dependent superstructures of chromatin, *J. Cell. Biol.*, **83** (1979), 403–427.
[23] P.M. Schwarz and J.C. Hansen, Formation and stability of higher order chromatin structures. Contributions of the histone octamer, *J. Biol. Chem.*, **269** (1994), 16284–16289.
[24] L.M. Carruthers and J.C. Hansen, The core histone N termini function independently of linker histones during chromatin condensation, *J. Biol. Chem.*, **275** (2000), 37285–37290.
[25] J.C. Hansen, Conformational dynamics of the chromatin fiber in solution: determinants, mechanisms, and functions, *Annu. Rev. Biophys. Biomol. Struct.*, **31** (2002), 361–392.

[26] G. Arya and T. Schlick, Role of histone tails in chromatin folding revealed by a mesoscopic oligonucleosome model, *Proc. Natl. Acad. Sci. USA*, **103** (2006), 16236–16241.

[27] L.M. Carruthers, J. Bednar, C.L. Woodcock, and J.C. Hansen, Linker histones stabilize the intrinsic salt-dependent folding of nucleosomal arrays: mechanistic ramifications for higher-order chromatin folding, *Biochemistry*, **37** (1998), 14776–14787.

[28] C.L. Woodcock, S.A. Grigoryev, R.A. Horowitz, and N. Whitaker, A chromatin folding model that incorporates linker variability generates fibers resembling the native structures, *Proc. Natl. Acad. Sci. USA*, **90** (1993), 9021–9025.

[29] S.H. Leuba, G. Yang, C. Robert, B. Samori, K. Van Holde, J. Zlatanova, and C. Bustamante, Three-dimensional structure of extended chromatin fibers as revealed by tapping-mode scanning force microscopy, *Proc. Natl. Acad. Sci. USA*, **91** (1994), 11621–11625.

[30] K. Van Holde and J. Zlatanova, What determines the folding of the chromatin fiber?, *Proc. Natl. Acad. Sci. USA*, **93** (1996), 10548–10555.

[31] T. Schalch, S. Duda, D.F. SARGENT, and T.J. Richmond, X-ray structure of a tetranucleosome and its implications for the chromatin fibre, *Nature*, **436** (2005), 138–141.

[32] V. Katritch, C. Bustamante, and W.K. Olson, Pulling chromatin fibers: Computer simulations of direct physical micromanipulations, *Journal of molecular biology*, **295** (2000), 29–40.

[33] D.A. Beard and T. Schlick, Computational modeling predicts the structure and dynamics of chromatin fiber, *Structure*, **9** (2001), 105–114.

[34] G. Wedemann and J. Langowski, Computer simulation of the 30-nanometer chromatin fiber, *Biophysical journal*, **82** (2002), 2847–2859.

[35] H. Wong, J.M. Victor, and J. Mozziconacci, An all-atom model of the chromatin fiber containing linker histones reveals a versatile structure tuned by the nucleosomal repeat length, *PLoS ONE*, **2** (2007), e877.

[36] S.H. Leuba, J. Zlatanova, and K. Van Holde, On the location of histones H1 and H5 in the chromatin fiber. Studies with immobilized trypsin and chymotrypsin, *Journal of molecular biology*, **229** (1993), 917–929.

[37] V. Graziano, S.E. Gerchman, D.K. Schneider, and V. Ramakrishnan, Histone H1 is located in the interior of the chromatin 30-nm filament, *Nature*, **368** (1994), 351–354.

[38] J. Zlatanova, S.H. Leuba, G. Yang, C. Bustamante, and K. Van Holde, Linker DNA accessibility in chromatin fibers of different conformations: a reevaluation, *Proc. Natl. Acad. Sci. USA*, **91** (1994), 5277–5280.

[39] J. Bednar, R.A. Horowitz, J. Dubochet, and C.L. Woodcock, Chromatin conformation and salt-induced compaction: three-dimensional structural information from cryoelectron microscopy, *J. Cell. Biol.*, **131** (1995), 1365–1376.

[40] J. Bednar, R.A. Horowitz, S.A. Grigoryev, L.M. Carruthers, J.C. Hansen, A.J. Koster, and C.L. Woodcock, Nucleosomes, linker DNA, and linker histone form a unique structural motif that directs the higher-order folding and compaction of chromatin, *Proc. Natl. Acad. Sci. USA*, **95** (1998), 14173–14178.

[41] P.B. Becker and W. Horz, ATP-dependent nucleosome remodeling, *Annu. Rev. Biochem.*, **71** (2002), 247–273.

[42] T. Tsukiyama, The in vivo functions of ATP-dependent chromatin-remodelling factors, *Nat. Rev. Mol. Cell. Biol.*, **3** (2002), 422–429.

[43] B.R. Cairns, Chromatin remodeling: insights and intrigue from single-molecule studies, *Nat. Struct. Mol. Biol.*, **14** (2007), 989–996.

[44] V.K. Gangaraju and B. Bartholomew, Mechanisms of ATP dependent chromatin remodeling, *Mutat. Res.*, **618** (2007), 3–17.

[45] P. Choudhary and P. Varga-Weisz, ATP-dependent chromatin remodelling: action and reaction, *Subcell. Biochem.*, **41** (2007), 29–43.
[46] K.J. Polach and J. Widom, Mechanism of protein access to specific DNA sequences in chromatin: a dynamic equilibrium model for gene regulation, *Journal of molecular biology*, **254** (1995), 130–149.
[47] J.D. Anderson, A. Thastrom and J. Widom, Spontaneous access of proteins to buried nucleosomal DNA target sites occurs via a mechanism that is distinct from nucleosome translocation, *Mol. Cell. Biol.*, **22** (2002), 7147–7157.
[48] G. Li, M. Levitus, C. Bustamante, and J. Widom, Rapid spontaneous accessibility of nucleosomal DNA, *Nat. Struct. Mol. Biol.*, **12** (2005), 46–53.
[49] M. Tomschik, H. Zheng, K. Van Holde, J. Zlatanova, and S.H. Leuba, Fast, long-range, reversible conformational fluctuations in nucleosomes revealed by single-pair fluorescence resonance energy transfer, *Proc. Natl. Acad. Sci. USA*, **102** (2005), 3278–3283.
[50] L. Kelbauskas, N. Chan, R. Bash, P. Debartolo, J. Sun, N. woodbury, and D. Lohr, Sequence-dependent variations associated with H2A/H2B depletion of nucleosomes, *Biophysical journal*, **94** (2008), 147–158.
[51] U.M. Muthurajan, Y.J. Park, R.S. Edayathumangalam, R.K. Suto, S. Chakravarthy, P.N. Dyer, and K. Luger, Structure and dynamics of nucleosomal DNA, *Biopolymers*, **68** (2003), 547–556.
[52] G. Li and J. Widom, Nucleosomes facilitate their own invasion, *Nat Struct. Mol. Biol.*, **11** (2004), 763–769.
[53] G. Meersseman, S. Pennings, and E.M. Bradbury, Mobile nucleosomes-a general behavior, *Embo. J.*, **11** (1992), 2951–2959.
[54] J.H. White, Self-linking and the Gauss integral in higher dimensions, *Am. J. Math.*, **91** (1969), 693–728.
[55] F.B. Fuller, The writhing number of a space curve, *Proc. Natl. Acad. Sci. USA*, **68** (1971), 815–819.
[56] F.H. Crick, Linking numbers and nucleosomes, *Proc. Natl. Acad. Sci. USA*, **73** (1976), 2639–2643.
[57] D.S. Horowitz and J.C. Wang, Torsional rigidity of DNA and length dependence of the free energy of DNA supercoiling, *Journal of molecular biology*, **173** (1984), 75–91.
[58] L.E. Ulanovsky and E.N. Trifonov, Superhelicity of nucleosomal DNA changes its double-helical repeat, *Cell. Biophys.*, **5** (1983), 281–283.
[59] M. Le Bret, Computation of the helical twist of nucleosomal DNA, *Journal of molecular biology*, **200** (1988), 285–290.
[60] M.Y. Tolstorukov, A.V. Colasanti, D.M. McCandlish, W.K. Olson, and V.B. Zhurkin, A novel roll-and-slide mechanism of DNA folding in chromatin: implications for nucleosome positioning, *Journal of molecular biology*, **371** (2007), 725–738.
[61] Y. Zivanovic, I. Goulet, B. Revet, M. Le Bret, and A. Prunell, Chromatin reconstitution on small DNA rings. II. DNA supercoiling on the nucleosome, *Journal of molecular biology*, **200** (1988), 267–290.
[62] Y. Zivanovic, I. Duband-Goulet, P. Schultz, E. Stofer, P. oudet, and A. Prunell, Chromatin reconstitution on small DNA rings. III. Histone H5 dependence of DNA supercoiling in the nucleosome, *Journal of molecular biology*, **214** (1990), 479–495.
[63] F. De Lucia, M. Alilat, A. Sivolob, and A. Prunell, Nucleosome dynamics. III. Histone tail-dependent fluctuation of nucleosomes between open and closed DNA conformations. Implications for chromatin dynamics and the linking number paradox. A relaxation study of mononucleosomes on DNA minicircles, *Journal of molecular biology*, **285** (1999), 1101–1119.
[64] R.T. Simpson and D.W. Stafford, Structural features of a phased nucleosome core particle, *Proc. Natl. Acad. Sci. USA*, **80** (1983), 51–55.

[65] A. Hamiche and A. Prunell, Chromatin reconstitution on small DNA rings. V. DNA thermal flexibility of single nucleosomes, *Journal of molecular biology*, **228** (1992), 327–337.

[66] A. Sivolob, C. Lavelle, and A. Prunell, Sequence-dependent nucleosome structural and dynamic polymorphism. Potential involvement of histone H2B N-terminal tail proximal domain, *Journal of molecular biology*, **326** (2003), 49–63.

[67] N. Conde e Silva, B.E. Black, A. Sivolob, J. Filipski, D.W. Cleveland, and A. Prunell, CENP-A-containing nucleosomes: Easier disassembly versus exclusive centromeric localization, *Journal of molecular biology*, **370** (2007), 555–573.

[68] D. Swigon, B.D. Coleman, and I. Tobias, The elastic rod model for DNA and its application to the tertiary structure of DNA minicircles in mononucleosomes, *Biophysical journal*, **74** (1998), 2515–2530.

[69] I. Tobias, D. Swigon, and B.D. Coleman, Elastic stability of DNA configurations. I. General theory, *Phys. Rev. E. Stat. Phys. Plasmas Fluids Relat. Interdiscip. Topics*, **61** (2000), 747–758.

[70] B.D. Coleman, D. Swigon, and I. Tobias, Elastic stability of DNA configurations. II. Supercoiled plasmids with self-contact, *Phys. Rev. E. Stat. Phys. Plasmas Fluids Relat. Interdiscip. Topics*, **61** (2000), 759–770.

[71] M. Alilat, A. Sivolob, B. Revet, and A. Prunell, Nucleosome dynamics. Protein and DNA contributions in the chiral transition of the tetrasome, the histone (H3-H4)2 tetramer-DNA particle, *Journal of molecular biology*, **291** (1999), 815–841.

[72] C. Lavelle and A. Prunell, Chromatin polymorphism and the nucleosome superfamily: A genealogy, *Cell cycle*, **6** (2007), 2113–2119.

[73] A. Sivolob, F. De Lucia, M. Alilat, and A. Prunell, Nucleosome dynamics. VI. Histone tail regulation of tetrasome chiral transition. A relaxation study of tetrasomes on DNA minicircles, *Journal of molecular biology*, **295** (2000), 55–69.

[74] M. Le Bret, Twist and writhing in short circular DNAs according to first-order elasticity, *Biopolymers*, **23** (1984), 1835–1867.

[75] T. Schlick and W.K. Olson, Supercoiled DNA energetics and dynamics by computer simulation, *Journal of molecular biology*, **223** (1992), 1089–1119.

[76] T. Schlick, W.K. Olson, T. Westcott, and J.P. Greenberg, On higher buckling transitions in supercoiled DNA, *Biopolymers*, **34** (1994), 565–597.

[77] J. Bednar, P. Furrer, A. Stasiak, J. Dubochet, E.H. Egelman, and A.D. Bates, The twist, writhe and overall shape of supercoiled DNA change during counterion-induced transition from a loosely to a tightly interwound superhelix. Possible implications for DNA structure in vivo, *Journal of molecular biology*, **235** (1994), 825–847.

[78] A. Sivolob, F. De Lucia, B. Revet, and A. Prunell, Nucleosome dynamics. II. High flexibility of nucleosome entering and exiting DNAs to positive crossing. An ethidium bromide fluorescence study of mononucleosomes on DNA minicircles, *Journal of molecular biology*, **285** (1999), 1081–1099.

[79] A. Sivolob and A. Prunell, Nucleosome dynamics V. Ethidium bromide versus histone tails in modulating ethidium bromide-driven tetrasome chiral transition. A fluorescence study of tetrasomes on DNA minicircles, *Journal of molecular biology*, **295** (2000), 41–53.

[80] A. Sivolob and A. Prunell, Nucleosome conformational flexibility and implications for chromatin dynamics, *Philosophical transactions*, **362** (2004), 1519–1547.

[81] V. Morales and H. Richard-Foy, Role of histone N-terminal tails and their acetylation in nucleosome dynamics, *Mol. Cell. Biol.*, **20** (2000), 7230–7237.

[82] A. Hamiche, V. Carot, M. Alilat, F. De Lucia, M.F. O'Donohue, B. Revet, and A. Prunell, Interaction of the histone (H3-H4)2 tetramer of the nucle-

osome with positively supercoiled DNA minicircles: Potential flipping of the protein from a left- to a right-handed superhelical form, *Proceedings of the National Academy of Sciences of the United States of America*, **93** (1996), 7588–7593.
[83] A. Hamiche and H. Richard-Foy, The switch in the helical handedness of the histone (H3-H4)2 tetramer within a nucleoprotein particle requires a reorientation of the H3-H3 interface, *J. Biol. Chem.*, **273** (1998), 9261–9269.
[84] S. Peterson, R. Danowit, A. Wunsch, and V. Jackson, NAP1 catalyzes the formation of either positive or negative supercoils on DNA on basis of the dimer-tetramer equilibrium of histones H3/H4, *Biochemistry*, **46** (2007), 8634–8646.
[85] F. Marc, K. Sandman, R. Lurz, and J.N. Reeve, Archaeal histone tetramerization determines DNA affinity and the direction of DNA supercoiling, *J. Biol. Chem.*, **277** (2002), 30879–30886.
[86] J.L. Baneres, J. Parello, J. Zaccai, and D. Svergun, A neutron scattering study of the histone sub-assemblies within the nucleosome protein core, *In ILL Millenium Symposium & European User Meeting, A.J. Dianoux, ed. (I.L.L. Grenoble, France)* (2001), 55–57.
[87] A. Bancaud, G. Wagner, E.S.N. Conde, C. Lavelle, H. Wong, J. Mozziconacci, M. Barbi, A. Sivolob, E. Le Cam, L. Mouawad, J.L. Viovy, J.M. Victor, and A. Prunell, Nucleosome chiral transition under positive torsional stress in single chromatin fibers, *Molecular cell*, **27** (2007), 135–147.
[88] I. Goulet, Y. Zivanovic, A. Prunell, and B. Revet, Chromatin reconstitution on small DNA rings. I, *Journal of molecular biology*, **200** (1988), 253–266.
[89] J. Mozziconacci and J.M. Victor, Nucleosome gaping supports a functional structure for the 30nm chromatin fiber, *J. Struct. Biol.*, **143** (2003), 72–76.
[90] J. Mozziconacci, C. Lavelle, M. Barbi, A. Lesne, and J.M. Victor, A physical model for the condensation and decondensation of eukaryotic chromosomes, *FEBS Lett.*, **580** (2006), 368–372.
[91] I. Duband-Goulet, V. Carot, A.V. Ulyanov, S. Douc-Rasy, and A. Prunell, Chromatin reconstitution on small DNA rings. IV. DNA supercoiling and nucleosome sequence preference, *Journal of molecular biology*, **224** (1992), 981–1001.
[92] H.M. Wu and D.M. Crothers, The locus of sequence-directed and protein-induced DNA bending, *Nature*, **308** (1984), 509–513.
[93] D. Angelov, J.M. Vitolo, V. Mutskov, S. Dimitrov, and J.J. Hayes, Preferential interaction of the core histone tail domains with linker DNA, *Proc. Natl. Acad. Sci. USA*, **98** (2001), 6599–6604.
[94] L. Hong, G.P. Schroth, H.R. Matthews, P. Yau, and E.M. Bradbury, Studies of the DNA binding properties of histone H4 amino terminus. Thermal denaturation studies reveal that acetylation markedly reduces the binding constant of the H4 "tail" to DNA, *J. Biol. Chem.*, **268** (1993), 305–314.
[95] K. Toth, N. Brun, and J. Langowski, Chromatin compaction at the mononucleosome level, *Biochemistry*, **45** (2006), 1591–1598.
[96] B.P. Chadwick and H.F. Willard, A novel chromatin protein, distantly related to histone H2A, is largely excluded from the inactive X chromosome, *J. Cell. Biol.*, **152** (2001), 375–384.
[97] Y. Bao, K. Konesky, Y.J. Park, S. Rosu, P.N. Dyer, D. Rangasamy, D.J. Tremethick, P.J. Laybourn, and K. Luger, Nucleosomes containing the histone variant H2A.Bbd organize only 118 base pairs of DNA, *Embo. J.*, **23** (2004), 3314–3324.
[98] C.M. Doyen, F. Montel, T. Gautier, H. Menoni, C. Claudet, M. Delacour-Larose, D. Angelov, A. Hamiche, J. Bednar, C. Faivre-Moskalenko, P. Bouvet, and S. Dimitrov, Dissection of the unusual structural and functional properties of the variant H2A.Bbd nucleosome, *Embo. J.*, **25** (2006), 4234–4244.

[99] D.K. Palmer, K. O'Day, M.H. Wener, B.S. Andrews, and R.L. Margolis, A 17-kD centromere protein (CENP-A) copurifies with nucleosome core particles and with histones, *J. Cell. Biol.*, **104** (1987), 805–815.

[100] K. Yoda, S. Ando, S. Morishita, K. Houmura, K. Hashimoto, K. Takeyasu, and T. Okazaki, Human centromere protein A (CENP-A) can replace histone H3 in nucleosome reconstitution in vitro, *Proc. Natl. Acad. Sci. USA*, **97** (2000), 7266–7271.

[101] K.A. Collins, S. Furuyama, and S. Biggins, Proteolysis contributes to the exclusive centromere localization of the yeast Cse4/CENP-A histone H3 variant, *Curr. Biol.*, **14** (2004), 1968–1972.

[102] O. Moreno-Moreno, M. Torras-Llort, and F. Azorin, Proteolysis restricts localization of CID, the centromere-specific histone H3 variant of Drosophila, to centromeres, *Nucleic acids research*, **34** (2006), 6247–6255.

[103] M.G. Schueler and B.A. Sullivan, Structural and functional dynamics of human centromeric chromatin, *Annual review of genomics and human genetics*, **7** (2006), 301–313.

[104] A. Sivolob and A. PrunelL, Linker histone-dependent organization and dynamics of nucleosome entry/exit DNAs, *Journal of molecular biology*, **331** (2003), 1025–1040.

[105] A. Bancaud, N. Conde e Silva, M. Barbi, G. Wagner, J.F. Allemand, J. Mozziconacci, C. Lavelle, V. Croquette, J.M. Victor, A. Prunell, and J.L. Viovy, Structural plasticity of single chromatin fibers revealed by torsional manipulation, *Nature structural & molecular biology*, **13** (2006), 444–450.

[106] T.R. Strick, J.F. Allemand, D. Bensimon, A. Bensimon, and V. Croquette, The elasticity of a single supercoiled DNA molecule, *Science*, **271** (1996), 1835–1837.

[107] S. Neukirch, Extracting DNA twist rigidity from experimental supercoiling data, *Physical review letters*, **93** (2004), 198107.

[108] C. Bouchiat and M. Mézard, Elasticity model of a supercoiled DNA molecule, *Physical review letters*, **80** (1998), 1556–1559.

[109] E. Ben-Haïm, A. Lesne, and J.M. Victor, Chromatin: a tunable spring at work inside chromosomes, *Phys. Rev. E Stat. Nonlin. Soft Matter Phys.*, **64** (2001), 051921.

[110] Y. Cui AND C. Bustamante, Pulling a single chromatin fiber reveals the forces that maintain its higher-order structure, *Proc. Natl. Acad. Sci. USA*, **97** (2000), 127–132.

[111] B.D. Brower-Toland, C.L. Smith, R.C. Yeh, J.T. Lis, C.L. Peterson, and M.D. Wang, Mechanical disruption of individual nucleosomes reveals a reversible multistage release of DNA, *Proc. Natl. Acad. Sci. USA*, **99** (2002), 1960–1965.

[112] C. Bouchiat, M.D. Wang, J. Allemand, T. Strick, S.M. Block, and V. Croquette, Estimating the persistence length of a worm-like chain molecule from force-extension measurements, *Biophysical journal*, **76** (1999), 409–413.

[113] A. Bertin, M. Renouard, J.S. Pedersen, F. Livolant, and D. Durand, H3 and H4 histone tails play a central role in the interactions of recombinant NCPs, *Biophysical journal*, **92** (2007), 2633–2645.

[114] R.C. Benedict, E.N. Moudrianakis, and G.K. Ackers, Interactions of the nucleosomal core histones: a calorimetric study of octamer assembly, *Biochemistry*, **23** (1984), 1214–1218.

[115] S. Neukirch, G.H.M. Van Der Heijden, and J.M.T. Thompson, Writhing instabilities of twisted rods: from infinite to finite lengths, *J. Mech. Phys. Solids*, **50** (2002), 1175–1191.

[116] D.J. Clark AND G. Felsenfeld, Formation of nucleosomes on positively supercoiled DNA, *Embo. J.*, **10** (1991), 387–395.

[117] J.E. Germond, B. Hirt, P. Oudet, M. Gross-Bellark, and P. Chambon, Folding of the DNA double helix in chromatin-like structures from simian virus 40, *Proc. Natl. Acad. Sci. USA*, **72** (1975), 1843–1847.

[118] R.T. Simpson, F. Thoma, and J.M. Brubaker, Chromatin reconstituted from tandemly repeated cloned DNA fragments and core histones: A model system for study of higher order structure, *Cell*, **42** (1985), 799–808.

[119] V.G. Norton, B.S. Imai, P. Yau, and E.M. Bradbury, Histone acetylation reduces nucleosome core particle linking number change, *Cell*, **57** (1989), 449–457.

[120] W. Keller, U. Muller, I. Eicken, I. Wendel, and H. Zentgraf, Biochemical and ultrastructural analysis of SV40 chromatin, *Cold Spring Harb. Symp. Quant. Biol.*, **42 Pt 1** (1978), 227–244.

[121] M.M. Garner, G. Felsenfeld, M.H. O'Dea, and M. Gellert, Effects of DNA supercoiling on the topological properties of nucleosomes, *Proc. Natl. Acad. Sci. USA*, **84** (1987), 2620–2623.

[122] A. Klug and L.C. Lutter, The helical periodicity of DNA on the nucleosome, *Nucleic acids research*, **9** (1981), 4267–4283.

[123] J.C. Wang, The path of DNA in the nucleosome, *Cell*, **29** (1982), 724–726.

[124] A. Prunell, A topological approach to nucleosome structure and dynamics: the linking number paradox and other issues, *Biophysical journal*, **74** (1998), 2531–2544.

[125] J.T. Finch, L.C. Lutter, D. Rhodes, R.S. Brown, B. Rushton, M. Levitt, and A. Klug, Structure of nucleosome core particles of chromatin, *Nature*, **269** (1977), 29–36.

[126] A. Klug and A.A. Travers, The helical repeat of nucleosome-wrapped DNA, *Cell*, **56** (1989), 10–11.

[127] J.H. White and W.R. Bauer, The helical repeat of nucleosome-wrapped DNA, *Cell*, **56** (1989), 9–10.

[128] J.J. Hayes, T.D. Tullius, and A.P. Wolffe, The structure of DNA in a nucleosome, *Proc. Natl. Acad. Sci. USA*, **87** (1990), 7405–7409.

[129] A. Prunell and A. Sivolob, Paradox lost: nucleosome structure and dynamics by the DNA minicircle approach *in Chromatin Structure and Dynamics: State-of-the-Art, Vol. 39 (eds. Zlatanova, J. & Leuba, S.H.) 45–73 (Elsevier, London, 2004)* (2004).

[130] L.C. Lutter, L. Judis, and R.F. Paretti, Effects of histone acetylation on chromatin topology in vivo, *Mol. Cell. Biol.*, **12** (1992), 5004–5014.

[131] A. Stein, DNA wrapping in nucleosomes. The linking number problem re-examined, *Nucleic acids research*, **8** (1980), 4803–4820.

[132] M. Shure and J. Vinograd, The number of superhelical turns in native virion SV40 DNA and minicol DNA determined by the band counting method, *Cell*, **8** (1976), 215–226.

[133] R.H. Morse and C.R. Cantor, Effect of trypsinization and histone H5 addition on DNA twist and topology in reconstituted minichromosomes, *Nucleic acids research*, **14** (1986), 3293–3310.

[134] A. Rodriguez-Campos, A. Shimamura, and A. Worcel, Assembly and properties of chromatin containing histone H1, *Journal of molecular biology*, **209** (1989), 135–150.

[135] L.A. Freeman and W.T. Garrard, DNA supercoiling in chromatin structure and gene expression, *Crit. Rev. Eukaryot. Gene. Expr.*, **2** (1992), 165–209.

[136] R.H. Morse and C.R. Cantor, Nucleosome core particles suppress the thermal untwisting of core DNA and adjacent linker DNA, *Proc. Natl. Acad. Sci. USA*, **82** (1985), 4653–4657.

[137] R.H. Morse, D.S. Pederson, A. Dean, and R.T. Simpson, Yeast nucleosomes allow thermal untwisting of DNA, *Nucleic acids research*, **15** (1987), 10311–10330.

[138] L.F. Liu and J.C. Wang, Supercoiling of the DNA template during transcription, *Proc. Natl. Acad. Sci. USA*, **84** (1987), 7024–7027.

[139] Y.P. Tsao, H.Y. Wu, and L.F. Liu, Transcription-driven supercoiling of DNA: direct biochemical evidence from in vitro studies, *Cell*, **56** (1989), 111–118.

[140] C. Lavelle, Transcription elongation through a chromatin template, *Biochimie*, **89** (2007), 516–527.

[141] F. Caron and J.O. Thomas, Exchange of histone H1 between segments of chromatin, *Journal of molecular biology*, **146** (1981), 513–537.

[142] Y.J. Jin and R.D. Cole, H1 histone exchange is limited to particular regions of chromatin that differ in aggregation properties, *J. Biol. Chem.*, **261** (1986), 3420–3427.

[143] L. Louters and R. Chalkley, Exchange of histones H1, H2A, and H2B in vivo, *Biochemistry*, **24** (1985), 3080–3085.

[144] M.A. Lever, J.P. Th'ng, X. Sun, and M.J. Hendzel, Rapid exchange of histone H1.1 on chromatin in living human cells, *Nature*, **408** (2000), 873–876.

[145] T. Misteli, A. Gunjan, R. Hock, M. Bustin, and D.T. Brown, Dynamic binding of histone H1 to chromatin in living cells, *Nature*, **408** (2000), 877–881.

[146] D.T. Brown, Histone H1 and the dynamic regulation of chromatin function, *Biochem. Cell. Biol.*, **81** (2003), 221–227.

[147] M. Bustin, F. Catez, and J.H. Lim, The dynamics of histone H1 function in chromatin, *Mol. Cell.*, **17** (2005), 617–620.

[148] W. An, Histone acetylation and methylation: combinatorial players for transcriptional regulation, *Subcell. Biochem.*, **41** (2007), 351–369.

[149] M.D. Shahbazian and M. Grunstein, Functions of site-specific histone acetylation and deacetylation, *Annu. Rev. Biochem.*, **76** (2007), 75–100.

[150] Y. Harada, O. Ohara, A. Takatsuki, H. Itoh, N. Shimamoto, and K. Kinosita, Jr., Direct observation of DNA rotation during transcription by Escherichia coli RNA polymerase, *Nature*, **409** (2001), 113–115.

[151] T. Ito, T. Ikehara, T. Nakagawa, W.L. Kraus, and M. Muramatsu, p300-mediated acetylation facilitates the transfer of histone H2A-H2B dimers from nucleosomes to a histone chaperone, *Genes. Dev.*, **14** (2000), 1899–1907.

[152] D. Reinberg and R.J. Sims, 3rd, de FACTo nucleosome dynamics, *J. Biol. Chem.*, **281** (2006), 23297–23301.

[153] J. Salceda, X. Fernandez, and J. Roca, Topoisomerase II, not topoisomerase I, is the proficient relaxase of nucleosomal DNA, *Embo. J.*, **25** (2006), 2575–2583.

[154] F. Kouzine, S. Sanford, Z. Elisha-Feil, and D. Levens, The functional response of upstream DNA to dynamic supercoiling in vivo, *Nat. Struct. Mol. Biol.*, **15** (2008), 146–154.

[155] C. Lavelle, DNA torsional stress propagates through chromatin fiber and participates in transcriptional regulation, *Nat. Struct. Mol. Biol.*, **15** (2008), 123–125.

THE MATHEMATICS OF DNA STRUCTURE, MECHANICS, AND DYNAMICS

DAVID SWIGON*

Abstract. A brief review is given of the main concepts, ideas, and results in the fields of DNA topology, elasticity, mechanics and statistical mechanics. Discussion includes the notions of the linking number, writhe, and twist of closed DNA, elastic rod models, sequence-dependent base-pair level models, statistical models such as helical worm-like chain and freely jointed chain, and dynamical simulation procedures. Experimental methods that lead to the development of the models and the implications of the models are also discussed. Emphasis is placed on illustrating the breadth of approaches and the latest developments in the field, rather than the depth and completeness of exposition.

Key words. DNA topology, elasticity, mechanics, statistical mechanics, stretching.

1. Introduction.

The discovery of DNA structure 55 years ago marked the beginning of a process that has transformed the foundations of biology and medicine, and accelerated the development of new fields, such as molecular biology or genetic engineering. Today, we know much about DNA, its properties, and function. We can determine the structure of short DNA fragments with picometer precision, find majority of the genes encoded in DNA, and we can manipulate, stretch and twist individual DNA molecules. We can utilize our knowledge of gene regulatory apparatus encoded in DNA to produce new microorganisms with unexpected properties. Yet, there are aspects of DNA function that defy our understanding, mostly because the molecule is just one, albeit essential, component of a complex cellular machinery.

From the very beginning, abstraction and modeling played a significant role in research on DNA, since the molecule could not be visualized by any available experimental methods. These models gave rise to mathematical concepts and techniques for study of DNA configurations at the macroscopic and mesoscopic levels, which are the subject of this short review. The paper begins with a brief description of DNA atomic-level structure, followed by a discussion of topological properties of DNA such as knotting, catenation, and the definitions of linking number and supercoiling. It continues with an outline of continuum and discrete models of DNA elasticity, focusing on local energy contributions and analysis of equilibrium states. Modeling of long range electrostatic interactions is described next, followed by the treatment of thermal fluctuations and statistical mechanics. The paper concludes with an outline of dynamical models of DNA, and

*Department of Mathematics, University of Pittsburgh, 301 Thackeray Hall, Pittsburgh, PA 15260 (swigon@pitt.edu). The work was supported by Institute for Mathematics and its Applications (IMA), Alfred P SLoan Fellowship and NSF Grant DMS 0516646.

C.J. Benham et al. (eds.), *Mathematics of DNA Structure, Function and Interactions*,
The IMA Volumes in Mathematics and its Applications 150,
DOI 10.1007/978-1-4419-0670-0_14,

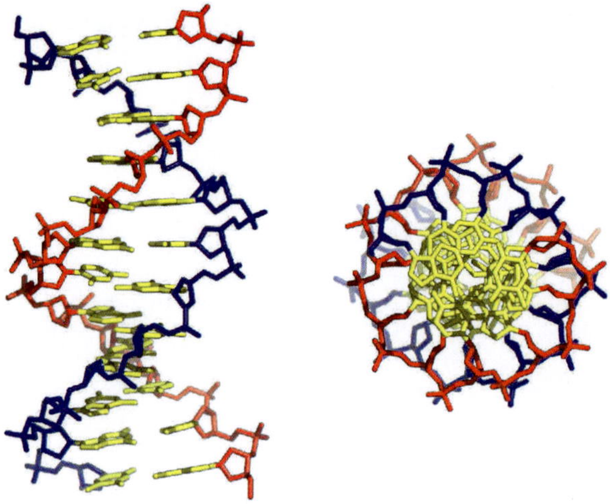

FIG. 1. *Side view (left) and a view along the axis (right) of DNA double helix in atomic level detail, showing the two DNA backbones (blue and red) and the base pairs (yellow).*

a discussion of future directions in DNA research. The analysis of DNA sequences, or modeling of the atomic-level structure and dynamics of DNA are not covered here.

2. Background. DNA is made up of two polymeric strands composed of monomers that include a nitrogenous base (A-adenine, C-cytosine, G-guanine, and T-thymine), deoxyribose sugar, and a phosphate group. The sugar and phosphate groups, which form the backbone of each strand, are located on the surface of DNA while the bases are on the inside of the structure (see Fig. 1). Weak hydrogen bonds between complementary bases of each strand (i.e., between A and T and between C and G) give rise to pairing of bases that holds the two strands together. The base pairs (bp) are flat and stack on top of each other like dominoes with centers separated by approximately 0.34 nm. In normal conditions each base pair is rotated relative to its predecessor by approximately 34°, giving rise to the familiar right-handed Watson-Crick double helix.

The chemical nature of the backbone gives each strand an orientation - one end is called the $5'$-end and the other the $3'$-end. In duplex DNA the two strands run antiparallel to each other. A closed DNA (also called a plasmid or ring) is formed when the ends of each strand are joined by a covalent bond. A prokaryotic organism, e.g., a bacterium, lacks nuclear structures and its entire genome is in the form of a single closed duplex DNA. Genomic DNA of a eukaryotic cell is contained within a nucleus and it is divided into a number of chromosomes.

The DNA of any organism must be folded and packed in a complicated fashion in order to fit inside a cell.[1] This is complicated by the fact that DNA resists bending and twisting deformations and also has a tendency to repel itself electrostatically. In addition to being compacted, portions of DNA must be accessible at various moments during the lifetime of the cell, so that the genes encoded in the DNA can be expressed and proteins produced when necessary. The effort to understand how DNA is packed and unpacked in cells, and how its mechanical properties influence the processes of transcription, replication and recombination, is one of the driving forces behind the development of mathematical models of DNA.

3. Topology. When Watson and Crick first proposed the double helical model for DNA [147], they remarked:

> "Since the two chains in our model are intertwined, it is essential for them to untwist if they are to separate. Although it is difficult at the moment to see how these processes occur without everything getting tangled, we do not feel that this objection would be insuperable."

The entanglement of DNA and Nature's ways of coping with it is the subject of DNA topology.

In the first approximation, a closed DNA molecule can be treated as a single closed curve in space. (The resistance of DNA to bending implies that this curve is rather smooth.) Because during regular deformation the bonds in DNA strands do not break, it is natural to consider the problems of DNA knotting and catenation.[2]

DNA plasmids can become catenated during DNA replication, a process in which the two strands of DNA are separated, each strand is complemented by one newly formed strand, and instead of a single plasmid one obtains two plasmids that are catenated in the same way the strands were linked in the original plasmid. Of course, it is crucial that during replication the catenation of the plasmids is removed so that they can be separated and placed one in each of the daughter cells. The enzymes that preform decatenation are called *type II topoisomerases* [146]. They operate by a *strand passage* mechanism in which two DNA segments are brought to a close contact, one of the segments is severed in such a way that both backbone chains of the molecule are broken, the second segment of DNA is passed through the gap in the first segment, and finally the severed segment is resealed.

[1]For example the DNA of *E. coli* is a closed DNA of circumference 1.58 mm that must fit inside a cell of diameter 1 μm. Human genome has more that 3 billion bp, i.e., a linear length of 1 m. Two copies of the genome must be packed inside every cell of human body, which range in size between 3 and 35 μm.

[2]As is customary in DNA research we here use the term *catenation* for linking of two DNA molecules and reserve the term *linking* for topological relation between two DNA strands.

DNA knotting rarely occurs naturally, but it has been achieved in a laboratory using the aforementioned topoisomerases and also DNA recombinases, enzymes that cut two DNA molecules at specific recognition sites and then switch and reconnect the ends. Because a given recombinase only forms knots of certain types, knot theory, and in particular tangle analysis, has been applied to the problem of determining the structure and function of these enzymes [55, 47, 141]. The changes in knot type resulting from strand passages have been classified and the probabilities of such passages have been estimated [46, 70]. Knotting also occurs in DNA closure experiments in which open (linear) DNA segments spontaneously cyclize to form closed DNAs. Since DNA thermally fluctuates, the probability of forming a knot can be related to the probability that a random configuration of a phantom DNA (i.e., a DNA allowed to pass through itself) has the topology of a knot (see Section 6). It was shown that in the limit of length going to infinity a randomly cyclized polygon will be knotted with probability 1 [48].

A closed DNA molecule can also be viewed as a collection of two continuous curves - the DNA strands. This is because the biochemical nature of the strands guarantees that during closure ceach strand of the DNA can only bind to itself. The axial curve of a closed DNA, which can be thought of as the curve passing through the centroids of the base pairs, is also a closed curve.

For any two closed curves $\mathcal{C}_1$ and $\mathcal{C}_2$ one can define a quantity, called the *linking number* Lk, that characterizes how the curves are interwound with each other. The linking number can be found by examining a generic projection of the two curves on a plane (a projection in which every crossing of one curve with the other is transversal). First, orientation is assigned to each curve and a sign to each crossing of one curve over the other, in accord with the convention shown in Fig. 2A.

The linking number Lk is then taken to be one half the sum of all signed crossings (see Fig. 2B and C); it is a topological invariant of the two curves, i.e., a number independent of homotopic deformations of the curves that do not pass one curve through the another. In DNA research it is customary to take $\mathcal{C}_1$ to be the axial curve of the molecule and $\mathcal{C}_2$ one of the backbone chains.

For differentiable curves, a formula for linking number in terms of a double integral was found by Gauss [42]

$$Lk(\mathcal{C}_1, \mathcal{C}_2) = \frac{1}{4\pi} \oint_{\mathcal{C}_1} \oint_{\mathcal{C}_2} \frac{\mathbf{t}_1(s_1) \times \mathbf{t}_2(s_2) \cdot [\mathbf{x}_1(s_1) - \mathbf{x}_2(s_2)]}{|\mathbf{x}_1(s_1) - \mathbf{x}_2(s_2)|^3} \, ds_2 ds_1, \quad (3.1)$$

where $\mathcal{C}_i$ is defined by giving its position $\mathbf{x}_i(s)$ in space as a function of the arc-length s, and $\mathbf{t}_i(s) = \mathbf{x}_i'(s) = d\mathbf{x}_i(s)/ds$.

There are two geometric properties of curves that are intimately related to the linking number. The first property, called the *writhe* Wr,

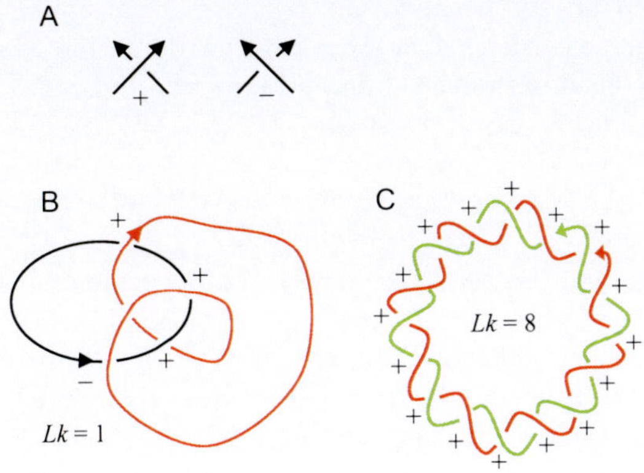

FIG. 2. *The linking number of two curves. A: A sign convention for crossings. B and C: Examples of calculation of Lk for two curves.*

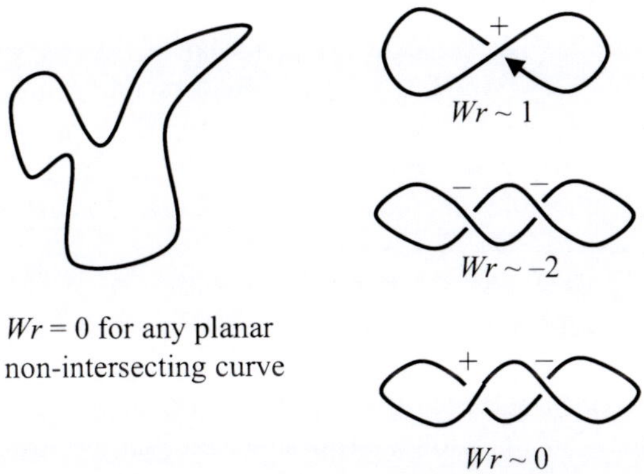

FIG. 3. *The writhe of a curve.*

characterizes the amount of chiral deformation of a single curve. To find Wr, one assigns orientation to the curve and computes the sum of signed crossings in a planar projection along every direction; Wr is equal to the average of such sums over all projections. Examples of curves with various values of Wr are shown in Fig. 3. For a closed differentiable curve $\mathcal{C}$ a formula for Wr analogous to (3.1) exists:

$$Wr(\mathcal{C}) = \frac{1}{4\pi} \oint_{\mathcal{C}} \oint_{\mathcal{C}} \frac{\mathbf{t}(s_1) \times \mathbf{t}(s_2) \cdot [\mathbf{x}(s_1) - \mathbf{x}(s_2)]}{|\mathbf{x}(s_1) - \mathbf{x}(s_2)|^3} \, ds_2 ds_1. \tag{3.2}$$

Alternative formulae relating Wr to the area swept by the vector $\mathbf{x}(s_1) - \mathbf{x}(s_2)$ on a unit sphere when traversing $\mathcal{C}$, or the difference in writhe of two closed curves can be found in [57, 1].

The second property, called the *twist* Tw, measures the winding of one curve about the other. The most familiar definition requires that the curves under consideration be differentiable; the twist of $\mathcal{C}_2$ about $\mathcal{C}_1$ is then

$$Tw(\mathcal{C}_2, \mathcal{C}_1) = \frac{1}{2\pi} \oint_{\mathcal{C}_1} [\mathbf{t}_1(s) \times \mathbf{d}(s)] \cdot \mathbf{d}'(s) ds \tag{3.3}$$

where $\mathbf{d}(s) = \mathbf{x}_2(\sigma(s)) - \mathbf{x}_2(s)$ is taken to be perpendicular to $\mathbf{t}_1(s)$.

Neither the writhe nor the twist are topological invariants. However, it follows from the results of Calugareanu [29] and White [149] that the linking number of two closed curves is the sum of the writhe of one curve and the twist of the second curve about the first:

$$Lk(\mathcal{C}_1, \mathcal{C}_2) = Wr(\mathcal{C}_1) + Tw(\mathcal{C}_2, \mathcal{C}_1). \tag{3.4}$$

This relation has important implications for a closed DNA molecule. Since in a closed duplex DNA Lk is invariant, any change in Tw, which may come about as a result of binding of DNA to proteins (such as histones) or intercalating molecules, will induce a corresponding opposite change in Wr. Alternatively, DNA mechanics tells us that if Lk is changed by cutting and resealing of DNA strands, that change will be partitioned into a change in Tw and a change in Wr of equal signs. In DNA research an increase in the magnitude of writhe, accompanied by an increase in the number of crossings of the molecule, is called *supercoiling*, and a molecule with high $|Wr|$ is known as *supercoiled DNA*.

Supercoiling is a characteristic deformation of a closed DNA that can be observed and quantified experimentally. Supercoiling can be either detrimental or beneficial to a cell, depending on its magnitude and circumstances. Each cell contains enzymes topoisomerases that regulate DNA supercoiling by constantly adjusting the linking number. Since the linking number of a closed DNA molecule remains constant during any deformation of the molecule that preserves chemical bonding, it can therefore be changed only by mechanisms in which chemical bonds are disrupted. There are two such mechanisms: (i) a relaxation, in which a bond in one of the backbone chains is broken, one end of the broken backbone is rotated about the other backbone by 360° and the broken bond is repaired, or (ii) a strand passage, described earlier, in which one segment of DNA is passed through a gap created in the second segment. Type I DNA topoisomerases use the first mechanism and hence change Lk by by ± 1, while type II topoisomerases use the second mechanism and change Lk by ± 2.

Natural questions arise, such as what is the configuration of supercoiled DNA with prescribed Lk, what is the probability of occurence of topoisomers or knot types, or how much time does it take for a segment of DNA to form a closed molecule. These questions can be answered with the help of theories of DNA elasticity, statistical mechanics and dynamics, described in subsequent sections.

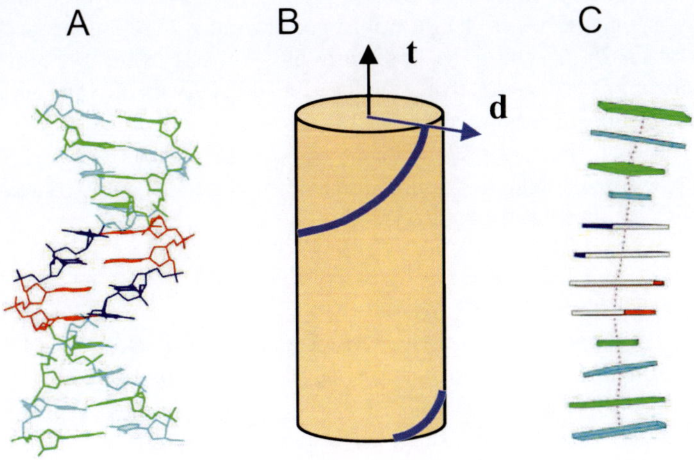

FIG. 4. *Schematic representations of DNA. A: a wireframe representation of the atomic level structure. B: continuum elastic rod. C: base-pair level description*

4. Elasticity. The elasticity of DNA is governed by interactions between the atoms of the molecule and by interactions of the molecule with the surrounding solvent. The primary interaction responsible for DNA bending stiffness is base stacking, a tendency of the flat hydrophobic nucleotides to aggregate in such a way as to minimize water accessible surface [115]. Such a stacking occurs even in the absence of backbone connections. The twisting rigidity of the molecule is due to the presence of two backbone polymeric chains. The elastic properties are signigficantly affected by electrostatic interaction between negatively charged phosphate groups in the backbone, which are strongly modulated by ionic properties of the surrounding solvent.

Continuum models. The simplest model of DNA deformability treats DNA as an *ideal elastic rod*, i.e., thin elastic body that is inextensible, intrinsically straight, transversely isotropic and homogeneous [12, 13]. The configuration of DNA is described by giving the position $\mathbf{x}(s)$ of its axial curve in space and its twist density $\Omega(s)$ as functions of the arc-length s, where $\Omega(s) = [\mathbf{t}_1(s) \times \mathbf{d}(s)] \cdot \mathbf{d}'(s)$ with $\mathbf{d}(s)$ a vector pointing from the axial curve to one of the backbones (see Figure 4A and Eq. (3.3)). The elastic energy of the rod is given by

$$\Psi = \frac{1}{2} \int_0^L A\kappa(s)^2 + C\left(\Omega(s) - \bar{\Omega}(s)\right)^2 ds \tag{4.1}$$

where $\kappa(s) = |\mathbf{t}'(s)|$ is the curvature of the axial curve and $\bar{\Omega}(s)$ is the twist density in a stress free state. The bending modulus A and the twisting modulus C characterize the elastic properties of DNA. The accepted "average" value of A for B-DNA under standard conditions is 50 kT·nm [66, 23] and C is between 25 kT·nm and 100 kT·nm [69, 121, 131], (here kT, an widely used unit of energy in molecular biology, is the product of Boltzmann constant k and absolute temperature T).

A rod with the energy (4.1) obeys the classical theory of Kirchhoff [75, 51], which implies that in equilibrium $\Delta\Omega = \Omega(s) - \bar{\Omega}(s)$ is constant and $\mathbf{t}(s)$ obeys a differential equation,

$$A(\mathbf{t} \times \mathbf{t}'') + C\Delta\Omega\mathbf{t}' = \mathbf{F} \times \mathbf{t} \tag{4.2}$$

with the constant F playing the role of a force. Solutions of (4.2) have been obtained in a closed form in terms of elliptic functions and integrals [82, 139].

Although each solution of (4.2) corresponds to an equilibrium configuration of the rod, from a practical point of view it is important to know which of these solutions are locally stable in the sense that any small perturbation of the configuration compatible with the boundary conditions leads to an increase in elastic energy. Stability theory for closed Kirchhoff elastic rods has been developed by a number of researchers using the framework of calculus of variations; necessary conditions (the slope of the graph of Lk versus Wr for a family of equilibrium configurations [88, 140, 41]), suficient conditions (the absence of conjugate points [96, 67]), or general observations about stability of rod configurations [84].

Bifurcation theory of straight rods subject to tension and twist is a classical subject [92, 3, 136, 105] and bifurcations of a closed rod with a given linking number have also been analyzed [155, 87, 49]. The general conclusion is that the straight or circular solution of (4.2) is stable for Lk smaller than a critical value, while other solutions of (4.2) can be stable only if $|Wr|$ is small and C/A is larger than a critical value that depends on the boundary conditions.[3] Experiments with steel wires, which have $C/A < 1$, confirm this result [137]. Consequently, the solutions of (4.2) cannot represent minimum energy configurations of supercoiled DNA with high $|Wr|$, because such configurations show self-contact, i.e., a contact between the surfaces of two distinct subsegments of the rod.

In any theory of rod configurations with self-contact, the forces exerted on the surface of DNA can be accounted for as external forces in the balance equations. The existence of a globally minimizing configuration for general nonlinearly elastic rods with self-contact has been demonstrated [62, 120]. In the case of an ideal elastic rod, segments of the rod between points of contact can be treated using Kirchhoff's theory, and by putting together explicit expressions for contact-free segments and balance equations for forces at the contact points one obtains a system of algebraic equations

[3]This critical value is 11/8 for closed rods subject to twisting.

that can be solved to obtain a configuration of DNA plasmid with self-contact [88, 74, 49, 39]. The ideal rod model with self-contact has been applied to the study of DNA supercoiling [41, 39], configurations of straight DNA subject to stretching and twisting [132], and configurations of DNA loops in mononucleosomes [133].

The ability to account for self-contact is critical if one intends to study equilibrium configurations of knotted DNA, for it has been shown that knotted contact-free equilibrium configurations of closed DNA have the topology of torus knots and are all unstable [84]; examples of such configurations can be found in [88, 49, 129]. Thus any stable configuration of a DNA knot shows self-contact; minimum energy configuration of a DNA plasmid with the topology of a trefoil knot as a function of Lk has been found [40].

Departures from ideality, such as intrinsic curvature, bending anisotropy, shearing, or coupling between modes of deformation can be treated using *special Cosserat theory of rods* (see, e.g., [2]). In that theory the configuration of the rod is described by giving, as functions of the arc-length s, its axial curve $\mathbf{x}(s)$ and an orthonormal triad $(\mathbf{d}_1(s), \mathbf{d}_2(s), \mathbf{d}_3(s))$, which is embedded in the cross-section of the rod in such a way that $\mathbf{d}_3$ is normal to the cross-section. The vector $\mathbf{d}_3(s)$ need not be parallel to $\mathbf{x}'(s)$ and hence the theory can describe rods with shear. The elastic energy is expressed in terms of the variables $(\kappa_1, \kappa_2, \kappa_3, \nu_1, \nu_2, \nu_3)$ describing local deformation of the rod, i.e.,

$$\Psi = \int_0^L W(\boldsymbol{\kappa} - \bar{\boldsymbol{\kappa}}, \boldsymbol{\nu} - \bar{\boldsymbol{\nu}})ds \qquad (4.3)$$

where

$$\mathbf{d}_i'(s) = \boldsymbol{\kappa}(s) \times \mathbf{d}_i(s) \qquad (4.4)$$

$$\nu_i(s) = \mathbf{x}_i'(s) \cdot \mathbf{d}_i(s). \qquad (4.5)$$

When this theory is applied to DNA research [93, 6, 58], it is usually assumed that DNA is inextensible and unshearable (i.e., $\mathbf{d}_3(s) = \mathbf{x}'(s)$), and shows no coupling; consequently the energy density is given by

$$2W(\boldsymbol{\kappa} - \bar{\boldsymbol{\kappa}}) = K_1(\kappa_1 - \bar{\kappa}_1)^2 + K_2(\kappa_2 - \bar{\kappa}_2)^2 + K_3(\kappa_3 - \bar{\kappa}_3)^2. \qquad (4.6)$$

Variational equations in the Cosserat theory are identical to the balance equations in the Kirchhoff theory:

$$\mathbf{F}' + \mathbf{f} = 0 \qquad (4.7)$$

$$\mathbf{M}' + \mathbf{x}' \times \mathbf{F} + \mathbf{m} = 0. \qquad (4.8)$$

These equations cannot be solved explicitly and therefore are usually integrated numerically. Accurate numerical schemes employ a parametrization

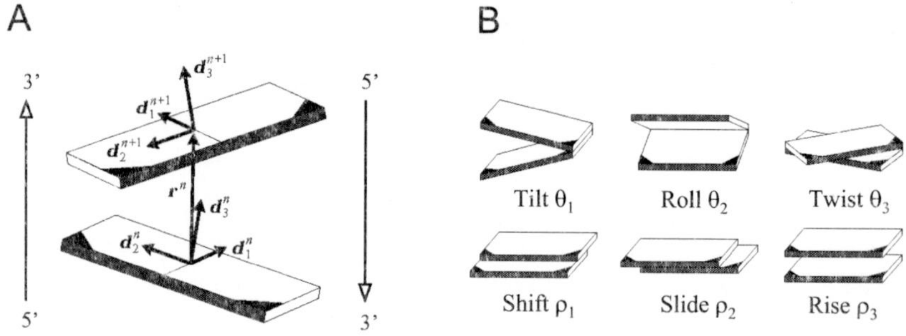

FIG. 5. *Parameters characterizing the base-pair step.*

for $(\mathbf{d}_1, \mathbf{d}_2, \mathbf{d}_3)$ using Euler angles or Euler parameters and reformulate the problem as a set of differential equations for these parameters [49]. The practical problem of computing DNA configurations using the Cosserat model requires one to determine the unstressed values $\bar{\boldsymbol{\kappa}}$ and elastic moduli K_1, K_2, K_3 for a given DNA sequence, which can be done, for example, by comparing computed equilibria with the results of a cyclization experiment [95]. Cosserat theory has been employed to show that intrinsically curved DNA circles and DNA segments with fixed ends can have multiple stable contact-free equilibrium configurations [58, 142, 68], and was also used to compute the structure of protein-induced DNA loops [7, 65].

Discrete models. Discrete models have been developed to model sequence-dependent elasticity of DNA in a way that closely resembles detailed DNA structure. The most common discrete models treat DNA as a collection of rigid subunits representing the base-pairs (see Figure 4C). This description has long been used by chemists to characterize DNA crystal structures [28, 107]. The DNA configuration is specified by giving, for each base pair, numbered by index n, its location $\mathbf{x}^n$ in space and its orientation described by an embedded orthonormal frame $(\mathbf{d}_1^n, \mathbf{d}_2^n, \mathbf{d}_3^n)$. The relative orientation and position of the base pair and its predecessor are specified by six kinematical variables $(\theta_1^n, \theta_2^n, \theta_3^n, \rho_1^n, \rho_2^n, \rho_3^n)$, termed, respectively, tilt, roll, twist, shift, slide, and rise (see Fig. 5). In the simplest, so-called *dinucleotide models*, the elastic energy Ψ is taken to be the sum of the base-pair step energies ψ^n , each of which is a function of the kinematical variables, i.e.,

$$\Psi = \sum_{n=1}^{N-1} \psi^n(\theta_1^n, \theta_2^n, \theta_3^n, \rho_1^n, \rho_2^n, \rho_3^n) \qquad (4.9)$$

TABLE 1
Sequence-dependent variability of DNA elastic properties.

Quantity	Range	Units
Intrinsic bending	$0.4 < \bar{\theta}_2 < 5.1$	*deg*
Bending anisotropy	$1.3 < F_{11}/F_{22} < 3.0$	
Twisting/bending ratio	$0.7 < F_{33}/F_{22} < 2.7$	
Twist-roll coupling	$0.1 < F_{23}/F_{22} < 0.6$	
Twist-stretch coupling	$-0.8 < G_{33} < -0.25$	$kT/(deg \cdot \text{Å})$
Shearing anisotropy	$0.7 < H_{22}/H_{11} < 2.8$	

where the function ψ^n depends on the base-pair composition of the nth step, and is commonly assumed to be a quadratic function

$$\psi^n = \frac{1}{2}\sum_{i=1}^{3}\sum_{j=1}^{3} F_{ij}^{\mathrm{XY}}\Delta\theta_i^n\Delta\theta_j^n + G_{ij}^{\mathrm{XY}}\Delta\theta_i^n\Delta\rho_j^n + H_{ij}^{\mathrm{XY}}\Delta\rho_i^n\Delta\rho_j^n. \tag{4.10}$$

Here XY is the nucleotide sequence (in the direction of the coding strand) of the nth base pair step, $\Delta\theta_i^n = \theta_i^n - \bar{\theta}_i^{\mathrm{XY}}$, $\Delta\rho_i^n = \rho_i^n - \bar{\rho}_i^{\mathrm{XY}}$ are the deviations of variables from their intrinsic values $\bar{\theta}_i^{\mathrm{XY}}$, $\bar{\rho}_i^{\mathrm{XY}}$, and F_{ij}^{XY}, G_{ij}^{XY}, H_{ij}^{XY} are the elastic moduli. A discrete version of the ideal elastic model can be obtained by taking

$$\bar{\theta}^{\mathrm{XY}} = \begin{pmatrix} 0 \\ 0 \\ 34^\circ \end{pmatrix}, \bar{\rho}^{\mathrm{XY}} = \begin{pmatrix} 0 \\ 0 \\ 0.34 \end{pmatrix} \mathrm{nm}, F^{\mathrm{XY}} = \begin{pmatrix} A & 0 & 0 \\ 0 & A & 0 \\ 0 & 0 & C \end{pmatrix}, \tag{4.11}$$

$$G^{\mathrm{XY}} = 0, \qquad H^{\mathrm{XY}} \to \infty. \tag{4.12}$$

Empirical estimates of intrinsic values and elastic moduli have been deduced from the averages and fluctuations of base-pair step parameters in high-resolution DNA protein complexes [108] after normalization so that the persistence length of mixed-sequence DNA matches observed values (circa 500Å). Departures from ideal behavior found by Olson and collaborators [108] and listed in Table 1 include intrinsic bending (in the roll variable), bending anisotropy, inhomogeneity in twisting to bending stiffness ratio, twist-roll coupling, twist-stretch coupling, and shearing anisotropy. Analysis of X-ray crystal structures and NMR experiments yields the most detailed information to date about DNA structure and flexibility. Other experimental methods, such as cyclization [43, 8], fluorescence resonance energy transfer (FRET) [111], gel mobility [20], or single-molecule stretching [27, 23] and twisting [24, 130], have been used to examine elastic behavior of longer segments in which the effects of individual base-pair steps are averaged over. The sequence-dependent nature of DNA deformability has been independently confirmed by research aimed to deduce DNA elastic properties from molecular dynamics simulations [17, 52].

For the dinucleotide model with energy (4.9) variational equations have been derived [38] and equilibrium configurations for plasmids of various compositions and end conditions have been found [38, 109], including (i) multiple equilibria of ligand-free DNA o-rings (plasmids that are circular when stress-free), (ii) minimum configuration of DNA o-rings with bound intercalating agents (iii) optimal distribution of intercalating agents that minimizes elastic energy of DNA o-rings, (iv) collapsed configurations of DNA o-rings subject to local overtwisting, (v) minimum energy configurations of intrinsically straight DNA plasmids with various distributions of twist-roll coupling, (vi) minimum energy of S-shaped DNA subject to local overtwisting. The theory has been extended to account for electrostatic repulsion and thermal fluctuations and applied to the study of minimum energy configurations and looping free energies of LacR-mediated DNA loops [134], and minimum energy configurations of free segments of promoter DNA bound to Class I and Class II CAP dependent transcription-activation complexes [86].

There have been suggestions that the local energy of DNA deformations may depend on the composition, or even the deformation, of more than just the immediate base-pair neighbors, for example

$$\Psi = \sum_{n=1}^{N-1} \psi^n(\theta^n, \rho^n, \ldots, \theta^{n+k}, \rho^{n+k}). \tag{4.13}$$

Trinucleotide and tetranucleotide models have been proposed to account for some DNA structural features [110], and they also seem to better represent averaged DNA properties extracted from molecular dynamics simulations [17, 52]; the mechanical theory of such models has not yet been constructed.

5. Electrostatics. DNA has a net negative charge that resides primarily at the phosphate groups on the DNA backbone (see Figure 6). Electrostatic interaction is an integral component of DNA response to deformations but its role in DNA is not completely understood, mainly because it is difficult to decouple such an effect from purely elastic local contributions. The effect of electrostatics is modulated by the ionic conditions of the solvent, such as its dielectric properties and the valence of counterions. The two most important effects of electrostatic repulsion appear to be the increase in DNA effective diameter [144, 118] and increase in DNA bending stiffness [10].

In accord with the classical theory of electrostatics, in the absence of counterions (charged particles in the solution) the electrostatic energy of DNA with M charged sites would be given by

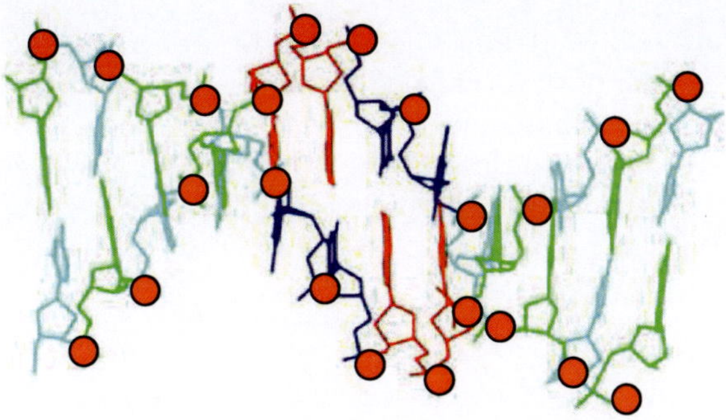

FIG. 6. *Negative charge on DNA is located at the phosphate groups (red).*

$$\Phi = \frac{(2\delta)^2}{4\pi\epsilon} \sum_{m=1}^{M-1} \sum_{n=m+1}^{M} \frac{1}{|\mathbf{r}^{mn}|} \tag{5.1}$$

where $\mathbf{r}^{mn} = \mathbf{x}^m - \mathbf{x}^n$ is the position vector connecting the charges m and n, δ is the elementary charge, and ϵ is the permittivity of water at 300K.

In the presence of counterions this long-range electrostatic interaction will be screened. Two main theories have been proposed to describe the effect of screening by monovalent counterions. The *Poisson-Boltzmann theory* replaces counterions by a continuous charge density and assumes that the this density is proportional to the Boltzmann factor of the electrostatic potential ϕ, which, after substituting in the classical equation of electrostatics, obeys the equation

$$\nabla(\epsilon(\mathbf{x})\nabla\phi(\mathbf{x})) = -4\pi\left(\rho(\mathbf{x}) + qe^{-\frac{q\phi(\mathbf{x})}{kT}}\right) \tag{5.2}$$

where ϵ is the dielectric, ρ is the charge density of DNA, q is the charge of counterions, and kT is Boltzmann constant times temperature. The electrostatic energy of DNA is then

$$\Phi = \frac{\delta}{2} \sum_{m=1}^{M} \phi(\mathbf{x}^m). \tag{5.3}$$

It was shown by Kirkwood [76] that the PB equation ignores the distinction between two different types of averages of the potential, which causes serious errors in the theory of strong electrolytes. Nonetheless, PB theory remains popular in studies of DNA at the atomic scale level [59, 19, 21, 138].

Alternative theory, proposed by Manning [94] and called the *counterion condensation theory*, separates the counterion distribution around

DNA into two parts: some counterions condense on the DNA and becomes immobile in all but one direction (along the DNA), the rest of the counterions remain mobile. The condensed portion of counterions neutralizes DNA charge to 24% of the original value, independent of the ionic strength. The weakened DNA charge can now be treated using Debye-Huckel theory (a linearized version of Poisson-Boltzmann theory) and yields, in place of (5.1) or (5.3), the following expression for DNA electrostatic energy:

$$\Phi = \frac{(2\delta)^2}{4\pi\epsilon} \sum_{m=1}^{M-1} \sum_{n=m+1}^{M} \frac{e^{-\kappa|\mathbf{r}^{mn}|}}{|\mathbf{r}^{mn}|} \tag{5.4}$$

where δ is now the net effective charge of $0.24\mathrm{e}^-$ and κ is the Debye screening parameter, which, for monovalent salt such as NaCl, depends on the molar salt concentration c as $\kappa = 0.329\sqrt{c}\text{Å}^{-1}$.

The counterion condensation theory has been included in some calculations of minimum energy configurations of DNA using continuum and discrete elastic models. The electrostatic energy gives rise to an additional term in the balance equation for forces, accounting for the force of repulsion between a DNA base pair and the rest of the molecule. For simplicity, the charges are usually assumed to be located in the centers of base-pairs, as opposed to the phosphate groups. The singularity in (5.4) makes it difficult to account for electostatics by a continuous charge density and hence, even in continuum models, the charges are generally assumed to be discrete and the resulting equations are solved numerically. The cases studied to date include superoiled configrations of DNA plasmids [148], the effect of electrostatics on LacR-induced DNA loops [6, 7], and the straghtening effect of electrostatics on intrinsically curved DNA segments [18].

Vologodskii and Cozzarelli have employed an alternative method to account for electrostatic repulsion of DNA, the so called *hard-core repulsion model* in which no energy is added to the elastic energy of DNA but configurations with intersegmental distance smaller than some effective DNA radius R are inadmissible [144]. They found that such a model yields accurate results in Monte Carlo simulations of the dependence of knotting probability on on ionic strength, in the sense that R can be calibrated for each ionic strength and with this calbibrated value their statistical model of DNA was able to predict correctly knotting probability for various types of experiments.

The effects of multivalent counterions are much more difficult to treat because such ions have the ability to interact with more than one charged phosphate group. They have been hypothesized to bridge DNA segments in DNA condensation or to participate in charge-neutralization induced DNA bending [72, 80].

6. Statistical mechanics. A long molecule of DNA in solution is subject to thermal fluctuations that perturb its shape away from the minimum energy configuration. Statistical mechanical theories of DNA account for fluctuations by assuming that each attainable configuration has a probability of occurence proportional to the Boltzmann factor of its total energy. Depending on the length and level of detail one seeks to describe, a fluctuating DNA can be treated using one of several polymer chain models: a freely-jointed chain, a worm-like chain, or a helical worm-like chain. In each of these models a DNA molecule is represented by a chain with vertices ${\mathbf{x}^n}_{n=1}^N$. The models are characterized by the dependence on N of the mean square end-to-end distance $< R^2 >=< |\mathbf{x}^N - \mathbf{x}^1|^2 >$, end-to-end (or radial) probability distribution function $\rho(R)$, or the closure (looping) probability $P(R = 0, \mathbf{t}^1 = \mathbf{t}^N)$.

The *freely-jointed chain* represents DNA as a chain of N rigid segments of length l, referred to as the Kuhn length, with uncorrelated orientations, i.e., $|\mathbf{x}^{n+1} - \mathbf{x}^n| = l$, $< (\mathbf{x}^{n+1} - \mathbf{x}^n) \cdot (\mathbf{x}^n - \mathbf{x}^{n-1}) >= 0$. (It corresponds to an unbiased random walk of equidistant steps in 3-space.) In the limit as $N \to \infty$, one finds that [53]:

$$< R^2 >\to Nl^2, \qquad \rho(R) \to \left(\frac{3}{2\pi N l^2}\right)^{3/2} \exp\left(-\frac{3R^2}{2Nl^2}\right) \tag{6.1}$$

provided that the orientation of the first segment is random, which implies that the root mean squared end-to-end distance scales with the square root of the length of the chain.

The *worm-like chain* accounts for bending rigidity of DNA. It can be derived in two ways - as a Kratky-Porod limit of a freely rotating chain with fixed angles between neighboring segments [81], or using Landau and Lifshitz method of averaging of configurations of an elastic rod with bending energy but no twisting energy (i.e., with C = 0 in (4.1)) [83]. In both cases one obtains the following expression:

$$< R^2 >= 2P\left(L - P(1-e^{-L/P})\right) \cong \begin{cases} L^2(1-L/3P) & \text{for } L << P \\ 2P(L-P) & \text{for } L >> P \end{cases} \tag{6.2}$$

where the persistence length P is a constant characterizing the stiffness of DNA; it is related to the bending rigidity A as $A = PkT$. The two limits of $< R^2 >$ in expression (6.2) tell us that a DNA that is much shorter than P behaves essentially as a stiff rod, while a DNA that is much larger than P behaves as a freely jointed chain with segments of length $l = 2P$.

The *helical worm-like chain* (HWLC) generalizes the worm-like chain model by accounting for the twisting deformation of DNA [123]. The partition function for HWLC is given by the path integral

$$Z(\mathbf{d}(L), \mathbf{t}(L), \mathbf{x}(L)|\mathbf{d}(0), \mathbf{t}(0), \mathbf{x}(0)) = \int_S \exp(-\Psi/kT) D\mathbf{x}(.) D\mathbf{d}(.) \tag{6.3}$$

with Ψ as in equation (4.1) and integration taken over the set S of all configurations $(\mathbf{x}(.), \mathbf{d}(.))$ with the specified end-conditions. For closed DNA plasmids these end-conditions are:

$$\mathbf{d}(L) = \mathbf{d}(0), \quad \mathbf{t}(L) = \mathbf{t}(0), \quad \mathbf{x}(L) = \mathbf{x}(0) \tag{6.4}$$

and an additional constraint of prescibed Lk is imposed on the contigurations in S which yields $Z = Z(Lk)$. This path integral cannot be evaluated explicitly but in various cases of interest approximate solutions have been obtained by asymptotic expansion [123, 99], Metropolis Monte-Carlo sampling [56, 90, 60], saddle-point expansion about the minimum energy configuration [156], or numerical integration on Euler motion group [34]. The integral (6.3) can also be evaluated using techniques developed for solving the Schrödinger equation [103, 22].

The HWLC theory has been employed in the study of DNA supercoiling and topoisomer distribution. Experimental results indicate that when DNA plasmids are reacted with type I topoisomerase [69], or are randomly formed by cyclization [125], one obtains a distribution of plasmid topoisomers that are identical apart from a difference in Lk. The resulting distribution of Lk is approximately Gaussian

$$P(Lk) \cong \exp(-G(Lk)/kT), \qquad G(Lk) = K(N)(Lk - N/h)^2, \tag{6.5}$$

where N is the plasmid size (in bp) and h is the helical repeat length (10.5 bp/turn). Theoretical predictions of this distribution [89] and the dependence of K on N by HWLC theory [79] were found to be in excellent agreement with experimental results. The shape of supercoiled configurations corresponding to high values of excess link $\Delta Lk = |Lk - N/h|$ was found to be of plectonemic nature with multiple terminal loops [119, 98].

The DNA cyclization experiment is one of the most sensitive methods for measuring DNA structural and elastic properties in solution [126, 43]. In the experiment identical linear DNA molecules with complementary free ends are reacted with an enzyme ligase that connects the free ends. The molecules can connect in two ways: (i) the two ends of a single molecule can join to form a cyclized molecule, or (ii) the ends of two molecules can dimerize to form a linear DNA segment of twice the length. The rates of cyclization and dimerization can be measured and their ratio, called the Jacobson-Stockmayer factor (or the J factor), can be plotted as a function of N to obtain the characteristic J curve [126, 124, 43, 8]. The J factor has been shown to be proportional to the probability of cyclization, which is an equilibrium quantity that can be computed using a HWLC model. Thus, material properties of DNA can be estimated by fitting the measured data with a computed J curve [123, 156, 90].

During a closure experiment DNA molecules may become knotted [122]. The probability of DNA knotting can be estimated using HWLC model [113], and the results are sensitive to DNA electrostatic repulsion, both in the magnitude of the screening and the treatment of electrostatic interaction [144]. DNA knotting is also produced by the action of topoisomerases of type II and type I (on nicked DNA), which has been used to elucidate the function of those enzymes. An important issue related to type II topoisomerases is that they are very efficient in removing knots, catenations, and supercoils well below the thermodynamic equilibrium [114], which is made possible by the fact that they utilize ATP, a source of energy, during their action. A definite mechanism of how a small enzyme manages to recognize a global property of DNA as a knot type has not yet been found although several hypotheses have been proposed [145, 152, 25]. DNA knotting probability within the confined volume of bacteriophage head has also been studied using FJC model [5, 101].

Another area of DNA research that has greatly benefitted from and stimulated the development of statistical modeling is the area of single-molecule DNA manipulation experiments. Single-molecule DNA stretching and twisting experiments represent breakthroughs in the study of DNA properties because they allow researchers to track time-dependent behavior of individual molecules as opposed to ensemble averaged quantities. In these experiments one end of DNA is attached to a fixed object, for example the microscope slide or a bead that is held by a pipette, while the other end is attached to a bead that is captured and manipulated by an optical or magnetic trap. By varying the distance between the beads experimenters can stretch the captured molecule, and by rotating the magnetic bead they can twist the molecule.

Results of DNA stretching experiments [128, 26] are in excellent accord with theoretical predictions using WLC and HWLC models in ranges of loading that preserve the duplex DNA structure. The dependence of force on extension for a torsionally relaxed molecule of length L is fitted very accurately by the formula [99]

$$\frac{FP}{kT} = \frac{1}{4}\left(1 - \frac{x}{L}\right)^{-2} - \frac{1}{4} + \frac{x}{L} \tag{6.6}$$

where P is the persistence length. DNA twisting experiments [131, 30] have also been found in close agreement with HWLC predictions [103, 102, 22]. When large forces and/or twist is imposed on DNA, the molecule changes its secondary structure into alternative structures - overstretched DNA [127, 97], melted duplex with separated strands [131], or Pauli structure with backbone on the inside and bases on the outside [130] - none of which are governed by the HWLC theory.

The HWLC theory is built on the simplest, ideal model of DNA elasticity. An equivalent statistical mechanical theory has been developed also for sequence-dependent base-pair level DNA model [61]. Various cases of interest for DNA with sequence dependent properties have been analyzed, such the statistics of polymer chains with intrinsic bends or elastic inhomogeneity [112], the looping free energy of LacR-mediated DNA loops [134, 6], the free energy of LacR loops in the presence of CAP [135], the effect of intrinsic curvature, anisotropy, or twist-roll coupling on ring closure probability [44].

7. Dynamics. Dynamical models of DNA have been constructed to help us uncover time-dependent features of DNA behavior, such as perturbation relaxation times, rates of transition between configurational states, rates of closure and loop formation, etc. The primary interactions controlling DNA dynamics are hydrodynamic resistance and thermal fluctuations.

Dynamical models of DNA can be divided into two groups - those based on theories of elastic rods and those based on theories of polymer dynamics. Dynamical theories of rods can be formulated within both Kirchhoff and special Cosserrat theories (see, e.g., [2]). Suppose that the rod is described by giving, as functions of the arc-length s and time t, its axial curve $\mathbf{x}(s,t)$ and an orthonormal triad $(\mathbf{d}_1(s,t), \mathbf{d}_2(s,t), \mathbf{d}_3(s,t))$ embedded in the cross-section. The equations of motion are given by

$$\dot{\mathbf{P}} = \mathbf{F}' + \mathbf{f} \tag{7.1}$$

$$\dot{\mathbf{R}} = \mathbf{M}' + \mathbf{x}' \times \mathbf{F} + \mathbf{m} \tag{7.2}$$

where dot stands for the time derivative, $\mathbf{P}$ and $\mathbf{R}$ are the linear and angular momentum of the cross-section, $\mathbf{F}$ and $\mathbf{M}$ are the contact forces and moments applied on the cross-section at s by material with arc-length greater than s, and $\mathbf{f}$ and $\mathbf{m}$ are the external forces and moments. The precise form of $\mathbf{P}$ and $\mathbf{R}$ and the constitutive equations for $\mathbf{F}$ and $\mathbf{M}$ depend on the type of rod under consideration and the approximations taken. The terms $\mathbf{f}$ and $\mathbf{m}$ account for hydrodynamic resistance.

When $\mathbf{f} = \mathbf{m} = 0$, exact solutions of (7.1)–(7.2) can be obtained for special motions, called traveling waves, in which the shape of the axial curve remains invariant and its apex is moving with constant velocity along the rod [4, 37, 50]. Other results include perturbation analysis of looping and ring collapse transitions [63, 64]. Hydrodynamic resistance was accounted for in numerical analyses of the formation of supercoiled states of overtwisted rings plasmids [77, 91]. Some researchers have used the solution of dynamical equations as a method for finding stable equilibrium configurations of DNA [6, 65]. With the exception of [6], the studies of DNA dynamics using continuum models published to date ignore thermal fluctuations but, nonetheless, yield useful information about the transition from circular to supercoiled DNA configurations.

In *polymer dynamics* models, a DNA molecule is replaced by a collection of rigid spheres of radius R with centers at $\{\mathbf{x}^n\}_{n=1}^N$ that are connected by elastic linkages simulating the bending and twisting rigidity of DNA. The total energy E of the chain is composed of stretching, bending, twisting, and electrostatic energy components. Because of high hydrodynamics resistance of the solvent, the molecule is assumed to move by diffusion which results in a Brownian type dynamics. At each time step Δt the positions of the beads change in accord with the following formula

$$\mathbf{x}^n(t+\Delta t) = \mathbf{x}^n(t) - \frac{\Delta t}{kT}\sum_m \mathbf{D}^{mn}(t)\nabla_{\mathbf{x}^n} E + \mathbf{R}^n(t) \tag{7.3}$$

where the Rotne-Prager diffusion tensor $\mathbf{D}^{mn}$ accounts for hydrodynamic coupling between beads m and n, and the random displacements $\mathbf{R}^n(t)$ obey

$$< \mathbf{R}^n(t) >= 0, \qquad < \mathbf{R}^n(t)\mathbf{R}^n(t)^T >= 2\Delta t\mathbf{D}^{mn}(t). \tag{7.4}$$

Brownian dynamics has been employed in the study of DNA tumbling and twisting, where computed results were compared with data coming from fluorescence depolarization experiments [91]. It has also been used to study the dynamics of DNA supercoiling [31, 78], and the dependence of site juxtaposition in DNA on the distance between sites [73, 71] and supercoiling density [32]. It was found, for example, that at low salt juxtaposition times are accelerated by a factor of 10 or more due to supercoiling [73]. DNA supercoiling was found [32] to occur on the timescale of 3–6 μs starting from a planar closed molecule, with an initial phase of 1–2 μs during which toroidal supercoiling appeared, followed by a conversion into a plectonemic supercoiling. The effect of intrinsic curvature on DNA supercoiling [33] and looping [100] has also been studied

8. Conclusion. This survey outlines the main mathematical results and models used by researchers to discuss DNA deformability and structure at the macroscopic level, covering a whole range of topological, geometrical, mechanical, electrostatic, statistical and dynamical models. There are few important topics in DNA research that do not naturally fit under the headings above, and one of them is the connection between DNA denaturation and supercoiling. It is known that although DNA molecule is stable under the conditions mimicking the intracellular environment, the base-pairing interaction can be disrupted, in a process called denaturation, as a result of a high temperature or mechanical deformation such as large untwisting, stretching, or bending. The energy required for DNA denaturation depends on base-pair composition and have been determined very accurately in calorimetric experiments. Craig Benham has used this information to develop a theory of stress-induced duplex destabilization [14, 15], which he used to compute the sites in a genome that would be most prone to denaturation due to supercoiling, and found that such sites coincide with transcription initiation regions [16].

The response of DNA to large deformations is still not well understood. Extensive stretching or twisting can induce the transition of DNA to alternative conformations with disrupted base-pairing (see Section 6), but it is not known whether such conformations play any biological role. DNA kinking - a higher order response to bending associated with disruption of base-stacking - has been proposed [154, 151, 116, 85, 54] as an explanation for unusually large cyclization probabilities of certain special DNA sequences [35, 36]. Furthermore, in addition to the well known B-DNA form there are other, alternative forms of DNA - A-DNA, C-DNA, Z-DNA, - which are induced by special experimental conditions (ethanol, high salt), and for which mechanical properties have not been explored, nor mechanical theories of transitions between these forms have been formulated.

A large area of DNA research is concerned with protein-DNA interactions. We are still far from complete understanding the principles of protein-DNA binding affinity and specificity, and the role of DNA deformations that many proteins impose on DNA. This problem requires the use of local atomic level description of DNA and proteins and is beyond the scope of this essay.

The modeling efforts described above have focused on the understanding of DNA physical properties. The ultimate goal of DNA modeling, however, is to address important biological problems such as the problem of DNA compaction, chromatin formation and remodeling, and the problem of the role of DNA deformability in replication, recombination, or regulation of transcription. The first steps in this direction are provided by methods that utilize current information about protein-DNA interaction (X-ray crystal data, binding affinity measurements, DNA footprinting, etc.) to compute the structures of complex multi-protein DNA assemblies [6, 142, 65, 134, 7, 86, 135, 45].

A new direction in simulation of DNA dynamics and mechanics is to move away from models tailored to the conditions *in vitro* (i.e., in the test tube) to models of DNA *in vivo* (i.e., inside of a living cell). One important difference here is that the DNA *in vivo* is subject to random interactions with a large number of DNA binding proteins, both sequence specific and non-specific, that bend and twist the molecule. The first examples of a reserch concerned with intracellular DNA modeling include the analysis of DNA stretching in the presence of randomly binding bending or stiffening agents [153], and the study of DNA cyclization in the presence of a nonspecifically binding bending protein HU [45].

Although this overview of various areas of DNA modeling is understandably sketchy and incomplete, it gives the reader an idea about the variety of areas of DNA research that benefit from the use of mathematics. Further information about specific areas can be found in numerous survey papers and books, e.g., DNA topology [9, 150], base-pair level DNA structure [28], detail DNA structure [104], DNA mechanics [11, 65], single-molecule DNA stretching [26] and twisting [130], DNA supercoiling [106, 117, 143], or DNA topoisomerase action [146]. Additional material is available online, and includes two lectures given by the author at the tutorial "Mathematics of Nucleic Acids" which has taken place at the IMA on September 15, 2007. The slides and videorecordings of these lectures can be found on the IMA website (www.ima.umn.edu).

Acknowledgements. The author wishes to express his thanks to Zuzana Swigonova for careful proofreading of the manuscript and numerous suggestions. Much of this work was written during a stimulating one-semester visit at the Institute for Mathematics and its Applications, University of Minnesota. Support by A.P. Sloan Fellowship and NSF grant DMS-05-16646 is also acknowledged.

REFERENCES

[1] J. Aldinger, I. Klapper, and M. Tabor, *Formulae for the calculation and estimation of writhe*, J. Knot Theory Ramifications, **4** (1995), pp. 343–372.

[2] S.S. Antman, *Nonlinear problems of elasticity*, Vol. 107 of Applied Mathematical Sciences, Springer, New York, second ed., 2005.

[3] S.S. Antman and C.S. Kenney, *Large buckled states of nonlinearly elastic rods under torsion, thrust, and gravity*, Arch. Rational Mech. Anal., **76** (1981), pp. 289–338.

[4] S.S. Antman and T.-P. Liu, *Travelling waves in hyperelastic rods*, Quart. Appl. Math., **36** (1979), pp. 377–399.

[5] J. Arsuaga, M. Vázquez, S. Trigueros, D. Sumners, and J. Roca, *Knotting probability of DNA molecules confined in restricted volumes: DNA knotting in phage capsids*, Proc. Natl. Acad. Sci. U.S.A., **99** (2002), pp. 5373–5377.

[6] A. Balaeff, C.R. Koudella, L. Mahadevan, and K. Schulten, *Modelling DNA loops using continuum and statistical mechanics*, Philos. Trans. R. Soc. Lond. Ser. A Math. Phys. Eng. Sci., **362** (2004), pp. 1355–1371.

[7] A. Balaeff, L. Mahadevan, and K. Schulten, *Modeling DNA loops using the theory of elasticity*, Phys. Rev. E (3), **73** (2006), pp. 031919, 23.

[8] A. Barbic and D.M. Crothers, *Comparison of analyses of DNA curvature*, J. Biomol. Struct. Dyn., **21** (2003), pp. 89–97.

[9] A.D. Bates and A. Maxwell, *DNA topology*, Oxford University Press, 1993.

[10] C. Baumann, S. Smith, V. Bloomfield, and C. Bustamante, *Ionic effects on the elasticity of single DNA molecules*, Proc. Natl. Acad. Sci. U.S.A., **94** (1997), pp. 6185–6190.

[11] C. Benham and S. Mielke, *DNA mechanics*, Annu Rev Biomed Eng, **7** (2005), pp. 21–53.

[12] C.J. Benham, *Elastic model of supercoiling*, Proc. Natl. Acad. Sci. U.S.A., **74** (1977), pp. 2397–2401.

[13] ———, *An elastic model of the large-scale structure of duplex DNA*, Biopolymers, **18** (1979), pp. 609–623.

[14] ———, *Theoretical analysis of heteropolymeric transitions in superhelical DNA molecules of specified sequence.*, J. Chem. Phys., **92** (1990), pp. 6294–6305.

[15] ———, *Energetics of the strand separation transition in superhelical DNA*, J. Mol. Biol., **225** (1992), pp. 835–847.

[16] ———, *Duplex destabilization in superhelical DNA is predicted to occur at specific transcriptional regulatory regions.*, J. Mol. Biol., **255** (1996), pp. 425–434.

[17] D.L. Beveridge, G. Barreiro, K.S. Byun, D.A. Case, T.E. Cheatham, S.B. Dixit, E. Giudice, F. Lankas, R. Lavery, J.H. Maddocks, R. Osman, E. Seibert, H. Sklenar, G. Stoll, K.M. Thayer, P. Varnai, and M.A. Young, *Molecular dynamics simulations of the 136 unique tetranucleotide sequences of DNA oligonucleotides. I. Research design and results on d(CpG) steps*, Biophys. J., **87** (2004), pp. 3799–3813.

[18] Y.Y. Biton, B.D. Coleman, and D. Swigon, *On bifurcations of equilibria of intrinsically curved, electrically charged, rod-like structures that model DNA molecules in solution*, J. Elasticity, **87** (2007), pp. 187–210.

[19] V.A. Bloomfield and I. Rouzina, *Use of Poisson-Boltzmann equation to analyze ion binding to DNA*, Meth. Enzymol., **295** (1998), pp. 364–378.

[20] A. Bolshoy, P. McNamara, R.E. Harrington, and E.N. Trifonov, *Curved DNA without A-A: experimental estimation of all 16 DNA wedge angles*, Proc. Natl. Acad. Sci. U.S.A., **88** (1991), pp. 2312–2316.

[21] A.H. Boschitsch and M.O. Fenley, *Hybrid boundary element and finite difference method for solving the nonlinear Poisson-Boltzmann equation*, J. Comput. Chem., **25** (2004), pp. 935–955.

[22] C. Bouchiat and M. Mezard, *Elasticity model of a supercoiled DNA molecule*, Phys. Rev. Lett., **80** (1998), pp. 1556–1559.

[23] C. Bouchiat, M. Wang, J. Allemand, T. Strick, S. Block, and V. Croquette, *Estimating the persistence length of a worm-like chain molecule from force-extension measurements*, Biophys. J., **76** (1999), pp. 409–413.

[24] Z. Bryant, M.D. Stone, J. Gore, S.B. Smith, N.R. Cozzarelli, and C. Bustamante, *Structural transitions and elasticity from torque measurements on DNA*, Nature, **424** (2003), pp. 338–341.

[25] G.R. Buck and E.L. Zechiedrich, *DNA disentangling by type-2 topoisomerases*, J. Mol. Biol., **340** (2004), pp. 933–939.

[26] C. Bustamante, Z. Bryant, and S.B. Smith, *Ten years of tension: single-molecule DNA mechanics*, Nature, **421** (2003), pp. 423–427.

[27] C. Bustamante, S.B. Smith, J. Liphardt, and D. Smith, *Single-molecule studies of DNA mechanics*, Curr. Opin. Struct. Biol., **10** (2000), pp. 279–285.

[28] C.R. Calladine and H.R. Drew, *Understanding DNA*, Academic Press, 1992.

[29] G. Călugăreanu, *Sur les classes d'isotopie des nœuds tridimensionnels et leurs invariants*, Czechoslovak Math. J., **11** (86) (1961), pp. 588–625.

[30] G. Charvin, J.F. Allemand, T.R. Strick, D. Bensimon, and V. Croquette, *Twisting DNA: single molecule studies*, Cont. Phys., **45** (2004), pp. 383–403.

[31] G. Chirico and J. Langowski, *Calculating hydrodynamic properties of DNA through a second-order Brownian dynamics algorithm*, Macromolecules, **25** (1992), pp. 769–775.

[32] ———, *Kinetics of DNA supercoiling studied by Brownian dynamics simulation*, Biopolymers, **34** (1994), pp. 211–225.

[33] ———, *Brownian dynamics simulations of supercoiled DNA with bent sequences*, Biophys. J., **71** (1996), pp. 955–971.

[34] G.S. Chirikjian and Y. Wang, *Conformational statistics of stiff macromolecules as solutions to partial differential equations on the rotation and motion groups*, Phys. Rev. E Stat. Phys. Plasmas. Fluids Relat. Interdiscip. Topics, **62** (2000), pp. 880–892.

[35] T. Cloutier and J. Widom, *Spontaneous sharp bending of double-stranded DNA*, Mol. Cell, **14** (2004), pp. 355–362.

[36] ———, *DNA twisting flexibility and the formation of sharply looped protein-DNA complexes*, Proc. Natl. Acad. Sci. U.S.A., **102** (2005), pp. 3645–3650.

[37] B.D. Coleman, E.H. Dill, M. Lembo, Z. Lu, and I. Tobias, *On the dynamics of rods in the theory of Kirchhoff and Clebsch*, Arch. Rational Mech. Anal., **121** (1992), pp. 339–359.

[38] B.D. Coleman, W.K. Olson, and D. Swigon, *Theory of sequence-dependent DNA elasticity*, J. Chem. Phys., **118** (2003), pp. 7127–7140.

[39] B.D. Coleman and D. Swigon, *Theory of supercoiled elastic rings with self-contact and its application to DNA plasmids*, J. Elasticity, **60** (2000), pp. 173–221 (2001).

[40] ———, *Theory of self-contact in Kirchhoff rods with applications to supercoiling of knotted and unknotted DNA plasmids*, Philos. Trans. R. Soc. Lond. Ser. A Math. Phys. Eng. Sci., **362** (2004), pp. 1281–1299.

[41] B.D. Coleman, D. Swigon, and I. Tobias, *Elastic stability of DNA configurations. II. Supercoiled plasmids with self-contact*, Phys. Rev. E (3), **61** (2000), pp. 759–770.

[42] R. Courant, *Differential and Integral Calculus*, Vol. 2, Blackie, London, 1936.

[43] D.M. Crothers, J. Drak, J.D. Kahn, and S.D. Levene, *DNA bending, flexibility, and helical repeat by cyclization kinetics*, Meth. Enzymol., **212** (1992), pp. 3–29.

[44] L. Czapla, D. Swigon, and W.K. Olson, *Sequence-dependent effects in the cyclization of short DNA*, J. Chem. Theory Comput., **2** (2006), pp. 685–695.

[45] ———, *Effects of the nucleoid protein HU on the structure, flexibility, and ring-closure properties of DNA deduced from monte-carlo simulations*, submitted to J. Mol. Biol. (2008).

[46] I.K. Darcy and D.W. Sumners, *A strand passage metric for topoisomerase action*, in KNOTS '96 (Tokyo), World Sci. Publ., River Edge, NJ, 1997, pp. 267–278.

[47] ———, *Rational tangle distances on knots and links*, Math. Proc. Cambridge Philos. Soc., **128** (2000), pp. 497–510.

[48] Y. Diao, J.C. Nardo, and Y. Sun, *Global knotting in equilateral random polygons*, J. Knot Theory Ram., **10** (2001), pp. 597–607.

[49] D.J. Dichmann, Y. Li, and J.H. Maddocks, *Hamiltonian formulations and symmetries in rod mechanics*, in Mathematical approaches to biomolecular structure and dynamics (Minneapolis, MN, 1994), Vol. 82 of IMA Vol. Math. Appl., Springer, New York, 1996, pp. 71–113.

[50] D.J. Dichmann, J.H. Maddocks, and R.L. Pego, *Hamiltonian dynamics of an elastica and the stability of solitary waves*, Arch. Rational Mech. Anal., **135** (1996), pp. 357–396.

[51] E.H. Dill, *Kirchhoff's theory of rods*, Arch. Hist. Exact Sci., **44** (1992), pp. 1–23.

[52] S.B. Dixit, D.L. Beveridge, D.A. Case, T.E. Cheatham, E. Giudice, F. Lankas, R. Lavery, J.H. Maddocks, R. Osman, H. Sklenar, K.M. Thayer, and P. Varnai, *Molecular dynamics simulations of the 136 unique tetranucleotide sequences of DNA oligonucleotides. II: sequence context effects on the dynamical structures of the 10 unique dinucleotide steps*, Biophys. J., **89** (2005), pp. 3721–3740.

[53] L.D. Doi and S.F. Edwards, *The Theory of Polymer Dynamics*, Clarendon Press, 1988.

[54] Q. Du, A. Kotlyar, and A. Vologodskii, *Kinking the double helix by bending deformation*, Nucleic Acids Res., **36** (2008), pp. 1120–1128.

[55] C. Ernst and D. W. Sumners, *A calculus for rational tangles: applications to DNA recombination*, Math. Proc. Cambridge Philos. Soc., **108** (1990), pp. 489–515.

[56] M.D. Frank-Kamenetskii, A.V. Lukashin, and A.V. Vologodskii, *Statistical mechanics and topology of polymer chains*, Nature, **258** (1975), pp. 398–402.

[57] F.B. Fuller, *The writhing number of a space curve*, Proc. Nat. Acad. Sci. U.S.A., **68** (1971), pp. 815–819.

[58] P.B. Furrer, R.S. Manning, and J.H. Maddocks, *DNA rings with multiple energy minima*, Biophys. J., **79** (2000), pp. 116–136.

[59] S. Gavryushov and P. Zielenkiewicz, *Electrostatic potential of B-DNA: effect of interionic correlations*, Biophys. J., **75** (1998), pp. 2732–2742.

[60] J.A. Gebe, S.A. Allison, J.B. Clendenning, and J.M. Schurr, *Monte Carlo simulations of supercoiling free energies for unknotted and trefoil knotted DNAs*, Biophys. J., **68** (1995), pp. 619–633.

[61] O. Gonzales and J.H. Maddocks, *Extracting parameters for base-pair level models of DNA from molecular dynamics simulations*, Theor. Chem. Acc., **106** (2001), pp. 76–82.

[62] O. Gonzalez, J.H. Maddocks, F. Schuricht, and H. von der Mosel, *Global curvature and self-contact of nonlinearly elastic curves and rods*, Calc. Var. Partial Differential Equations, **14** (2002), pp. 29–68.

[63] A. Goriely and M. Tabor, *Nonlinear dynamics of filaments. IV. Spontaneous looping of twisted elastic rods*, R. Soc. Lond. Proc. Ser. A Math. Phys. Eng. Sci., **454** (1998), pp. 3183–3202.

[64] ———, *The nonlinear dynamics of filaments*, Nonlinear Dynam., **21** (2000), pp. 101–133.

[65] S. Goyal, N.C. Perkins, and C.L. Lee, *Nonlinear dynamics and loop formation in Kirchhoff rods with implications to the mechanics of DNA and cables*, J. Comput. Phys., **209** (2005), pp. 371–389.
[66] P.J. Hagerman, *Flexibility of DNA*, Annu Rev Biophys Biophys Chem, **17** (1988), pp. 265–286.
[67] K.A. Hoffman, *Methods for determining stability in continuum elastic-rod models of DNA*, Philos. Trans. R. Soc. Lond. Ser. A Math. Phys. Eng. Sci., **362** (2004), pp. 1301–1315.
[68] K.A. Hoffman, R.S. Manning, and J.H. Maddocks, *Link, twist, energy, and the stability of DNA minicircles*, Biopolymers, **70** (2003), pp. 145–157.
[69] D.S. Horowitz and J.C. Wang, *Torsional rigidity of DNA and length dependence of the free energy of DNA supercoiling*, J. Mol. Biol., **173** (1984), pp. 75–91.
[70] X. Hua, D. Nguyen, B. Raghavan, J. Arsuaga, and M. Vazquez, *Random state transitions of knots: A first step towards modeling unknotting by type II topoisomerases*, Topology Appl., **154** (2007), pp. 1381–1397.
[71] J. Huang, T. Schlick, and A. Vologodskii, *Dynamics of site juxtaposition in supercoiled DNA*, Proc. Natl. Acad. Sci. U.S.A., **98** (2001), pp. 968–973.
[72] N. Hud and J. Plavec, *A unified model for the origin of DNA sequence-directed curvature*, Biopolymers, **69** (2003), pp. 144–158.
[73] H. Jian, T. Schlick, and A. Vologodskii, *Internal motion of supercoiled DNA: Brownian dynamics simulations of site juxtaposition*, J. Mol. Biol., **284** (1998), pp. 287–296.
[74] F. Julicher, *Supercoiling transitions of closed DNA*, Phys. Rev. E, **49** (1994), pp. 2429–2435.
[75] G. Kirchhoff, *Über das Gleichgewicht und die Bewegung eines unendlich dünen elastischen Stabes*, J. Reine angew. Math. (Crelle), **56** (1859), pp. 285–313.
[76] J.G. Kirkwood, *On the theory of strong electrolyte solutions*, J. Chem. Phys., **2** (1934), pp. 767–781.
[77] I. Klapper, *Biological applications of the dynamics of twisted elastic rods*, J. Comput. Phys., **125** (1996), pp. 325–337.
[78] K. Klenin, H. Merlitz, and J. Langowski, *A Brownian dynamics program for the simulation of linear and circular DNA and other wormlike chain polyelectrolytes*, Biophys. J., **74** (1998), pp. 780–788.
[79] K.V. Klenin, A.V. Vologodskii, V.V. Anshelevich, A.M. Dykhne, and M.D. Frank-Kamenetskii, *Computer simulation of DNA supercoiling*, J. Mol. Biol., **217** (1991), pp. 413–419.
[80] K.M. Kosikov, A.A. Gorin, X.J. Lu, W.K. Olson, and G.S. Manning, *Bending of DNA by asymmetric charge neutralization: all-atom energy simulations*, J. Am. Chem. Soc., **124** (2002), pp. 4838–4847.
[81] O. Kratky and G. Porod, *Röntgenuntersuchung gelöster fadenmoleküle*, Rec. Trav. Chim. Pays-Bas., **68** (1949), pp. 1106–1123.
[82] L.D. Landau and E.M. Lifshitz, *Theory of Elasticity*, Pergamon Press, Oxford, 1959.
[83] ———, *Statistical Physics*, Butterworth-Heinemann, 1984.
[84] J. Langer and D.A. Singer, *Knotted elastic curves in* $\mathbf{R}^3$, J. London Math. Soc. (2), **30** (1984), pp. 512–520.
[85] F. Lankas, R. Lavery, and J. Maddocks, *Kinking occurs during molecular dynamics simulations of small DNA minicircles*, Structure, **14** (2006), pp. 1527–1534.
[86] C.L. Lawson, D. Swigon, K.S. Murakami, S.A. Darst, H.M. Berman, and R.H. Ebright, *Catabolite activator protein: DNA binding and transcription activation*, Curr. Opin. Struct. Biol., **14** (2004), pp. 10–20.
[87] M. Le Bret, *Catastrophic variation of twist and writhing of circular DNAs with constraint?*, Biopolymers, **18** (1979), pp. 1709–1725.

[88] ———, *Twist and writhing in short circular DNAs according to first-order elasticity*, Biopolymers, **23** (1984), pp. 1835–1867.

[89] S. Levene and D. Crothers, *Topological distributions and the torsional rigidity of DNA. A Monte Carlo study of DNA circles*, J. Mol. Biol., **189** (1986), pp. 73–83.

[90] S. D. Levene and D. M. Crothers, *Ring closure probabilities for DNA fragments by Monte Carlo simulation*, J. Mol. Biol., **189** (1986), pp. 61–72.

[91] S. Lim, A. Ferent, X.S. Wang, and C.S. Peskin, *Dynamics of a closed rod with twist and bend in fluid*, preprint.

[92] A.E.H. Love, *Treatise on the mathematical theory of elasticity*, Cambridge University Press, 1927.

[93] J.H. Maddocks, *Bifurcation theory, symmetry breaking and homogenization in continuum mechanics descriptions of DNA. Mathematical modelling of the physics of the double helix*, in A celebration of mathematical modeling, Kluwer Acad. Publ., Dordrecht, 2004, pp. 113–136.

[94] G.S. Manning, *The molecular theory of polyelectrolyte solutions with applications to the electrostatic properties of polynucleotides*, Q. Rev. Biophys., **11** (1978), pp. 179–246.

[95] R.S. Manning, J.H. Maddocks, and J.D. Kahn, *A continuum rod model of sequence-dependent DNA structure*, J. Chem. Phys., **105** (1996), pp. 5626–5646.

[96] R.S. Manning, K.A. Rogers, and J.H. Maddocks, *Isoperimetric conjugate points with application to the stability of DNA minicircles*, R. Soc. Lond. Proc. Ser. A Math. Phys. Eng. Sci., **454** (1998), pp. 3047–3074.

[97] J.F. Marko, *DNA under high tension: Overstretching, undertwisting, and relaxation dynamics*, Phys. Rev. E, **57** (1998), pp. 2134–2149.

[98] J.F. Marko and E.D. Siggia, *Fluctuations and supercoiling of DNA*, Science, **265** (1994), pp. 506–508.

[99] ———, *Stretching DNA*, Macromolecules, **28** (1995), pp. 8759–8770.

[100] H. Merlitz, K. Rippe, K.V. Klenin, and J. Langowski, *Looping dynamics of linear DNA molecules and the effect of DNA curvature: a study by Brownian dynamics simulation*, Biophys. J., **74** (1998), pp. 773–779.

[101] C. Micheletti, D. Marenduzzo, E. Orlandini, and D. Summers, *Knotting of random ring polymers in confined spaces*, J. Chem. Phys., **124** (2006), p. 64903.

[102] J. Moroz and P. Nelson, *Torsional directed walks, entropic elasticity, and DNA twist stiffness*, Proc. Natl. Acad. Sci. U.S.A., **94** (1997), pp. 14418–14422.

[103] J.D. Moroz and P. Nelson, *Torsional directed walks, entropic elasticity, and DNA twist stiffness*, Proc. Natl. Acad. Sci. U.S.A., **94** (1997), pp. 14418–14422.

[104] S. Neidle, *Principles of nucleic acid structure*, Elsevier, 2007.

[105] S. Neukirch and M.E. Henderson, *Classification of the spatial equilibria of the clamped elastica: symmetries and zoology of solutions*, J. Elasticity, **68** (2002), pp. 95–121 (2003). Dedicated to Piero Villaggio on the occasion of his 70th birthday.

[106] W.K. Olson, *Simulating DNA at low resolution*, Curr. Opin. Struct. Biol., **6** (1996), pp. 242–256.

[107] W.K. Olson, M. Bansal, S.K. Burley, R.E. Dickerson, M. Gerstein, S.C. Harvey, U. Heinemann, X.J. Lu, S. Neidle, Z. Shakked, H. Sklenar, M. Suzuki, C.S. Tung, E. Westhof, C. Wolberger, and H.M. Berman, *A standard reference frame for the description of nucleic acid base-pair geometry*, J. Mol. Biol., **313** (2001), pp. 229–237.

[108] W.K. Olson, A.A. Gorin, X.J. Lu, L.M. Hock, and V.B. Zhurkin, *DNA sequence-dependent deformability deduced from protein-DNA crystal complexes*, Proc. Natl. Acad. Sci. U.S.A., **95** (1998), pp. 11163–11168.

[109] W.K. Olson, D. Swigon, and B.D. Coleman, *Implications of the dependence of the elastic properties of DNA on nucleotide sequence*, Philos Transact A Math Phys Eng Sci, **362** (2004), pp. 1403–1422.

[110] M.J. Packer, M.P. Dauncey, and C.A. Hunter, *Sequence-dependent DNA structure: tetranucleotide conformational maps*, J. Mol. Biol., **295** (2000), pp. 85–103.

[111] L.J. Parkhurst, K.M. Parkhurst, R. Powell, J. Wu, and S. Williams, *Time-resolved fluorescence resonance energy transfer studies of DNA bending in double-stranded oligonucleotides and in DNA-protein complexes*, Biopolymers, **61** (2001), pp. 180–200.

[112] C. Rivetti, C. Walker, and C. Bustamante, *Polymer chain statistics and conformational analysis of DNA molecules with bends or sections of different flexibility*, J. Mol. Biol., **280** (1998), pp. 41–59.

[113] V. Rybenkov, N. Cozzarelli, and A. Vologodskii, *Probability of DNA knotting and the effective diameter of the DNA double helix*, Proc. Natl. Acad. Sci. U.S.A., **90** (1993), pp. 5307–5311.

[114] V.V. Rybenkov, C. Ullsperger, A.V. Vologodskii, and N.R. Cozzarelli, *Simplification of DNA topology below equilibrium values by type II topoisomerases*, Science, **277** (1997), pp. 690–693.

[115] W. Saenger, *Principles of Nucleic Acid Structure*, Springer-Verlag, New York and Berlin, 1984.

[116] S. Sankararaman and J. Marko, *Formation of loops in DNA under tension*, Phys Rev E Stat Nonlin Soft Matter Phys, **71** (2005), p. 021911.

[117] T. Schlick, *Modeling superhelical DNA: recent analytical and dynamic approaches*, Curr. Opin. Struct. Biol., **5** (1995), pp. 245–262.

[118] T. Schlick, B. Li, and W.K. Olson, *The influence of salt on DNA energetics and dynamics*, Biophys. J., **67** (1994), pp. 2146–2166.

[119] T. Schlick, W. Olson, T. Westcott, and J. Greenberg, *On higher buckling transitions in supercoiled DNA*, Biopolymers, **34** (1994), pp. 565–597.

[120] F. Schuricht, *Global injectivity and topological constraints for spatial nonlinearly elastic rods*, J. Nonlin. Science, **12** (2002), pp. 423–444.

[121] J.M. Schurr, B.S. Fujimoto, P. Wu, and S.L., *Fluorescence studies of nucleic acids: Dynamics, rigidities, and structures*, in Biochemical Applications, Vol. 3 of Topics in Fluorescence Spectroscopy, Plenum Press, New York, 1992.

[122] S. Shaw and J. Wang, *Knotting of a DNA chain during ring closure*, Science, **260** (1993), pp. 533–536.

[123] J. Shimada and H. Yamakawa, *Ring-closure probabilities for twisted wormlike chains*, Macromolecules, **17** (1984), pp. 689–698.

[124] D. Shore and R. Baldwin, *Energetics of DNA twisting. I. Relation between twist and cyclization probability*, J. Mol. Biol., **170** (1983), pp. 957–981.

[125] ———, *Energetics of DNA twisting. II. Topoisomer analysis*, J. Mol. Biol., **170** (1983), pp. 983–1007.

[126] D. Shore, J. Langöwski, and R. L. Baldwin, *DNA flexibility studied by covalent closure of short fragments into circles*, Proc. Natl. Acad. Sci. USA, **78** (1981), pp. 4833–4837.

[127] S. Smith, Y. Cui, and C. Bustamante, *Overstretching B-DNA: the elastic response of individual double-stranded and single-stranded DNA molecules*, Science, **271** (1996), pp. 795–799.

[128] S. Smith, L. Finzi, and C. Bustamante, *Direct mechanical measurements of the elasticity of single DNA molecules by using magnetic beads*, Science, **258** (1992), pp. 1122–1126.

[129] E.L. Starostin, *Three-dimensional shapes of looped DNA*, Meccanica, **31** (1996), pp. 235–271.

[130] T. Strick, J. Allemand, V. Croquette, and D. Bensimon, *Twisting and stretching single DNA molecules*, Prog. Biophys. Mol. Biol., **74** (2000), pp. 115–140.

[131] T.R. Strick, J.F. Allemand, D. Bensimon, A. Bensimon, and V. Croquette, *The elasticity of a single supercoiled DNA molecule*, Science, **271** (1996), pp. 1835–1837.

[132] D. Swigon, *Configurations with Self-Contact in the Theory of the Elastic Rod Model for DNA*, Doctoral Dissertation, Rutgers University, 1999.

[133] D. Swigon, B. Coleman, and I. Tobias, *The elastic rod model for DNA and its application to the tertiary structure of DNA minicircles in mononucleosomes*, Biophys. J., **74** (1998), pp. 2515–2530.

[134] D. Swigon, B.D. Coleman, and W.K. Olson, *Modeling the Lac repressor-operator assembly: the influence of DNA looping on Lac repressor conformation*, Proc. Natl. Acad. Sci. U.S.A., **103** (2006), pp. 9879–9884.

[135] D. Swigon and W.K. Olson, *Mesoscale modeling of multi-protein-DNA assemblies: the role of the catabolic activator protein in Lac-repressor-mediated looping*, submitted to Int. J. Nonlin. Mech. (2007).

[136] J.M.T. Thompson and A.R. Champneys, *From helix to localized writhing in the torsional post-buckling of elastic rods*, in Localization and solitary waves in solid mechanics, Vol. 12 of Adv. Ser. Nonlinear Dynam., World Sci. Publ., River Edge, NJ, 1999, pp. 111–132.

[137] J.M.T. Thompson, G.H.M. van der Heijden, and S. Neukirch, *Supercoiling of DNA plasmids: mechanics of the generalized ply*, R. Soc. Lond. Proc. Ser. A Math. Phys. Eng. Sci., **458** (2002), pp. 959–985.

[138] A.V. Tkachenko, *Electrostatic effects in DNA stretching*, Phys. Rev. E Stat. Nonlin. Soft. Matter. Phys., **74** (2006), p. 041801.

[139] I. Tobias, B.D. Coleman, and W.K. Olson, *The dependence of DNA tertiary structure on end conditions: Theory and implications for topological transitions*, J. Chem. Phys., **101** (1994), pp. 10990–10996.

[140] I. Tobias, D. Swigon, and B.D. Coleman, *Elastic stability of DNA configurations. I. General theory*, Phys. Rev. E (3), **61** (2000), pp. 747–758.

[141] M. Vazquez and D.W. Sumners, *Tangle analysis of Gin site-specific recombination*, Math. Proc. Cambridge Philos. Soc., **136** (2004), pp. 565–582.

[142] E. Villa, A. Balaeff, L. Mahadevan, and K. Schulten, *Multiscale method for simulating protein-DNA complexes*, Multiscale Model. Simul., **2** (2004), pp. 527–553 (electronic).

[143] A. Vologodskii and N. Cozzarelli, *Conformational and thermodynamic properties of supercoiled DNA*, Annu Rev Biophys Biomol Struct, **23** (1994), pp. 609–643.

[144] A. Vologodskii and N. Cozzarelli, *Modeling of long-range electrostatic interactions in DNA*, Biopolymers, **35** (1995), pp. 289–296.

[145] A.V. Vologodskii, W. Zhang, V.V. Rybenkov, A.A. Podtelezhnikov, D. Subramanian, J.D. Griffith, and N.R. Cozzarelli, *Mechanism of topology simplification by type II DNA topoisomerases*, Proc. Natl. Acad. Sci. U.S.A., **98** (2001), pp. 3045–3049.

[146] J. Wang, *DNA topoisomerases*, Annu. Rev. Biochem., **65** (1996), pp. 635–692.

[147] J.D. Watson and F.H. Crick, *Molecular structure of nucleic acids; a structure for deoxyribose nucleic acid*, Nature, **171** (1953), pp. 737–738.

[148] T.P. Westcott, I. Tobias, and W.K. Olson, *Modeling self-contact forces in the elastic theory of DNA supercoiling*, J. Chem. Phys., **107** (1997), pp. 3967–3980.

[149] J.H. White, *Self-linking and the Gauss integral in higher dimensions*, Amer. J. Math., **91** (1969), pp. 693–728.

[150] ———, *An introduction to the geometry and topology of DNA structure*, in Mathematical methods for DNA sequences, CRC, Boca Raton, FL, 1989, pp. 225–253.

[151] P. Wiggins, R. Phillips, and P. Nelson, *Exact theory of kinkable elastic polymers*, Phys. Rev. E Stat. Nonlin. Soft. Matter. Phys., **71** (2005), p. 021909.

[152] J. Yan, M.O. Magnasco, and J.F. Marko, *A kinetic proofreading mechanism for disentanglement of DNA by topoisomerases*, Nature, **401** (1999), pp. 932–935.

[153] J. Yan and J. Marko, *Effects of DNA-distorting proteins on DNA elastic response*, Phys. Rev. E Stat. Nonlin. Soft. Matter. Phys., **68** (2003), p. 011905.

[154] ———, *Localized single-stranded bubble mechanism for cyclization of short double helix DNA*, Phys. Rev. Lett., **93** (2004), p. 108108.

[155] E.E. Zajac, *Stability of two planar loop elasticas*, J. Appl. Mech., **29** (1962), pp. 136–142.

[156] Y. Zhang and D.M. Crothers, *Statistical mechanics of sequence-dependent circular DNA and its application for DNA cyclization*, Biophys. J., **84** (2003), pp. 136–153.

PARADOX REGAINED: A TOPOLOGICAL COUPLING OF NUCLEOSOMAL DNA WRAPPING AND CHROMATIN FIBRE COILING

ANDREW TRAVERS*

The folding and unfolding of the chromatin fibre is a fundamental control point for the regulation of eukaryotic transcription. Although recent efforts have elucidated many of the mechanistic elaborations that regulate this process, the underlying mechanical basis of the folding transitions is poorly understood. Here I present a novel solution to the so-called 'linking number paradox' problem (Finch et al., 1977) and show that this solution implies that the chromatin fibre acts a tunable coil. The folding/unfolding process is essentially a topological transition in which the wrapping of DNA around the nucleosome core particle is directly coupled to degree of compaction of the coil.

Early studies established that the folding and unfolding of the 30 nm chromatin fibre is exquisitely sensitive to variation in ionic strength (Butler and Thomas, 1980; Dimitrov et al., 1990; Thoma et al., 1979) and that these transitions are reflected in the state of coiling of the fibre (Thoma et al., 1979; Widom, 1992). However, for infinitely long, or even for long unconstrained fibres, a change in coiling is a topological transition, an aspect of fibre function that has so far received little attention. Because there is little data pertaining to the topology of chromatin fibres as such, a starting point for the discussion of fibres must be the topology of the nucleosome core particle itself. This latter topic is in itself controversial. The nub of the problem is the so-called 'linking number paradox' first identified thirty years ago as a mismatch between the change in linking number on nucleosome formation ($\Delta L = \sim -1$) (Germond et al., 1975; Keller et al., 1978) and the number (n) of superhelical turns of DNA wrapped around the nucleosome (~ -1.5 — -1.7) (Finch et al., 1977). Two main solutions have been proposed to reconcile these two numbers - that the surface helical repeat of the wrapped DNA changes from 10.5 bp to 10 bp on nucleosome formation (Klug and Lutter, 1981; White et al., 1988) and that the entering and leaving duplexes cross each other disproportionately in opposite senses (De Lucia et al., 1999).

The original experiments on SV40 chromatin studied a closed topological system. Subsequent experiments in which core nucleosomes were reconstituted on minicircles refined the value of the change in ΔL on nucleosome formation to -1.1 to -1.2 (Zivanovic et al., 1990). However, the DNA sequence repeat, which is equivalent to the surface helical repeat, of

*MRC Laboratory of Molecular Biology, Hills Road, Cambridge, CB2 0QH, U.K. (aat@mrc-lmb.cam.ac.uk).

C.J. Benham et al. (eds.), *Mathematics of DNA Structure, Function and Interactions*,
The IMA Volumes in Mathematics and its Applications 150,
DOI 10.1007/978-1-4419-0670-0_15, © Springer Science+Business Media, LLC 2009

wrapped DNA is $\sim$ 10.2 bp and so cannot account fully for the linking number paradox (Satchwell et al., 1986; Travers and Klug, 1987). For a cylindrical surface the change in linking number due to this component is given by the equation:

$\Delta L = \Delta STw + n$ where STw is the surface twist (White et al., 1988).

Assuming that in a core particle 127 bp of DNA are wrapped in -1.63 superhelical turns (Richmond and Davey, 2003), and the helical repeat of DNA in solution is 10.4 bp (Wang, 1979), then

$$\Delta L = (127/10.2 - 127/10.4) - 1.63 = -1.39.$$

There is thus a 'missing' component of ΔL of $\sim +0.2$ to $+0.3$. A more defined closed system – a crystal structure of a tetranucleosome, in effect a mini-fibre – indicates a function for this component (Schalch et al., 2005). In this crystal structure four core nucleosomes with a 20 bp separation between each nucleosome crystallise as two stacks in which nucleosome 1 stacks on nucleosome 3 and nucleosome 2 on nucleosome 4. Importantly these stacks are twisted relative to each other by -71.5^o and thus in a longer fibre would be coiled relative to each other (Schalch et al., 2005). Assuming conservation of Lk such that this twisting or coiling compensates for the intrinsic topology of the core particles, the twisting must have the opposite sense to the topological contribution by the core particles. For a cylindrical surface the two derivations of ΔL are equivalent, i.e. $\Delta L = \Delta STw + n = \Delta Tw + Wr$ where Tw is the intrinsic twist of DNA and Wr is the writhe. On nucleosome formation the change in intrinsic twist is negligible or unaltered (Zivanovic et al., 1988) and therefore in this situation

$$Wr = \Delta STw + n = -1.39.$$

The relative twisting of the two nucleosome stacks in the crystal structure is not strictly a property of intrinsic twist of the DNA itself but is a topological quantity which can be regarded as twist. Taking this into account and approximating to a twist angle of 72°

$$\Delta L = +0.2 + (-1.39) = -1.19.$$

This calculated value is within the best experimentally determined values of ΔL and shows that, at least for the tetranucleosome structure ΔL has a very similar value to that determined for SV40 chromatin and for single nucleosome core particles (Germond et al., 1975; Keller, 1975; Norton et al., 1989; Zivanovic et al., 1988). Nevertheless it is important to note that the experimental values for ΔL, DNA wrapping in the core particle and the helical repeat of DNA in solution are not precise and may vary according to the methodology used. Nevertheless this calculation identifies a topological component in the tetranucleosome crystal structure that is provided by the core particle and suggests that, in this structure at least, it is not necessary to invoke variable crossings of the entering and leaving DNA.

Although the crystal structure of a tetranucleosome indicates the 'missing' ΔL is reflected in the relative twisting of the two nucleosome stacks it does not pinpoint the origin of this topological component. Two considerations are important. First in the original experiments with SV40 chromatin the DNA between nucleosomes was relaxed with topoisomerase I (Keller, 1975) and therefore the 'missing' component must represent a direct or imposed constraint by the histone octamer. Second the crystal structures of the core particle reveal only the contacts of the 'core' domains of the histones with DNA and not most contacts of histone tails (Luger et al., 1997; Richmond and Davey, 2003). There is however, direct evidence that some histone tails bind to DNA outside the 147 bp wrapped around the core histones (Angelov et al., 2001). Further, electron microscopy of isolated core particles shows that although the trajectories of the entering and leaving DNA duplexes are variable in many cases these duplexes bend away from the particle in the opposite sense to the negatively wrapped DNA (Hamiche et al., 1996) (Figure 1a). I propose that this configuration, which is equivalent to the 'open' configuration defined by De Lucia et al. (1999), is the origin of the 'missing' positive component of ΔL. In practice, it may be an average of an ensemble of configurations, that, at least in part, is stabilised by interactions with the histone tails and may also be imposed, in part, by electrostatic repulsion between closely approaching duplexes.

If the 'missing' component of ΔL is indeed related to both the entering and leaving trajectories of the core particle DNA and also to the coiling of the fibre changes in these trajectories should result directly in changes in compaction of the fibre. Two factors that alter the compaction of the fibre are the association of linker histone and the acetylation of core histone tails. These result in respectively an increase and a decrease in the packing density of the fibre (Butler and Thomas, 1980; Hizume et al., 2005; Thoma et al., 1979; Tse et al., 1998). Binding of a canonical linker histone both to core particles and within fibres result in the formation of a protein-DNA 'stem' protruding approximately at the dyad of the core particle (Bednar et al., 1995; Hamiche et al., 1996). This structure correlates with an increase in DNA wrapping around the core histones from 1.65–1.7 superhelical turns to 1.8–1.9 turns (as determined by electron microscopy) and an accentuation of the bends in the entering and leaving DNA without inducing crossings of the two duplexes (Bednar et al., 1995). The sense of these bends relative to the wrapping is positive and thus is significantly greater than the corresponding bends in the core particle (Figure 1a). This implies that the compensating left-handed coiling in the fibre will also be greater and thus that the fibre will be more compact. This argument also justifies the ad hoc assumption made elsewhere of the functional role of linker histones (Wu et al., 2007). In contrast to nucleosomes containing

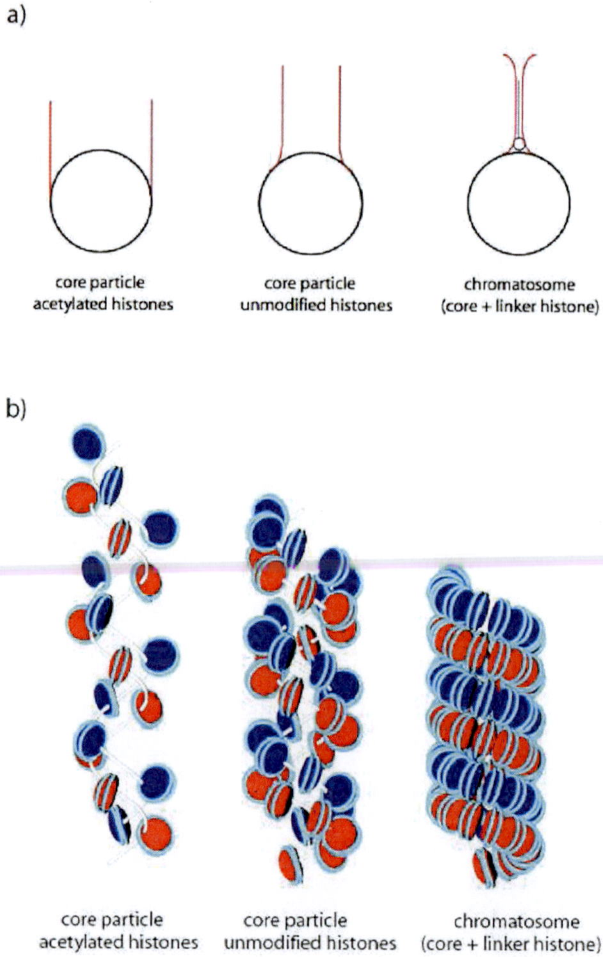

FIG. 1. *a) Proposed variation of exit/entry angle of nucleosomal DNA comparing an unmodified core particle (centre) with a particle containing hyperacetylated histones (centre) or a linker histone (right).*
b) The corresponding variation in coiling of the fibre. The figure represents a generalised 30 nm fibre with the paths of the two nucleosome coils shown. The precise trajectory of the linker DNA between successive nucleosomes depends on linker length and is not shown.

linker histone there are no direct determinations of the extent of wrapping in nucleosomes containing acetylated histones. However hyperacetylation would be expected to release any constraints imposed on the entering and leaving duplexes by the histone tails and so would decrease the positive bend in these duplexes (Figure 1a). In concert with electrostatic effects involving the acetylation of lysine 16 of histone H4 (Shogren-Knaak et al.,

2006) this would result in a decrease in coiling associated with an 'opening up' of the fibre (Figure 1b).

Linker histone binding increases (Stein, 1980; Zivanovic et al., 1990) and histone acetylation reduces the experimentally determined values of ΔL on nucleosome formation (Bauer et al., 1994; De Lucia et al., 1999; Norton et al., 1989). These changes are correlated with the amount of negative DNA wrapping around the core particle and would have the straightforward effect of decreasing and increasing respectively the amount of DNA between adjacent octamers. But changes in DNA wrapping would also have topological consequences. Although the extent to which the experimental determinations reflect the trajectories of the exiting and entering DNA duplexes is uncertain, increasing or decreasing the negative wrapping of DNA by the core histones would, in a closed nucleosome array, be expected to decrease and increase respectively the positive topological component available for coiling. This counterbalance between the negative wrapping component and a positive exit/entry component would mean that the changes in form associated with coiling and uncoiling of a fibre would be accompanied by little or no overall change in ΔL per nucleosome in a fibre as observed (Keller et al., 1978) and so would not require direct facilitation by topoisomerases.

I have argued that the coiling of the chromatin fibre is directly related to the positive bending of DNA as it enters and leaves the nucleosome and, in turn, the extent of this bending correlates with the extent of wrapping around the histone octamer. This relation implies that, for a given average linker length, the chromatin fibre acts as a tunable coil, with the extent of coiling, and hence of compaction, being determined by the interactions of the linker histones and the core histone tails with DNA in the immediate vicinity of the particle. In other words, within a fibre each nucleosome makes an individual topological contribution to coiling. Since the number of nucleosomes/helical turn of a fibre can vary the degree of compaction and the form of the fibre may depend of this parameter (Athey et al., 1990; Robinson et al., 2006). For example, whereas the canonical 30 nm fibre contains $\sim$ 6 nucleosomes/11 nm (Gerchman and Ramakrishnan, 1987; Ghirlando et al., 2004; Ghirlando and Felsenfeld, 2008) those reconstituted on arrays of optimally-spaced nucleosome positioning signals in vitro may contain between 11 and 16 nucleosomes/11 nm (Robinson et al., 2006). Thus per turn there will be a greater contribution to coiling from the nucleosomes for the more compact relative to the canonical fibres. This is consistent with the interpretation that the more compact fibres can, for short linker lengths, assume a more coiled form of the fibre. This more coiled form could be a supercoiled helical-ribbon rather than a relaxed crossed-linker form or a transitional helical form more akin to the A-form of DNA than to the crossed-linker form comparable to B-form DNA (Wu et al., 2007). This view also reconciles apparently disparate structural interpretations of canonical and compact fibres (Athey et al.,

1990; Robinson et al., 2006; Schalch et al., 2005) and validates the observation that the dimensions of compact fibres reconstituted in vitro can be fitted by a 2-start model in which a more coiled form relaxes to a less coiled form (Wu et al., 2007). All the relevant observations are consistent with a simple structural transition of 2-start chromatin helices. It is important to note that the coiling imparted by the configuration of individual nucleosomes is a topological quantity while the compaction depends on the configuration assumed by the resultant coils. For example, toroidal and plectonemic modes of coiling made differ substantially in compaction for equivalent amounts of coiling.

The tunable coil concept is fully consistent with experimental observations relating to the dependence of fibre compaction and of changes in linear dichroism on ionic strength in vitro (Butler and Thomas, 1980; Dimitrov et al., 1990). However, it places strong constraints on the connectivity between spatially adjacent nucleosomes in a fibre. Although it readily explains folding and unfolding in the context of a 2-start helical model, i.e. crossed-linker and helical ribbon forms, it is not simply compatible with 1-start models and their variants because, apart from the difficulty in reconciling 'stem' structures with 1-start models, this class of model requires tight DNA bending or coiling, either positive or negative, between adjacent nucleosomes in a helical stack and consequently any change in wrapping of nucleosomal DNA will result in the relative rotation of adjacent nucleosomes rather than a change in coiling. However, the connectivity of the 30 nm fibre is currently an unresolved issue. While fibres lacking linker histone can adopt a 2-start form (Dorigo et al., 2004; Schalch et al., 2005), compact fibres containing linker histone have been proposed to adopt a 1-start interdigitated form (Robinson et al., 2006), which is equivalent to a 3- or 4-start helix of stacked nucleosomes.

In addition to the trajectories of the leaving and entering DNAs the length of the linker DNA between nucleosomes can also directly affect the extent of coiling by altering the relative rotation of one nucleosome relative to its neighbour on a linear map (Widom, 1992; Wu et al., 2007). In general the relative rotation will be conserved only if this length changes by an integral number of helical turns. These changes affect both the ability of nucleosomes to stack on each other in a folded fibre and the average topological contribution of each nucleosome (Wu et al., 2007). For optimal linker lengths, as discussed above, the positive contribution of the exiting/entering nucleosomal DNA would generate a left-handed coil. However, substantial departures from optimal linker lengths of integral, or possibly also half-integral, numbers of helical turns could result in a twist deficit in the linker DNA that would reverse the sense of coiling of the fibre.

In addition, the individual contributions of nucleosomes to coiling could also, in principle, facilitate the formation of higher-order structures beyond the 30 nm fibre. If coiling transitions were not completely topologically neutral such that, for example, the 30 nm fibre structure assumed did not match the total topological nucleosomal contributions, then any imbalance could be absorbed by a further coiling of the 30 nm fibre on itself, either as a plectoneme or as a toroid. Importantly since these structures would contain many nucleosomes the actual contribution per nucleosome would be small relative to that driving the coiling of the 30 nm fibre. Such a mechanism would also ensure that in an actively transcribed gene high DNA compaction could be maintained in the regions lacking RNA polymerase.

The arguments presented above assume that a fibre acts as a closed topological domain. But in the nucleus a fibre of a particular packing density may be connected to and directly abut on a nucleosome-free region or a fibre of different packing density. Unless the structural transitions within the fibre are completely topologically neutral, i.e. overall there is zero net change in linking number, these transitions could directly affect the structures of neighbouring regions. Such effects would be blocked by relaxation of any excess superhelicity - either positive or negative. The topoisomerase II in chromatin matrix- or scaffold-attachment regions (Gasser et al., 1986) could perform this function and thus act primarily to maintain the topological status of chromatin fibres in adjoining chromatin domains. Indeed, topoisomerase II has been shown to play an active role in fibre compaction (Hizume et al., 2007).

In summary, in any consideration of chromatin structure it is crucial to consider the chromatin fibre itself as a topological nanomachine whose configuration is closely correlated with function.

Acknowledgements. I am most grateful to Ernesto Di Mauro for suggesting the term 'tunable' and especially to the organisers of a recent workshop on the Mathematics of DNA Structure, Function and Interactions held at the Institute of Mathematics and its Applications at the University of Minnesota in Minneapolis for bringing this problem to my attention again.

REFERENCES

Angelov D., Vitolo J.M., Mutskov V., Dimitrov S., and Hayes J.J. (2001). Preferential interaction of the core histone tail domains with linker DNA. Proc. Nat. Acad. Sci. USA **98**: 6599–6604.

Athey B.D., Smith M.F., Rankert D.A., Williams S.P., and Langmore J.P. (1990). The diameters of frozen-hydrated chromatin fibers increase with DNA linker length: evidence in support of variable diameter models for chromatin. J. Cell Biol. **111**: 795–806.

Bauer W.R., Hayes J.J., White J.H., and Wolffe A.P. (1994). Nucleosome structural changes due to acetylation. J. Mol. Biol. **236**: 685–690.

Bednar J., Horowitz R.A., Dubochet J., and Woodcock C.L. (1995). Chromatin conformation and salt-induced compaction: three-dimensional structural information from cryoelectron microscopy. J. Cell Biol. **131**: 1365–1376.

BUTLER P.J. AND THOMAS J.O. (1980). Changes in chromatin folding in solution. J. Mol. Biol. **140**: 505–529.

DE LUCIA F., ALILAT M., SIVOLOB A., AND PRUNELL A. (1999). Nucleosome dynamics. III. Histone tail-dependent fluctuation of nucleosomes between open and closed DNA conformations. Implications for chromatin dynamics and the linking number paradox. A relaxation study of mononucleosomes on DNA minicircles. J. Mol. Biol. **285**: 1101–1119.

DIMITROV S.I., MAKAROV V.L., AND PASHEV I.G. (1990). The chromatin fiber: structure and conformational transitions as revealed by optical anisotropy studies. J. Biomol. Struct. Dynam. **8**: 23–35.

DORIGO B., SCHALCH T., KULANGARA A., DUDA S., SCHROEDER R.R., AND RICHMOND T.J. (2004) Nucleosome arrays reveal a two-start organisation of the chromatin fiber. Science **306**: 1571–1573.

FINCH J.T., LUTTER L.C., RHODES D., BROWN R.S., RUSHTON B., LEVITT M., AND KLUG A. (1977). Structure of nucleosome core particles of chromatin. Nature **269**: 29–36.

GASSER S.M., LAROCHE T., FALQUET J., BOY DE LA TOUR E., AND LAEMMLI U.K. (1986). Metaphase chromosome structure. Involvement of topoisomerase II. J. Mol. Biol. **188**: 613–629.

GERCHMAN S.E. AND RAMAKRISHNAN V. (1987). Chromatin higher-order structure studied by neutron scattering and scanning transmission electron microscopy. Proc. Nat. Acad. Sci. USA **84**: 7802–7806.

GERMOND J.E., HIRT B., OUDET P., GROSS-BELLARD M., AND CHAMBON P. (1975). Folding of the DNA double helix in chromatin-like structures from simian virus 40. Proc. Nat. Acad. Sci. USA **72**: 1843–1847.

GHIRLANDO R. AND FELSENFELD G. (2008). Hydrodynamic studies on defined heterochromatin fragments support a 30 nm fiber having 6 nucleosomes per turn. J. Mol. Biol. epub. 3 Jan. 2008.

GHIRLANDO R., LITT M.D., PRIOLEAU M.N., RECILLAS-TARGA F., AND FELSENFELD G. (2004). Physical properties of a genomic condensed chromatin fragment. J. Mol. Biol. **336**: 597–605.

HAMICHE A., SCHULTZ P., RAMAKRISHNAN V., OUDET P., AND PRUNELL A. (1996). Linker histone-dependent DNA structure in linear mononucleosomes. J. Mol. Biol. **257**: 30–42.

HIZUME K., ARAKI S., YOSHIKAWA K., AND TAKEYASU K. (2007). Topoisomerase II, scaffold component, promotes chromatin compaction in vitro in a linker-histone H1-dependent manner. Nucleic Acids Res. **35**: 2787–2799.

HIZUME K., YOSHIMURA S.H., AND TAKEYASU K. (2005). Linker histone H1 per se can induce three-dimensional folding of chromatin fiber. Biochemistry **44**: 12978–12989.

KELLER W., MÜLLER U., EICKEN I., WENDEL I., AND ZENTGRAF H. (1978). Biochemical and ultrastructural analysis of SV40 chromatin. Cold Spring Harb. Symp. Quant. Biol. **42**: 227–244.

KLUG A. AND LUTTER L.C. (1981). The helical periodicity of DNA on the nucleosome. Nucleic Acids Res. **9**: 4267–4283.

LUGER K., MADER A.W., RICHMOND R.K., SARGENT D.F., AND RICHMOND T.J. (1997). Crystal structure of the nucleosome core particle at 2.8 A resolution. Nature **389**: 251–260.

NORTON V.G., IMAI B.S., YAU P., AND BRADBURY E.M. (1989). Histone acetylation reduces nucleosome core particle linking number change. Cell **57**: 449–457.

RICHMOND T.J. AND DAVEY C.A. (2003). The structure of DNA in the nucleosome core. Nature **423**: 145–150.

ROBINSON P.J., FAIRALL L., HUYNH V.A., AND RHODES D. (2006). EM measurements define the dimensions of the "30-nm" chromatin fiber: evidence for a compact, interdigitated structure. Proc. Nat. Acad. Sci. USA **103**: 6506–6511.

Satchwell S.C., Drew H.R., and Travers A.A. (1986). Sequence periodicities in chicken nucleosome core DNA. J. Mol. Biol. **191**: 659–675.

Schalch T., Duda S., Sargent D.F., and Richmond T.J. (2005). X-ray structure of a tetranucleosome and its implications for the chromatin fibre. Nature **436**: 138–141.

Shogren-Knaak M., Ishii H., Sun J.M., Pazin M.J., Davie J.R., and Peterson C.L. (2006). Histone H4-K16 acetylation controls chromatin structure and protein interactions. Science **311**: 844–847.

Stein A. (1980). DNA wrapping in nucleosomes. The linking number problem reexamined. Nucleic Acids Res. **8**: 4803–4820.

Thoma F., Koller T., and Klug A. (1979). Involvement of histone H1 in the organization of the nucleosome and of the salt-dependent superstructures of chromatin. J. Cell Biol. **83**: 403–427.

Travers A.A. and Klug A. (1987). The bending of DNA in nucleosomes and its wider implications. Phil. Trans. Roy. Soc. (London) B **317**: 537–561.

Tse C., Sera T., Wolffe A.P., and Hansen J.C. (1998). Disruption of higher-order folding by core histone acetylation dramatically enhances transcription of nucleosomal arrays by RNA polymerase III. Mol. Cell. Biol. **18**: 4629–4638.

Wang J.C. (1979). Helical repeat of DNA in solution. Proc. Nat. Acad. Sci. USA **76**: 200–203.

White J.H., Cozzarelli N.R., and Bauer W.R. (1988). Helical repeat and linking number of surface-wrapped DNA. Science **241**: 323–327.

Widom J. (1992). A relationship between the helical twist of DNA and the ordered positioning of nucleosomes in all eukaryotic cells. Proc. Nat. Acad. Sci. USA **89**: 1095–1099.

Wu C., Bassett A., and Travers A.A. (2007). A variable topology for the '30 nm' chromatin fibre. EMBO Rep. **8**: 1129–1134.

Zivanovic Y., Duband-Goulet I., Schultz P., Stofer E., Oudet P., and Prunell A. (1990). Chromatin reconstitution on small DNA rings. III. Histone H5 dependence of DNA supercoiling in the nucleosome. J. Mol. Biol. **214**: 479–495.

Zivanovic Y., Goulet I., Revet B., Le Bret M., and Prunell A. (1988). Chromatin reconstitution on small DNA rings. II. DNA supercoiling on the nucleosome. J. Mol. Biol. **200**: 267–290.

STATISTICAL-MECHANICAL ANALYSIS OF ENZYMATIC TOPOLOGICAL TRANSFORMATIONS IN DNA MOLECULES

ALEXANDER VOLOGODSKII*

Abstract. There are two classes of enzymes, DNA topoisomerases and site-specific recombinases, which change DNA topology. Many details of the enzyme action remain unknown, and therefore different models of the reactions require critical testing. A great help in such testing comes from computer simulation. The computational approach, described in the review, allows simulating the distribution of the reaction products for a chosen model of the enzyme action. Comparing the simulated distribution with corresponding experimental data serves as a model test. The major principles and assumptions of the approach, which is based on the simulation of an equilibrium set of DNA conformations, are discussed. The general consideration is illustrated by two specific examples, models of type II DNA topoisomerases and tyrosine family of site-specific recombination.

Key words. DNA topology, Simulation of DNA conformations, DNA topoisomerases, Site-specific recombination.

AMS(MOS) subject classifications. 92-08 Computational methods.

1. Introduction. There are two very important classes of enzymes, type II DNA topoisomerases [1] and site-specific recombinases [2], which can change topology of circular DNA molecules. During their catalytic acts these enzymes introduce temporary breaks into DNA backbone making topological changes possible (Fig. 1).

These enzymes interact, in general, with two or more DNA sites which are separated along the DNA contour, so formation of the reaction complex (synaptic complex) is associated with formation of two or more DNA loops. If the loops are large enough, they can adopt many different conformations. In these cases the reaction products can have various topologies. An example of different outcomes of a topological transformation is shown in Fig. 2.

It is important to have detailed pictures and clear understanding the action of these enzymes. Such information is very difficult to obtain using only experimental methods, however. Even X-ray structure of a synaptic complex does not give all the information about the reaction pathway, and various implicit methods are used to address the problem. In this situation very valuable help can be obtained from computer simulation. We can simulate the distribution of the reaction products over different topologies for any chosen model of the reaction. The product distribution can be also determined experimentally, and by comparing the simulated and experi-

*Department of Chemistry, New York University, 31 Washington Place, New York, NY 10003 (alex.vologodskii@nyu.edu). The work on this review was funded by the National Institutes of Health grant GM54215 to the author.

C.J. Benham et al. (eds.), *Mathematics of DNA Structure, Function and Interactions*,
The IMA Volumes in Mathematics and its Applications 150,
DOI 10.1007/978-1-4419-0670-0_16,

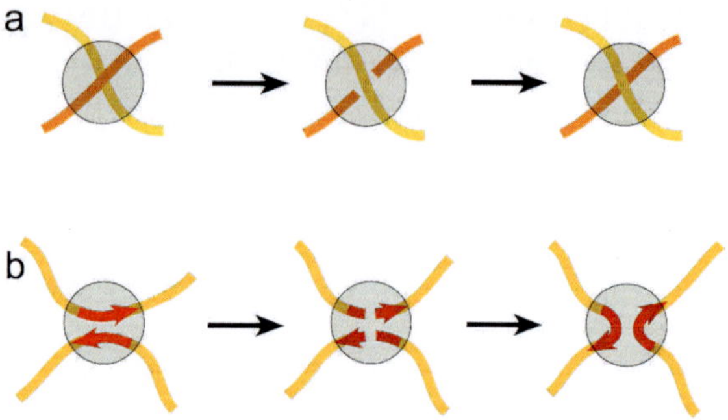

FIG. 1. *Diagrams of DNA enzymatic transformations. The enzyme and double stranded DNA are shown by the semitransparent disk and by the yellow lines, correspondingly.*
(a) Strand passing reaction catalyzed by type II topoisomerases. The enzymes make a transient double-stranded break in a "gate" segment (G segment) that allows passage of another segment (T segment) of the same or another DNA molecule. The topoisomerases reseal the break after the strand-passing.
(b) DNA rearrangement by tyrosine family of site-specific recombinases. Both specific DNA sites, shown by red arrows, aligned in the synaptic complex in the antiparallel orientation. The sites are cut and rejoined as it shown on the diagram.

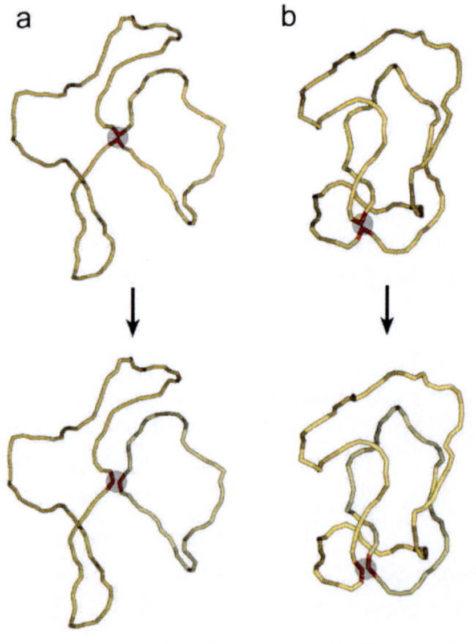

FIG. 2. *Topological outcomes of the enzymatic transformation. The example shows site-specific recombination in unknotted circular DNA with two specific sites oriented head-to-tail. The reaction product is two DNA circles which can be unlinked (a) or form a link (b). The product topology depends on the looping in the substrate DNA.*

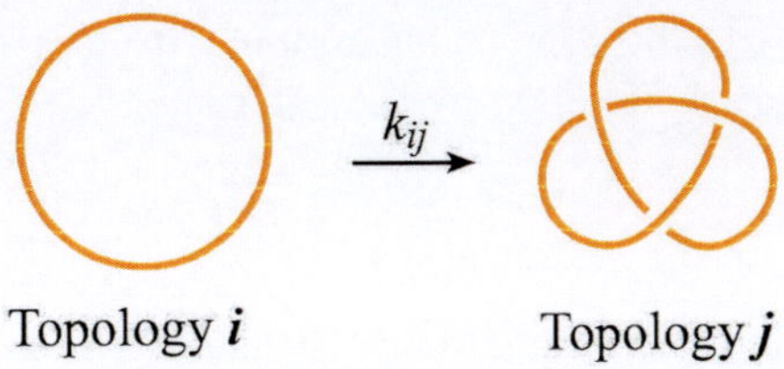

FIG. 3. *The rate constant of a topological transformation. The diagram shows the transformation from unknotted circular molecule to one with the topology of trefoil.*

mental distributions we can test the model. If the simulated and measured distributions are in agreement, we can usually perform additional tests of the model by checking experimentally its specific assumptions. We can simulate the distribution of the reaction products over different topologies for any chosen model of the reaction. The product distribution can be also determined experimentally, and by comparing the simulated and experimental distributions we can test the model. If the simulated and measured distributions are in agreement, we can usually perform additional tests of the model by checking experimentally its specific assumptions.

In this review we first consider the simulation approach in general terms. Then we analyze key elements of the simulation procedure. Finally, we consider two examples describing applications of the approach to specific problems.

2. General formulation. We can use computer simulation to calculate the rates of conversion of the substrate DNA with topology i into product with topology j, k_{ij} (Fig. 3). The value of k_{ij} is proportional to the probability of juxtaposition, and correspondingly the synaptic complex formation, of the DNA sites which would give topology j, under condition that the initial topology is i, p_{ij}:

$$k_{ij} = \alpha p_{ij}. \tag{2.1}$$

The coefficient between k_{ij} and p_{ij}, α, depends on the enzyme properties but does not depend on i and j. Eq. (2.1) assumes that the synaptic complex formation is not a diffusion-limited process and is specified by the equilibrium distribution of DNA conformations. In other words, many juxtapositions of the DNA sites, participating in the reaction, precede the synaptic complex formation. This is a common assumption in the analysis of protein-assisted DNA looping. Recently the assumption has been proven for enhancer-promoter complex formation [3].

The values of k_{ij} directly specify the distribution of reaction products formed from the substrate with topology i at early moments of the reaction, when only a small fraction of the substrate DNA has been transformed into molecules with other topologies. The values of k_{ij} can be also used to estimate the steady state distribution of topological forms, if such distribution is created by continuous action of the enzyme. In the latter case the distribution is specified by the equations

$$\frac{dc_i}{dt} = \sum_j k_{ji}c_j - c_i \sum_j k_{ij} = 0, \tag{2.2}$$

where c_i is the concentration of DNA molecules with topology i. By definition of the steady state the values of c_i do not change with time t. We will consider later examples which use either initial or steady state distributions of the reaction products.

Thus, to calculate the distribution of the reaction products we need to calculate p_{ij}. Let us now consider this calculation in detail. It is important that p_{ij} is specified by conformational distribution of molecules in topological state i and by a chosen reaction model only. Indeed, there is no randomness in the reaction pathway inside the complex (we do not consider here processive reactions which can be performed a random number of times on the same synaptic complex). Calculation of p_{ij} can be divided into three steps:

1) Generation a conformational set which corresponds to the equilibrium conformational ensemble of the substrate circular DNA with topology i, possibly with bound protein(s) which can introduce local conformational changes into this DNA;

2) Selection of DNA conformations where mutual separation and orientation of certain DNA segments correspond to those in the synaptic complex;

3) Calculation of the reaction outcome for all selected conformations and converting these data into distribution p_{ij}.

Below we consider these three steps separately.

3. DNA model and the simulation procedure. To generate an equilibrium conformational set for the substrate circular DNA with bound protein(s) we need, first of all, to choose a DNA model, appropriate for this kind of simulation. This should be the simplest model which properly describes DNA conformational properties on the scale of a few nanometers, a typical size of the DNA-protein complex. Such model is well known, it is the discrete wormlike chain [4–6]. A circular DNA molecule composed of n Kuhn statistical lengths is modeled as a closed chain consisting of kn rigid segments that are cylinders of equal length l and diameter d. The number of straight segments per Kuhn length, k, is a computational parameter of our choice (Fig. 4). The bending elastic energy of the chain, E_b, is computed as

$$E_b = \frac{1}{2}k_BTg\sum_{i=1}^{kn}\theta_i^2, \tag{3.1}$$

where the summation extends over all the joints between the elementary segments, θ_i is the angular displacement of segment i relative to segment

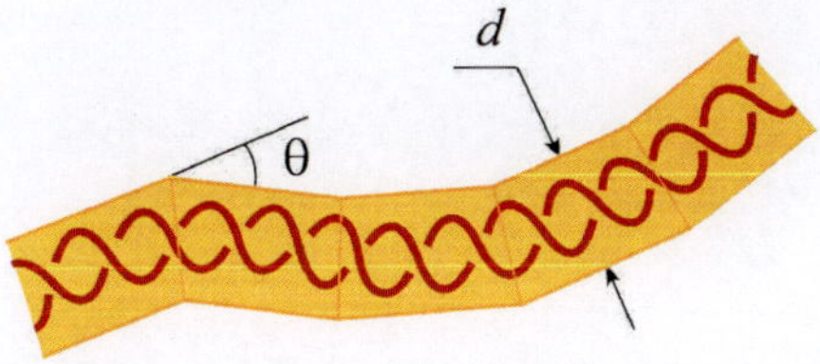

FIG. 4. *The model of double-stranded DNA. The length of the cylinders (shown by yellow) can vary, although it usually equals 30 or 15 base pairs of the double helix (1/5 or 1/10 of DNA persistence length). The DNA molecule (red) is shown to illustrate the scale of the model cylinders.*

$i-1$, and g is the bending rigidity constant, k_BT is the Boltzmann temperature factor. If the value of k is sufficiently large ($k \geq 10$), the bending constant g is proportional to the DNA persistence length, a [7]:

$$g = a/l. \tag{3.2}$$

More precise relation between g, l and a is described in ref. [5].

Proper choice of k, which is equal to $2a/l$, is very important here. The computer time needed for a simulation increases approximately as $(kn)^2$. It is therefore necessary to choose the value of k so that it is large enough to ensure reliable results but small enough to keep the computational time reasonable. This kind of polymer models cannot describe conformational properties on a scale smaller than the length of one straight segment, l. Taking into account the size of DNA-protein complexes we conclude that k cannot be smaller than 10. Therefore, one cannot use the freely-jointed chain or a lattice model for this kind of problems, as it was done in some studies [8, 9].

The diameter of impenetrable cylindrical segments of model chain accounts both for the DNA geometrical diameter and for the electrostatic repulsion between the segments. The quantitative definition of d is based on the concept of the second virial coefficient [10]. It was shown that approximation of the electrostatic interaction by this hard core potential and by the corresponding Debye-Hückel potential give very similar results in Monte Carlo simulations of DNA equilibrium properties, with the exception of conformations of supercoiled DNA at low concentration of monovalent ions (≤ 0.02 M) [11]. The value of d strongly affects DNA topological properties [12, 13], and therefore its proper setting is extremely important. For ionic conditions close to the physiological d equals 5 nm [14, 15], that is 5% of the DNA statistical length.

The model's features specified above are sufficient to simulate nicked circular DNA molecules, which do not maintain torsional stress. If both strands of the molecule are closed, the chain energy also depends on DNA twist, Tw. To use the model in this case one can express the displacement of chain twist from its equilibrium value, ΔTw, by the equation:

$$\Delta Tw = \Delta Lk - Wr, \tag{3.3}$$

where writhe, Wr, is a property of the chain axis alone [16], and ΔLk is the linking number difference of the simulated DNA.The value of ΔLk should be considered here as a simulation parameter [17], while Wr is calculated for each conformation of the model chain [18]. Hence, in this model, the torsional energy, E_t, is defined by the conformation of the DNA axis and may be expressed as

$$E_t = (2\pi^2 C/L)(\Delta Lk - Wr)^2, \tag{3.4}$$

where C is the torsional rigidity constant, and L is the DNA length.

Proteins, which are bound with one or more DNA segments during the assembling of the reaction complex, can introduce local distortions into the DNA molecule. Such distortions have to be included into the model. Typically, the bound proteins introduce local bends into the bounding sites, increase the volume of the sites, and make them more rigid. The first effect is the most important one, so it must be taken into account properly. There is no way to determine the bending rigidity of the complex, so it seems reasonable to assume that it is absolutely rigid. The volume increase does not affect global properties of the model chain as long as only few proteins are bound with it, so it can be neglected in most cases. We did not account for this effect in the applications described here.

Sampling the equilibrium ensemble of the chain conformations is performed by Metropolis Monte Carlo procedure [17, 19]. In some cases, however, the probability of conformations suitable for the synaptic complex formation is very low, so unbiased sampling of the equilibrium ensemble does not provide a sufficient number of such conformations for the statistical analysis of the reaction outcomes. To enrich the sampling by conformations with the properly juxtaposed sites, a biased sampling procedure, based on adding an artificial potential between specific sites, can be used [20, 21]. Although the potential disturbs the whole conformational ensemble, it does not affect the conformational distribution among the states with juxtaposed sites, since the energy of such conformations is change by the same value.

The simulation procedure allows passing one segment through another, so it does not prevent from topology change in the model chain. To keep the desired topology i over the simulation run one has to check the chain topology for each trial conformation of the Metropolis procedure. It can be done by calculating the Alexander polynomial, $\Delta(t)$ [19, 22]. It is sufficient in the most cases to calculate $\Delta(t)$ at points $t = -1$ and $t = -2$. The values of $\Delta(-1)$ and $\Delta(-2)$ distinguish the great majority of all 166 knots which can be drawn with less than 11 intersections on their projection [23].

The definition of juxtaposition depends on the reaction model. Correspondingly, the procedure of testing the simulated conformations on the presence of properly juxtaposed sites can hardly be described in general terms. For each conformation with the juxtaposed sites we need to generate the reaction product and determine its topology. The topology determination is based on the calculation of the Alexander polynomial for one chain or for two chains [19], if the reaction produces two circular molecules (see Fig. 2).

The distribution p_{ij} is calculated as

$$p_{ij} = n_{ij}/N_i, \tag{3.5}$$

where n_{ij} is the number of outcomes with topology j, N_i is the total number of conformations in the generated conformational set with topology i.

Successive conformations of the model chains obtained in the Metropolis procedure may be strongly correlated, so the total number of steps in the simulation, M, should be many times larger than the correlation length, m_c. This condition is always satisfied for relaxed circular molecules where m_c has the order of 10, while M exceeds 10^8. The value of m_c is much larger for supercoiled chains, somewhere between 10^6 and 10^7 steps, depending on the superhelix density of the model chain, σ. The correlation length is further increased by permanent bends at certain places of the chain, used in the simulation, since such bends have a tendency to be localized at the superhelix apices and suppress slithering motion in the superhelix. Still, even in these cases the simulation runs can be made much longer than m_c. What makes the issue of correlation really critical is the artificial potential between specific sites. Special care is required when choosing the potential to avoid locking the chain in a particular set of conformations. We use analysis of the simulated trajectories as a major way to address the issue.

From numerous simulations of supercoiled DNA we know that typical conformations of such molecules represent branched superhelices (see Fig. 9), and variation in the number of braches is the slowest relaxation process [17]. Thus, choosing the potential we need to check that in each simulation run there are many stretches of conformations where specific sites are not juxtaposed and these stretches are sufficiently long for the branch number relaxation.

An artificial potential was used in our simulations of site-specific recombination [21]. We were able to find the potential parameters to satisfy the above condition for model chains with σ of -0.05, but failed to do it for σ of -0.06.

4. Type II DNA topoisomerases. It was found that type II DNA topoisomerases actively reduce the fractions of knotted and linked circular DNA molecules below thermodynamic equilibrium values [24]. In particular, it was found that the steady state fraction of trefoil knots created by topo II (the only knots observed in experiments) is many times lower than the equilibrium fraction of these knots in the same DNA molecules and solution conditions. It was suggested, in an attempt to explain this surprising finding, that the enzymes create a sharp bend in G segment,

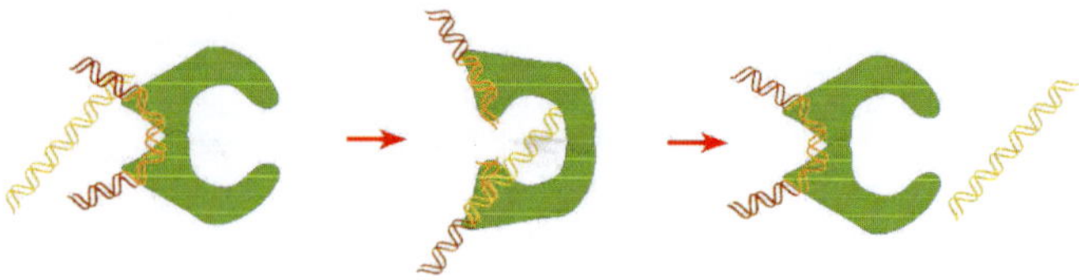

FIG. 5. *The model of type II topoisomerase action. The enzyme (shown by green) bends the DNA G segment (orange) into a hairpin. The entrance gate of the enzyme for the T segment (yellow) is inside the hairpin. Thus, the T segment can pass through the G segment only from inside to outside the hairpin.*

the first bound DNA segment [25]. If the enzymes create such bend, they have to have a specific orientation relative to the bend. It is known that the type II topoisomerases can harness the energy of ATP hydrolysis and promote passage of T segment in one direction relative to themselves [26–28]. Thus, the complex with bent G segment can provide a unidirectional passage of the T segment from inside to outside the hairpin formed by G segment (Fig. 5). This directionality of strand passage is only local, because the hairpin can have any orientation relative to the DNA chain. Surprisingly though, the quantitative analysis of the model showed that it leads to a large decrease of the steady state fraction of knots and catenanes compared with the equilibrium levels [25]. The basic elements of the analysis are described below.

The DNA model described above was used to simulate equilibrium set of DNA conformations with the G segment-bound protein. In accordance with the model, four elementary segments of the model chain form a rigid hairpin which does not change its conformation during the Metropolis sampling (Fig. 6). Since any segment of the chain could serve as a T segment, there was no way to use a biased sampling of the conformational space. Therefore, large sets of conformations (up to $2{\cdot}10^9$) were simulated to obtain statistically reliable estimations of the required parameters p_{ij}.

The experiments were performed for nicked circular DNA molecules, so there was no need to account for DNA torsional energy in the simulations. Only unknotted DNA molecules (denoted by 0) and trefoils (denoted by 3) were observed in the experimental study, so Eq. (2.2) are reduced to a single equation

$$p_{03}c_0 - p_{30}c_3 = 0. \tag{4.1}$$

Correspondingly, we needed to calculate only p_{03} and p_{30}. It is sufficient to calculate $\Delta(-1)$ to distinguish between trefoils and unknotted contours.

The chain conformations having a segment juxtaposed with the protein-bound G segment were identified by the inspection of constructed sets. Segment i was considered to be juxtaposed with the G segment, if

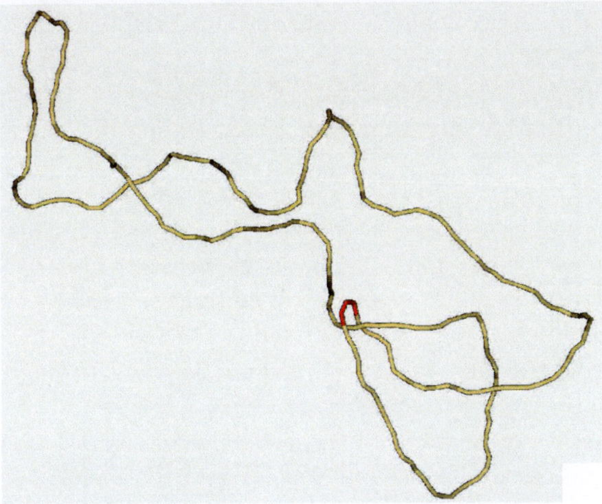

FIG. 6. *Typical simulated conformation of a knotted DNA with a hairpin-like G segment created by the topoisomerase binding (red). Another segment of the 3 kb model chain is inside the hairpin in this conformation, so the conformation was selected from the generated set as capable to form the reaction complex. The straight segments of the model chain correspond to 15 base pairs of the double helix (5 nm).*

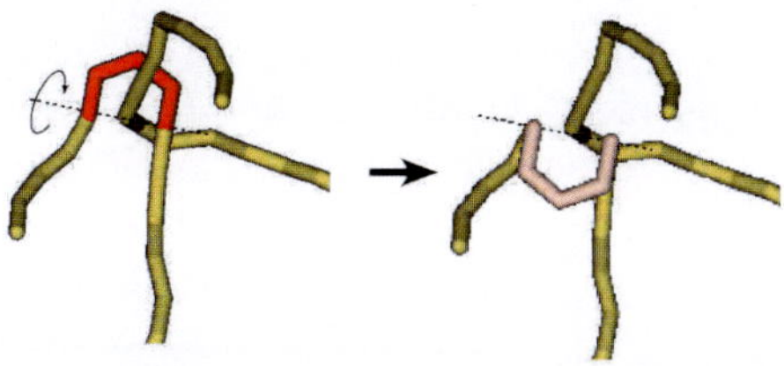

FIG. 7. *Determination of the chain topology after the strand passage. For each conformation with properly juxtaposed sites the G segment (red) was rotated by a 180° around the axis connecting its ends. The new orientation of the hairpin is shown by pink on the right side. The rest of the chain was immobilized during the rotation.*

1) The distance between the center of segment i and the ends of the G segment is smaller than ρ_o (5 nm for the case shown in Fig. 6);

2) The angle between segment i and the G segment plane is larger than ϕ_o (60° for the majority of the simulations).

For each selected conformation the strand-passage was modeled to determine the product topology. It was achieved by local deformation of the chain, as diagramed in Fig. 7. Visual inspection of numerous examples showed that the deformation does provide passage of the T segment through the G segment. Although the deformations can interfere with other segments which can occasionally be in the close vicinity to G segment, the probability of such event is very low.

The simulations performed for DNA molecules 7,000 base pairs in length showed that for the chain with a hairpin-like G segment the steady-state fraction of knots, c_3/c_0, is reduced by factor 15 compared with the results for DNA with a straight G segment [25]. This is a surprisingly large effect because the hairpin occupies only a small portion of the model chain. Presence of such small hairpin in the circular chain does not change the equilibrium probability of knots by more than 10% [25]. Two factors determine the effect:

1) The unidirectional transport of the T segment through the hairpin-like G segment;

2) An increased, by factor 3–5, probability to have a potential T segment inside the hairpin-like G segment for knotted molecules [25].

Thus, the simulation showed that the model is capable to explain large simplification of DNA topology by type II topoisomerases. Further support for the model came from the analysis of the G segment conformation. According to the model, the segment has to be bent upon the protein binding. Solution experiments and electron microscopy data confirmed the assumption quantitatively [25], and recently strong DNA bend was established by X-ray analysis of the DNA-protein complex [29].

5. Site-specific recombinases of the tyrosine family. Site-specific recombination is utilized by a great variety of organisms for different genomic rearrangements [2, 30, 31]. *In vitro* studies of the recombination usually use circular plasmid substrates harboring a pair of specific target sites. In the case of the tyrosine family of site-specific recombinases, the substrate can carry the target sites in either head-to-head (inversion substrate) or head-to-tail (deletion substrate) orientations. The recombination products from deletion substrates are two smaller circles which can be unlinked, or form torus links with various linking numbers (this linking number is denoted by Ca to distinguish it from the linking number of DNA complementary strands). Similarly, the products from the inversion substrates can be unknotted circles, or torus knots of various complexity [32]. The topological complexity of the products, obtained from the supercoiled substrate DNA, varies strongly depending on a particular recombination system of the tyrosine family. This large difference in the topological complexity of products remained a puzzle before it was solved in the recent study based on computer simulation of the reaction outcomes [21].

Two simplest and well studied systems of the tyrosine family, Cre and Flp, were used by Du et al. as models [21]. The structures of their synaptic complexes and reaction pathways had been well understood in the earlier studies [33–35]. Thus, only assembling of the synaptic complexes had to be simulated to determine the product distributions. This assembling occurs through collision of preassembled halves of the synaptic complex [36–39].

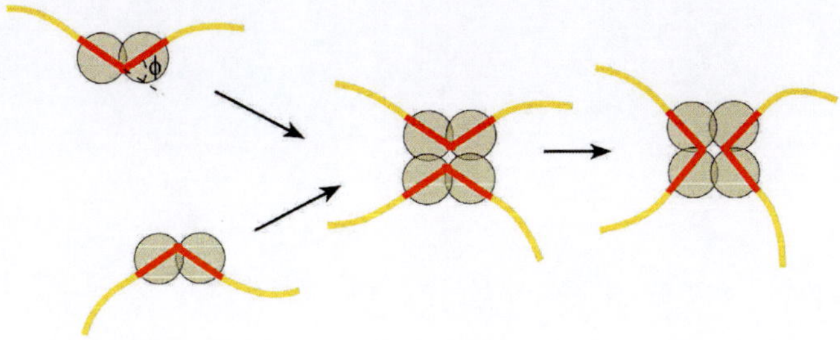

FIG. 8. *The model for the assembly of the synaptic complex by tyrosine family recombinases. A key element of the model is that formation of the recombinase-bound DNA partners is a necessary first step in synapsis. DNA structure in the 'pre-synaptic' halves is assumed to be bent by angle ϕ. The value of ϕ can be different from the value of the corresponding angle in the mature synaptic complex.*

Correspondingly, the synapsis was simulated as a collision of two properly oriented halves of the complex (Fig. 8). Each half was assumed to consist of a specific site bound by two recombinase monomers. It was also assumed that the protein-bound specific sites are bent by angle ϕ. This was consistent with the gel electrophoresis data for the isolated halves [40, 41] and with the structure of complete synaptic complexes [33–35]. However, the bend angles observed in the X-ray structures of the synapses [33–35] and those present in the pre-synaptic halves need not be identical.

The same discrete wormlike chain was used to simulate the synapsis. In this case the model chain had both bending and torsional rigidity,

making possible simulation of supercoiled DNA molecules. The bent conformations of the specific sites were assumed to be rigid during the simulation procedure. The value of the bent angle, ϕ, was considered as the simulation parameter. Typical simulated conformations of the chain are shown in Fig. 9 (panels A and B).

The equilibrium sampling, which corresponds to the described model, does not provide sufficient number of chain conformations with properly juxtaposed specific sites. To increase the number of such conformations a biased sampling was used, based on adding an artificial potential to the model chain. The potential depends on the mutual conformation of the recombination sites:

$$U = A \cdot \left[\left(\frac{r_0}{r} \right)^{2q} - 2 \left(\frac{r_0}{r} \right)^{q} \right] \cdot \exp \left(- \frac{\varphi^2 + \theta^2 + \varsigma^2}{2\sigma^2} \right), \tag{5.1}$$

where r, φ, θ, and ς are variables whose definitions are explained in the legend to Figure 10.

Metropolis sampling with this additional potential produces alternating stretches of two kinds of conformations, with close distance between the specific sites and with large separation between them. The juxtapositions in a single stretch are not independent and mainly give identical recombi-

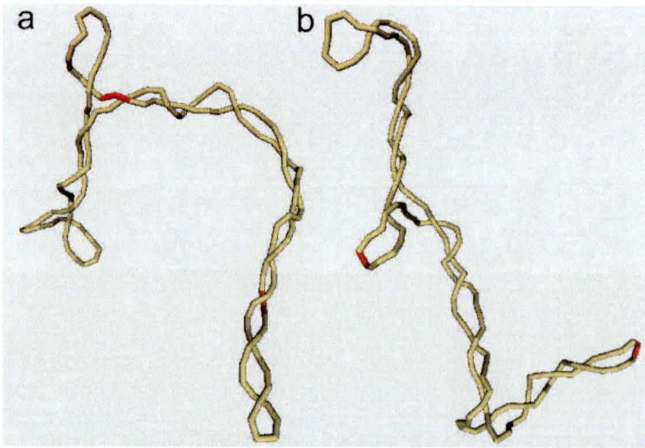

FIG. 9. *Typical simulated conformations of supercoiled molecules 4400 base pairs in length with protein-induced bends in the specific sites (shown by red). Strongly bent sites tend to be localized in the superhelix apices (b).*

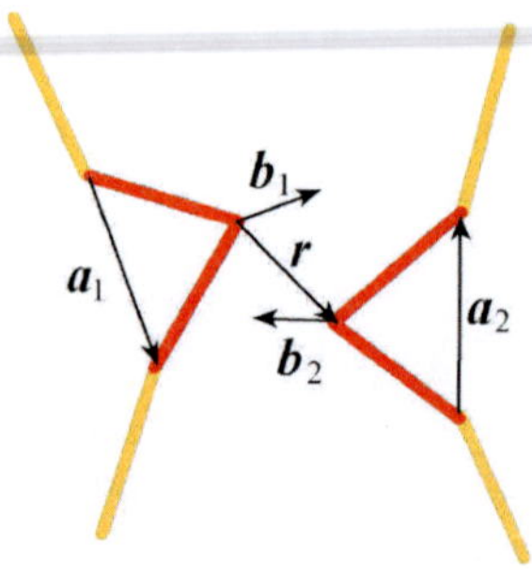

FIG. 10. *Monte Carlo simulations of synapsis. The diagram shows variables that define the mutual orientation of the recombination sites (shown by red). The variables are used in the potential specified by Eq. (5.1) and for the definition of the synapsis. Vectors* $\mathbf{b}_1$ *and* $\mathbf{b}_2$ *are perpendicular to* $\mathbf{a}_1$ *and* $\mathbf{a}_2$*, respectively, and are in the planes of the sites. The angles which appear in Eq. (5.1) are defined as:* $\varphi = \arccos\left[\frac{-(\mathbf{a}_1\mathbf{a}_2)}{a_1 a_2}\right]$, $\theta = \arccos\left[\frac{-(\mathbf{b}_1\mathbf{b}_2)}{b_1 b_2}\right]$ *and* $\varsigma = \arccos\left[\frac{(\mathbf{b}_1\mathbf{r})}{b_1 r}\right]$.

nation products. Therefore, it was important to choose the values of the potential parameters, A, q, r_0, and σ, so that many independent stretches of conformations with juxtaposed sites appeared in a simulation run [21].

The specific sites were considered properly juxtaposed if r was less than $0.8\,l$ and angles ϕ and θ were smaller than 20°. It was tested that further tightening this conditions does not change the simulation results.

For all conformations with properly juxtaposed specific sites the recombination reaction was modeled. Calculation of the Alexander polynomial for two circular contours [19, 42], $\Delta(t, s)$, for $s = -1$, $t = -1$ was used to determine topology of the recombination products. This topologi-

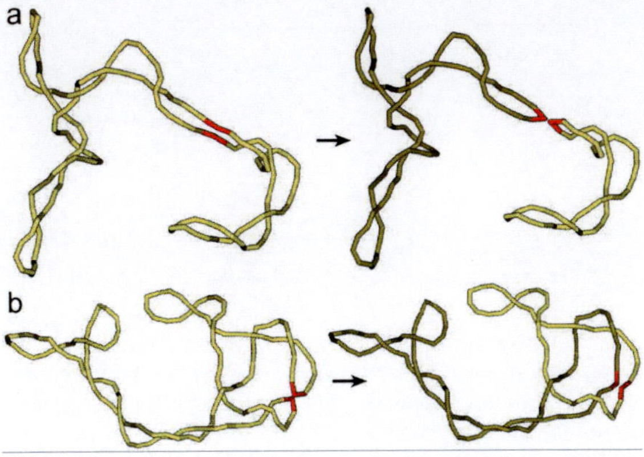

FIG. 11. *Monte Carlo simulation of the site-specific recombination.*
(a) A conformation of the model chain with the recombination sites (shown by red) juxtaposed within the same superhelix branch. This type of juxtaposition usually occurs if the bend angle, ϕ, in the sites is small. The recombination products are unlinked in this case.
(b) A conformation with juxtaposed sites located at superhelix apices. This type of juxtaposition is typical for large values of ϕ. The synapsis traps a few supercoils in such cases and the recombination products form torus links. The value of Ca equals 2 for the shown case.

cal invariant is powerful enough to distinguish the majority of the simplest links [23].

The simulation showed that the bend of the protein-bound recombination sites before their juxtaposition determines the distribution of *Ca* of the recombination products formed from directly oriented sites (or the distribution of knots for inversely oriented sites). If the bend angle, ϕ, is small, the collisions of the complex halves mainly occur in one branch of supercoiled DNA (Fig. 11a). The recombination event gives unlinked circles in this case. When ϕ increasess the bent DNA segments tend to be located at superhelix apices where the double helix has to be bent in any case (see Fig. 9b). This type of synapsis traps DNA supercoils and recombination products form torus links (Fig. 11b). This theoretical finding immediately suggests an explanation why the topological complexity of recombination products is different for different recombination systems, since the value of ϕ depends on a particular system [43, 44].

The simulated distributions of *Ca* obtained for different values of ϕ are presented in Fig. 12.

This simulation finding was tested experimentally by establishing the correlation between bend angle in the complex halves and the topological complexity of recombination products [21]. The difference in the product topology is strongly pronounced for the Cre/*lox*P and Flp/*FRT* systems which have striking similarities in their overall reaction mechanism (see review [2]). Action by Cre on head-to-tail target sitesproduces

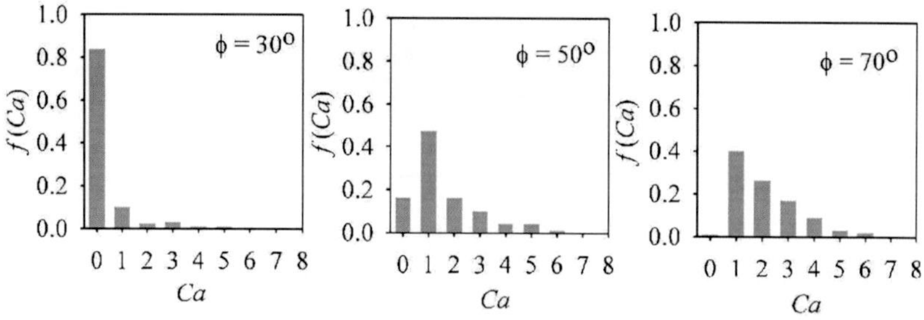

FIG. 12. *Computed distributions of recombination products for different values of the bend angle in the recombination sites. The fractions of products with different topologies, were computed for supercoiled substrate DNA 4200 bp in length with specific sites separated by 1400 bp. The distributions of the product linking number, Ca, are shown for DNA containing two target sites in direct orientation. The fraction of unlinked circles corresponds to Ca = 0. The DNA superhelix density in all these simulations was equal to* -0.05.

mainly unlinked circles, while that of Flp yields multiply linked catenanes [21, 45–47]. Du et al. measured the bent angles in the complex halves in solution, using the cyclization of short DNA fragments carrying the specific sites and bound proteins [21]. It was found, in full agreement with theoretical expectations, that the value of ϕ is close to 35° for Cre system and close to 80° for Flp. These values of bend angles have to provide the experimentally observed difference in topological complexity of recombination products. Thus, the established correlation between topological complexity of the recombination products and the bend angle gave very solid support to the explanation of the effect obtained in the computer simulation.

Acknowledgement. The author thanks the late Nick Cozzarelli for many stimulating discussions of topoisomerases and site-specific recombinases.

REFERENCES

[1] J.C. WANG, *Moving one DNA double helix through another by a type II DNA topoisomerase: the story of a simple molecular machine*, Q. Rev. Biophys. **31**: 107–144, 1998.

[2] N.D.F. GRINDLEY, K.L. WHITESON, AND P.A. RICE, *Mechanisms of site-specific recombination*, Ann. Rev. Biochem. **75**: 567–605, 2006.

[3] Y.S. POLIKANOV, V.A. BONDARENKO, V. TCHERNAENKO, Y.I. JIANG, L.C. LUTTER, A. VOLOGODSKII, AND V.M. STUDITSKY, *Probability of the site juxtaposition determines the rate of protein-mediated DNA looping*, Biophys. J. **93**: 2726–2731, 2007.

[4] J.A. SCHELLMAN, *Flexibility of DNA*, Biopolymers **13**: 217–226, 1974.

[5] M.D. FRANK-KAMENETSKII, A.V. LUKASHIN, V.V. ANSHELEVICH, AND A.V. VOLOGODSKII, *Torsional and bending rigidity of the double helix from data on small DNA rings*, J. Biomol. Struct. Dyn. **2**: 1005–1012, 1985.

[6] A. Vologodskii, in *Simulation of equilibrium and dynamic properties of large DNA molecules*, ed. by Lankas, F. and Sponer, J. Springer, Dordrecht, The Netherlands, pp. 579–604, 2006.

[7] L. Landau and E. Lifshitz, in *Statistical physics*, ed. by Elsevier, Oxford, 1980.

[8] Y. Burnier, C. Weber, A. Flammini, and A. Stasiak, *Local selection rules that can determine specific pathways of DNA unknotting by type II DNA topoisomerases*, Nucl. Acids Res. **35**: 5223–5231, 2007.

[9] Z. Liu, J.K. Mann, E.L. Zechiedrich, and H.S. Chan, *Topological Information embodied in local juxtaposition geometry provides a statistical mechanical basis for unknotting by type II DNA topoisomerases*, J. Mol. Biol. **361**: 268–285, 2006.

[10] D. Stigter, *Interactions of highly charged colloidal cylinders with applications to double-stranded DNA*, Biopolymers **16**: 1435–1448, 1977.

[11] A.V. Vologodskii and N. R. Cozzarelli, *Modeling of long-range electrostatic interactions in DNA*, Biopolymers **35**: 289–296, 1995.

[12] M. Le Bret, *Monte Carlo computation of supercoiling energy, the sedimentation constant, and the radius of gyration of unknotted and knotted circular DNA*, Biopolymers **19**: 619–637, 1980.

[13] K.V. Klenin, A.V. Vologodskii, V.V. Anshelevich, A.M. Dykhne, and M.D. Frank-Kamenetskii, *Effect of excluded volume on topological properties of circular DNA*, J. Biomol. Struct. Dyn. **5**: 1173–1185, 1988.

[14] A.A. Brian, H.L. Frisch, and L.S. Lerman, *Thermodynamics and equilibrium sedimentation analysis of the close approach of DNA molecules and a molecular ordering transition*, Biopolymers **20**: 1305–1328, 1981.

[15] V.V. Rybenkov, N.R. Cozzarelli, and A.V. Vologodskii, *Probability of DNA knotting and the effective diameter of the DNA double helix*, Proc. Natl. Acad. Sci. USA **90**: 5307–5311, 1993.

[16] F.B. Fuller, *The writhing number of a space curve*, Proc. Natl. Acad. Sci. USA **68**: 815–819, 1971.

[17] A.V. Vologodskii, S.D. Levene, K.V. Klenin, M.D. Frank-Kamenetskii, and N.R. Cozzarelli, *Conformational and thermodynamic properties of supercoiled DNA*, J. Mol. Biol. **227**: 1224–1243, 1992.

[18] K. Klenin and J. Langowski, *Computation of writhe in modeling of supercoiled DNA*, Biopolymers **54**: 307–317, 2000.

[19] A. Vologodskii, in *Monte Carlo simulation of DNA topological properties*, ed. by Monastryrsky, M. Springer, Berlin - Heidelberg - New York, pp. 23–41 2007.

[20] I. Grainge, S. Pathania, A. Vologodskii, R. Harshey, and M. Jayaram, *Symmetric DNA sites are functionally asymmetric within Flp and Cre site-specific DNA recombination synapses*, J. Mol. Biol. **320**: 515–527, 2002.

[21] Q. Du, A. Livshits, A. Kwiatek, M. Jayaram, and A. Vologodskii, *Protein-induced local DNA bends regulate global topological complexity of recombination products*, J. Mol. Biol. **368**: 170–182, 2007.

[22] M.D. Frank-Kamenetskii, A.V. Lukashin, and M.D. Vologodskii, *Statistical mechanics and topology of polymer chains*, Nature **258**: 398–402, 1975.

[23] D. Rolfsen, *Knots and Links* Publish or Perish, Inc., Berkeley, CA, 1976.

[24] V.V. Rybenkov, A.V. Vologodskii, and N.R. Cozzarelli, *The effect of ionic conditions on DNA helical repeat, effective diameter, and free energy of supercoiling*, Nucl. Acids Res. **25**: 1412–1418, 1997.

[25] A.V. Vologodskii, W.Zhang, V.V. Rybenkov, A.A. Podtelezhnikov, D. Subramanian, J.D. Griffith, and N.R. Cozzarelli, *Mechanism of topology simplification by type II DNA topoisomerases*, Proc. Natl. Acad. Sci. USA **98**: 3045–3049, 2001.

[26] J. Roca, J.M. Berger, S.C. Harrison, and J.C. Wang, *DNA transport by a type II topoisomerase – direct evidence for a two-gate mechanism*, Proc. Natl. Acad. Sci. USA **93**: 4057–4062, 1996.

[27] J. Roca and J. C. Wang, *The capture of a DNA double helix by an ATP-dependent protein clamp: A key step in DNA transport by topo II DNA topoisomerases*, Cell **71**: 833–840, 1992.

[28] J. Roca and J. Wang, *DNA transport by a type II DNA topoisomerase – evidence in favor of a two-gate mechanism*, Cell **77**: 609–616, 1994.

[29] K.C. Dong and J.M. Berger, *Structural basis for gate-DNA recognition and bending by type II A topoisomerases.*, Nature **450**: 1201–1205, 2007.

[30] H.A. Nash, in *Site-specific recombination: integration, excition, resolution and inversion of defined DNA segments*, ed. by Neidhardt, F.C. American Society for Microbiology, Washington DC, pp. 2330–2376 1996.

[31] D.N. Gopaul and G.D. Van Duyne, *Structure and mechanism in site-specific recombination*, Curr. Opin. Struct. Biol. **9**: 14–20, 1999.

[32] R. Kanaar and N.R. Cozzarelli, *Roles of supercoiled DNA structure in DNA transactions*, Curr. Opion. Struct. Biol. **2**: 369–379, 1992.

[33] F. Guo, D.N. Gopaul, and G.D. Van Duyne, *Asymmetric DNA bending in the Cre-loxP site-specific recombination synapse*, Proc. Natl. Acad. Sci. USA **96**: 7143–7148, 1999.

[34] Y. Chen, U. Narendra, L.E. Iype, M.M. Cox, and P.A. Rice, *Crystal structure of a Flp recombinase-Holliday junction complex: assembly of an active oligomer by helix swapping*, Mol. Cell **6**: 885–897, 2000.

[35] T. Biswas, H. Aihara, M. Radman-Livaja, D. Filman, A. Landy, and T. Ellenberger, *A structural basis for allosteric control of DNA recombination by [lambda] integrase*, Nature **435**: 1059–1066, 2005.

[36] R.H. Hoess and K. Abremski, *Interaction of the bacteriophage P1 recombinase Cre with the recombining site loxP*, Proc. Natl. Acad. Sci. USA **81**: 1026–1029, 1984.

[37] A. Mack, B. Sauer, K. Abremski, and R. Hoess, *Stoichiometry of the Cre recombinase bound to the lox recombining site*, Nucl. Acids Res. **20**: 4451–4455, 1992.

[38] B.J. Andrews, L.G. Beatty, and P.D. Sadowski, *Isolation of intermediates in the binding of the FLP recombinase of the yeast plasmid 2-micron circle to its target sequence*, J. Mol. Biol. **193**: 345–358, 1987.

[39] L. Ringrose, V. Lounnas, L. Ehrlich, F. Buchholz, R. Wade, and A.F. Stewart, *Comparative kinetic analysis of FLP and cre recombinases: mathematical models for DNA binding and recombination*, J. Mol. Biol. **284**: 363–384, 1998.

[40] C.J. Schwartz and P.D. Sadowski, *FLP protein of 2 mu circle plasmid of yeast induces multiple bends in the FLP recognition target site*, J. Mol. Biol. **216**: 289–298, 1990.

[41] L. Lee, L.C. Chu, and P.D. Sadowski, *Cre induces an asymmetric DNA bend in its target loxP site*, J. Biol. Chem. **278**: 23118–23129, 2003.

[42] A.V. Vologodskii, A.V. Lukashin, and M.D. Frank-Kamenetskii, *Topological interaction between polymer chains*, Sov. Phys. JETP **40**: 932–936, 1975.

[43] H.A. Nash, *Bending and supercoiling of DNA at the attachment site of bacteriophage l.*, Trends Biochem. Sci. **15**: 222–227., 1990.

[44] Y. Voziyanov, S. Pathania, and M. Jayaram, *A general model for site-specific recombination by the integrase family recombinases*, Nucl. Acids Res. **27**: 930–941, 1999.

[45] K. Abremski, R. Hoess, and N. Sternberg, *Studies on the properties of P1 site-specific recombination: Evidence for topologically unlinked products following recombination*, Cell **32**: 1301–1311, 1983.

[46] R.H. Hoess and K. Abremski, *Mechanism of strand cleavage and exchange in the Cre-lox site-specific recombination system*, J. Mol. Biol. **181**: 351–362, 1985.

[47] L.G. Beatty, D. Babineau-Clary, C. Hogrefe, and P.D. Sadowski, *FLP Site-specific Recombinase of Yeast 2-um Plasmid: Topological Features of the Reaction*, J. Mol. Biol. **188**: 529–544, 1986.

LIST OF WORKSHOP PARTICIPANTS

IMA Workshop on Mathematics of DNA Structure, Function, and Interactions

September 16-21, 2007

- Pranav Agarwal, Department of Electrical Engineering, University of Minnesota
- Tanuj Aggarwal, Department of Electrical Engineering, University of Minnesota
- Ramzi Alsallaq, Institute of Molecular Biophysics, Florida State University
- Douglas N. Arnold, School of Mathematics, University of Minnesota
- F. Javier Arsuaga, Department of Mathematics, San Francisco State University
- Nathan A. Baker, Department of Biochemistry and Molecular Biophysics, Washington University School of Medicine
- Daniel J. Bates, Institute for Mathematics and its Applications, University of Minnesota
- Peter W. Bates, Department of Mathematics, Michigan State University
- John Baxter, Institute for Mathematics and its Applications, University of Minnesota
- Craig John Benham, UC Davis Genome Center, University of California, Davis
- Meredith Betterton, Department of Physics, University of Colorado
- Yermal Sujeet Bhat, Institute for Mathematics and its Applications, University of Minnesota
- John Bida, Biochemistry Department, Mayo Clinic
- Betul Bilgin, University of Minnesota
- Victor Bloomfield, Department of Biochemistry, University of Minnesota
- Erik Boczko, Department of Biomedical Informatics, Vanderbilt University
- Richard J. Braun, Department of Mathematical Sciences, University of Delaware
- Dorothy E. Buck, Department of Mathematics, Imperial College London
- Gregory Buck, Department of Mathematics, Saint Anselm College
- Maria-Carme T. Calderer, School of Mathematics, University of Minnesota

C.J. Benham et al. (eds.), *Mathematics of DNA Structure, Function and Interactions*,
The IMA Volumes in Mathematics and its Applications 150,
DOI 10.1007/978-1-4419-0670-0, © Springer Science+Business Media, LLC 2009

- Hannah Callender, Institute for Mathematics and its Applications, University of Minnesota
- Shi-Jie Chen, Department of Physics, University of Missouri
- Gregory S. Chirikjian, Department of Mechanical Engineering, Johns Hopkins University
- Bernard D. Coleman, School of Engineering, Rutgers University
- Vincent Croquette, Departement de Physique, École Normale Supérieure
- John Crow, University of Minnesota
- Feng Cui, National Institutes of Health
- Jeremy Curuksu, School of Engineering and Science, Jacobs University
- Luke Czapla, Department of Chemistry and Chemical Biology, Rutgers University
- Isabel K. Darcy, Department of Mathematics, University of Iowa
- Melanie DeVries, Department of Mathematics, University of Iowa
- Yuanan Diao, Department of Mathematics and Statistics, University of North Carolina at Charlotte
- Claus Ernst, Department of Mathematics, Western Kentucky University
- Marcia O. Fenley, Department of Physics, Florida State University
- Laura Finzi, Department of Physics, Emory University
- Anant Godbole, Department of Mathematics, East Tennessee State University
- Jason E. Gower, Institute for Mathematics and its Applications, University of Minnesota
- Steve Harvey, Department of Chemistry and Biochemistry, Georgia Institute of Technology
- Christine E. Heitsch, School of Mathematics, Georgia Institute of Technology
- Milena Hering, Institute for Mathematics and its Applications, University of Minnesota
- Peter Hinow, Institute for Mathematics and its Applications, University of Minnesota
- Xia Hua, Department of Mathematics, Massachusetts Institute of Technology
- Richard D. James, Department of Aerospace Engineering and Mechanics, University of Minnesota
- Makkuni Jayaram, Section of Molecular Genetics and Microbiology, University of Texas at Austin
- Tiefeng Jiang, Department of Statistics, University of Minnesota
- Jason D. Kahn, Department of Chemistry and Biochemistry, University of Maryland
- George Karypis, Department of Computer Science and Engineering, University of Minnesota

- Alex Kasman, Department of Mathematics, College of Charleston
- Christopher Kauffman, Department of Computer Science, University of Minnesota
- Christine A. Kelley, Department of Mathematics, Ohio State University
- Soojeong Kim, Department of Mathematics, University of Iowa
- Debra Knisley, Department of Mathematics, East Tennessee State University
- Mark Kon, Department of Mathematics and Statistics, Boston University
- Christian E. Laing, Department of Mathematics, Chemistry, New York University,
- Jörg Langowski, Division of Biophysics of Macromolecules Deutsches Krebsforschungszentrum (Cancer Research)(DKFZ)
- Richard Lavery, Laboratoire de Biochimie Théorique, Centre National de la Recherche Scientifique (CNRS)
- Stephen D. Levene, Department of Molecular and Cellular Biology, University of Texas at Dallas
- Anton Leykin, Institute for Mathematics and its Applications, University of Minnesota
- David M.J. Lilley, Cancer Research UK Nucleic Acid Structure Research Group, The University of Dundee
- Sookkyung Lim, Department of Mathematical Sciences, University of Cincinnati
- Chun-Chi Lin, Department of Mathematics, National Taiwan Normal University
- Maggie Linak, CEMS, University of Minnesota
- Roger Lui, Department of Mathematical Sciences, Worcester Polytechnic Institute
- Laura Lurati, Institute for Mathematics and its Applications, University of Minnesota
- James Maher, Mayo Clinic
- Jennifer Mann, Department of Mathematics, University of Texas
- John F. Marko, Department of Physics and Astronomy and Department of Biochemistry, Molecular Biology and Cell Biology, Northwestern University
- Kyle McQuisten, Department of Mathematics, University of Iowa
- Ezra Miller, School of Mathematics, University of Minnesota
- Willard Miller, Jr., School of Mathematics, University of Minnesota
- Kenneth C. Millett, Department of Mathematics, University of California, Santa Barbara
- Hyeyoung Moon, Department of Mathematics, University of Iowa
- Maria Giovanna Mora, Department of Mathematics, International School for Advanced Studies (SISSA/ISAS)

- Tolkynay Myrzakul, Physics Department, Kazakh Al-Farabi State National University
- Junalyn Navarra-Madsen, Department of Mathematics and Computer Science, Texas Woman's University
- Timothy Newman, Department of Physics, Arizona State University
- Olalla Nieto Faza, Department of Chemistry, University of Minnesota
- Duane Nykamp, School of Mathematics, University of Minnesota
- Wilma K. Olson, Department of Chemistry and Chemical Biology, Rutgers University
- Isamu Onishi, Department of Mathematical and Life Sciences, Hiroshima University
- John Oprea, Department of Mathematics, Cleveland State University
- Hans G. Othmer, Department of Mathematics, University of Minnesota
- Jinhae Park, Department of Mathematics, Purdue University
- Ariel Prunell, Genome Biology, Institut Jacques Monod
- Teresita Ramirez-Rosas, Department of Mathematics, University of California, Santa Barbara
- Graham L. Randall, Department of Molecular Virology and Microbiology, Baylor College of Medicine
- Huzefa Rangwala, Department of Computer Science, University of Minnesota
- Eric J. Rawdon, Department of Mathematics, University of St. Thomas
- Dale Rolfsen, Department of Mathematics, University of British Columbia
- Ioulia Rouzina, Biochemistry, Molecular Biology and Biophysics, University of Minnesota
- George Rublein, Department of Mathematics, College of William and Mary
- Deena Schmidt, Institute for Mathematics and its Applications, University of Minnesota
- Chehrzad Shakiban, Institute of Mathematics and its Application, University of Minnesota
- Koya Shimokawa, Department of Mathematics, Saitama University
- David Snyder, Department of Mathematics, Texas State University at San Marcos
- Andrzej Stasiak, Laboratoire d'Analyse Ultrastructurale, University of Lausanne
- Andrew M. Stein, Institute for Mathematics and its Applications, University of Minnesota

- De Witt L. Sumners, Department of Mathematics, Florida State University
- Vladimir Sverak, School of Mathematics, University of Minnesota
- David Swigon, Department of Mathematics, University of Pittsburgh
- Irwin Tobias, Department of Chemistry and Chemical Biology, Rutgers University
- Andrew Travers, LBPA, Insitut d'Alembert, ENS de Cachan
- Erkan Tüzel Institute for Mathematics and its Applications, University of Minnesota
- Lawrence Varela, Computer Science Department, San Francisco State University
- Tanya Vassilevska, Center for Applied Scientific Computing, Lawrence Livermore National Laboratory
- Mariel Vazquez, Department of Mathematics, San Francisco State University
- Alexander Vologodskii, Department of Chemistry, New York University
- Haiyan Wang, Department of Mathematical Sciences and Applied Computing, Arizona State University
- Zhian Wang, Institute for Mathematics and its Applications, University of Minnesota
- Annika Wedemeier, Biophysics of Macromolecules Department, German Cancer Research Center
- Guowei Wei, Department of Mathematics, Michigan State University
- Shimon Weiss, Department of Chemistry and Biochemistry, University of California, Los Angeles
- Jonathan Widom, Department of Chemistry, Northwestern University
- Peng Wu, Rockefeller University
- Zhijun Wu, Department of Mathematics, Iowa State University
- Yanji Xu, Minnesota Supercomputing Institute
- Lynn Zechiedrich, Department of Molecular Virology and Microbiology, Baylor College of Medicine
- Hongchao Zhang, Institute for Mathematics and its Applications, University of Minnesota
- Guohui Zheng, Department of Chemistry and Chemical Biology, Rutgers University
- Victor B. Zhurkin, Laboratory of Cell Biology, National Cancer Institute

Continued from page ii

1997–1998 Emerging Applications of Dynamical Systems
1998–1999 Mathematics in Biology
1999–2000 Reactive Flows and Transport Phenomena
2000–2001 Mathematics in Multimedia
2001–2002 Mathematics in the Geosciences
2002–2003 Optimization
2003–2004 Probability and Statistics in Complex Systems: Genomics, Networks, and Financial Engineering
2004–2005 Mathematics of Materials and Macromolecules: Multiple Scales, Disorder, and Singularities
2005-2006 Imaging
2006-2007 Applications of Algebraic Geometry
2007-2008 Mathematics of Molecular and Cellular Biology
2008-2009 Mathematics and Chemistry
2009-2010 Complex Fluids and Complex Flows
2010-2011 Simulating Our Complex World: Modeling, Computation and Analysis

IMA SUMMER PROGRAMS

1987 Robotics
1988 Signal Processing
1989 Robust Statistics and Diagnostics
1990 Radar and Sonar (June 18–29)
New Directions in Time Series Analysis (July 2–27)
1991 Semiconductors
1992 Environmental Studies: Mathematical, Computational, and Statistical Analysis
1993 Modeling, Mesh Generation, and Adaptive Numerical Methods for Partial Differential Equations
1994 Molecular Biology
1995 Large Scale Optimizations with Applications to Inverse Problems, Optimal Control and Design, and Molecular and Structural Optimization
1996 Emerging Applications of Number Theory (July 15–26)
Theory of Random Sets (August 22–24)
1997 Statistics in the Health Sciences
1998 Coding and Cryptography (July 6–18)
Mathematical Modeling in Industry (July 22–31)
1999 Codes, Systems, and Graphical Models (August 2–13, 1999)
2000 Mathematical Modeling in Industry: A Workshop for Graduate Students (July 19–28)
2001 Geometric Methods in Inverse Problems and PDE Control (July 16–27)
2002 Special Functions in the Digital Age (July 22–August 2)

2003 Probability and Partial Differential Equations in Modern Applied Mathematics (July 21–August 1)
2004 n-Categories: Foundations and Applications (June 7–18)
2005 Wireless Communications (June 22–July 1)
2006 Symmetries and Overdetermined Systems of Partial Differential Equations (July 17–August 4)
2007 Classical and Quantum Approaches in Molecular Modeling (July 23-August 3)
2008 Geometrical Singularities and Singular Geometries (July 14-25)
2009 Nonlinear Conservation Laws and Applications (July 13-31)

IMA "HOT TOPICS/SPECIAL" WORKSHOPS

- Challenges and Opportunities in Genomics: Production, Storage, Mining and Use, April 24–27, 1999
- Decision Making Under Uncertainty: Energy and Environmental Models, July 20–24, 1999
- Analysis and Modeling of Optical Devices, September 9–10, 1999
- Decision Making under Uncertainty: Assessment of the Reliability of Mathematical Models, September 16–17, 1999
- Scaling Phenomena in Communication Networks, October 22–24, 1999
- Text Mining, April 17–18, 2000
- Mathematical Challenges in Global Positioning Systems (GPS), August 16–18, 2000
- Modeling and Analysis of Noise in Integrated Circuits and Systems, August 29–30, 2000
- Mathematics of the Internet: E-Auction and Markets, December 3–5, 2000
- Analysis and Modeling of Industrial Jetting Processes, January 10–13, 2001
- Special Workshop: Mathematical Opportunities in Large-Scale Network Dynamics, August 6–7, 2001
- Wireless Networks, August 8–10 2001
- Numerical Relativity, June 24–29, 2002
- Operational Modeling and Biodefense: Problems, Techniques, and Opportunities, September 28, 2002
- Data-driven Control and Optimization, December 4–6, 2002
- Agent Based Modeling and Simulation, November 3–6, 2003
- Enhancing the Search of Mathematics, April 26–27, 2004
- Compatible Spatial Discretizations for Partial Differential Equations, May 11–15, 2004
- Adaptive Sensing and Multimode Data Inversion, June 27–30, 2004
- Mixed Integer Programming, July 25–29, 2005

- New Directions in Probability Theory, August 5–6, 2005
- Negative Index Materials, October 2–4, 2006
- The Evolution of Mathematical Communication in the Age of Digital Libraries, December 8–9, 2006
- Math is Cool! and Who Wants to Be a Mathematician?, November 3, 2006
- Special Workshop: Blackwell-Tapia Conference, November 3–4, 2006
- Stochastic Models for Intracellular Reaction Networks, May 11–13, 2008
- Multi-Manifold Data Modeling and Applications, October 27–30, 2008
- Mixed-Integer Nonlinear Optimization: Algorithmic Advances and Applications, November 17–21, 2008
- Higher Order Geometric Evolution Equations: Theory and Applications from Microfluidics to Image Understanding, March 23–26, 2009
- Career Options for Women in Mathematical Sciences, April 2-4, 2009
- MOLCAS, May 4-8, 2009
- IMA Interdisciplinary Research Experience for Undergraduates, June 29-July 31, 2009

SPRINGER LECTURE NOTES FROM THE IMA:

The Mathematics and Physics of Disordered Media
Editors: Barry Hughes and Barry Ninham
(Lecture Notes in Math., Volume 1035, 1983)

Orienting Polymers
Editor: J.L. Ericksen
(Lecture Notes in Math., Volume 1063, 1984)

New Perspectives in Thermodynamics
Editor: James Serrin
(Springer-Verlag, 1986)

Models of Economic Dynamics
Editor: Hugo Sonnenschein
(Lecture Notes in Econ., Volume 264, 1986)